T0181718

CISM COURSES AND LECTURES

Series Editors:

The Rectors
Manuel Garcia Velarde - Madrid
Jean Salençon - Palaiseau
Wilhelm Schneider - Wien

The Secretary General
Bernhard Schrefler - Padua

Executive Editor
Carlo Tasso - Udine

The series presents lecture notes, monographs, edited works and
proceedings in the field of Mechanics, Engineering, Computer Science
and Applied Mathematics.
Purpose of the series is to make known in the international scientific
and technical community results obtained in some of the activities
organized by CISM, the International Centre for Mechanical Sciences.

CISM COURSES AND LECTURES

Series Editors:

The Rectors
Manuel Garcia Velarde - Madrid
Jean Salençon - Palaiseau
Wilhelm Schneider - Wien

The Secretary General
Bernhard Schrefler - Padua

Executive Editor
Carlo Tasso - Udine

The series presents lecture notes, monographs, edited works and
proceedings in the field of Mechanics, Engineering, Computer Science
and Applied Mathematics.
Purpose of the series is to make known in the international scientific
and technical community results obtained in some of the activities
organized by CISM, the International Centre for Mechanical Sciences.

INTERNATIONAL CENTRE FOR MECHANICAL SCIENCES

COURSES AND LECTURES - No 461

DEGRADATIONS AND INSTABILITIES
IN GEOMATERIALS

a DIGA RTN sponsored course
EDITED BY

FELIX DARVE
INSTITUT NATIONAL POLYTECHNIQUE DE GRENOBLE

IOANNIS VARDOULAKIS
NATIONAL TECHNICAL UNIVERSITY OF ATHENS

Springer-Verlag Wien GmbH

The publication of this volume was co-sponsored and co-financed by the UNESCO Venice Office - Regional Bureau for Science in Europe (ROSTE) and its content corresponds to a CISM Advanced Course supported by the same UNESCO Regional Bureau.

This volume contains 208 illustrations

This work is subject to copyright.
All rights are reserved,
whether the whole or part of the material is concerned
specifically those of translation, reprinting, re-use of illustrations,
broadcasting, reproduction by photocopying machine
or similar means, and storage in data banks.
© 2004 by Springer-Verlag Wien
Originally published by Springer-Verlag Wien New York in 2004
SPIN 11006794

In order to make this volume available as economically and as rapidly as possible the authors' typescripts have been reproduced in their original forms. This method unfortunately has its typographical limitations but it is hoped that they in no way distract the reader.

ISBN 978-3-211-21936-2   ISBN 978-3-7091-2768-1 (eBook)
DOI 10.1007/978-3-7091-2768-1

PREFACE

Degradation phenomena in geomaterials are linked to cyclic loading, ageing, weathering, internal erosion, capillary effects, etc. They can be treated today successfully within the frame of constitutive theories, which account for elasto-thermo-visco-plastic or damage effects, anisotropy, non-coaxiality and they are often extended so as to incorporate incremental non-linearity, non-local, second gradient and/or Cosserat effects. In addition the existing continuum chemo-mechanics framework can be utilized by assuming specific expressions for the chemical potentials and by identifying the relevant chemo-stress effects on the basis of the dominant physico-chemical mechanisms for concrete and rocks.

Nowadays it is recognized that there is a large variety of bifurcations and instabilities of material and/or of geometric nature, which are leading to various modes of failure, strictly inside the domain in stress-space which is bounded by the empirical "Mohr-Coulomb" failure criterion, but also inside the domain defined by the localization condition of plastic-strain or damage. The existence of such instabilities has been conjectured from the theory of non-associated elasto-plasticity and from the "controllability" theory. Some of these modes of collapse can be called "diffuse failure modes" and they seem to be the leading mechanisms in certain types of slope failures, debris flows and submarine slides in deltaic zones, which can be simulated presently successfully by the finite element method.

Once a localized or diffuse failure mechanism has been triggered, under the influence of its dead load a mass of soil may undergo a peculiar phase change and may start moving in a fluid-like manner. Analytical and numerical models of flow-slides can be classified in full 3D-formulations and in 1D-depth integrated or shallow-water models.

In recent years the use of scaled-down 1g laboratory experiments as well as 2D-discrete element numerical models are proposed for analyzing granular flows and also rockfalls. It is today possible to reproduce sliding and overturning of a group of blocks, impacts with fragmentation on a rock slope and the interaction of debris flows with a sheltering structure.

This book contains the lectures related to the classes given during the CISM course on "Degradations and Instabilities in Geomaterials", which was held at the CISM Centre in Udine on June 16-20, 2003.

The first three chapters of this book are devoted to the central question of constitutive relations for geomaterials. After an introduction on the notion of constitutive relation, a classification of the various existing models and an overview on incrementally non-linear relations, the basics of elasto-plasticity, with its fundamental concepts, and that of damage mechanics (including non-local models) are presented and discussed.

Localization of deformation leads to a change of scale of the problem, so that phenomena occurring at the scale of the grain cannot be ignored anymore in the mechanical modeling process of the macroscopic behavior of the material. Thus in order to describe correctly localization phenomena it appears necessary nowadays to resort to continuum models with micro-structure, such as Cosserat- or Gradient models. Chapter 4 serves as an introduction, where the simplest linear elasticity theory with microstructure is discussed. Chapter 5 deals directly with the issues of 2nd grade elasto-plasticity and localization.

Chapters 6 and 7 focus on the application of Hill's sufficient condition of stability to geomaterials The notion of "controllability", essentially related to Hill's condition, is introduced and some applications to slope stability are given. Both the continuous and discrete forms of second order work are considered and applied to determine the unstable stress domain of sands and to analyze granular avalanches. The discrete element method (which is basically an example of Molecular Dynamics methods) is then considered with its application to the discrete modeling of landslides.

Chapter 9 investigates the chemo-mechanical damage due to calcium leaching and discusses the evolution of the internal length with ageing of concretes.

The two last chapters consider the question of modeling of landslides by the finite element method. The initiation of failure (localized or diffuse) is first considered in an hydro-mechanical framework, then the propagation phase is described to tackle phenomena like flow-slides, avalanches, mudflows and debris flows.

The CISM School on "Degradations and Instabilities in Geomaterials", is linked to the European Research and Training Network, Degradation and Instabilities in Geomaterials with Application to Hazard Mitigation (DIGA)[1], which is run in the framework of the Human Potential Program (HPRN-CT-2002-00220). The teachers and authors of the present book are all DIGA mentors. At this point we want to acknowledge the EU support of all the DIGA young researchers, who participated in this CISM-DIGA School.

DIGA and other related activities on research and training of young researchers are done within the frame of the Alliance of Laboratories in Europe for Research and

---

[1] http://diga.mechan.ntua.gr/

Technology (ALERT-Geomaterials Network)[2]. ALERT-Geomaterials is a network of European laboratories that was created in 1989 as a pioneering effort to develop an European School of thought in the field of the Mechanics of Geomaterials. ALERT-Geomaterials now includes 20 European Universities or Organizations, which are most active in the field of numerical modeling of geomaterials and geo-structures. Its main research areas are the following:

- *micromechanics and constitutive modeling of geomaterials,*
- *failure, strain localization and instabilities,*
- *large scale computations for geomaterials and geo-structures,*
- *integrity of geo-structures and inverse analysis in geo-mechanics,*
- *environmental geo-mechanics and durability of geomaterials.*

Finally we would like to thank warmly the Authors not only for their precious scientific contributions to this book, but also for their swiftness for preparing their manuscripts !

With great pleasure we want to acknowledge the Centre International des Sciences Mécaniques (CISM), and its General Secretary Professor Bernhard Schrefler in particular, for his invaluable support and the encouragement for scientific excellence in the field of Geomechanics.

Félix Darve and Ioannis Vardoulakis

---

[2] http://alert.epfl.ch/

CONTENTS

Fundamentals of constitutive equations for geomaterials

Félix Darve [‡] and Guillaume Servant [‡]

[‡] Institut National Polytechnique de Grenoble
Laboratoire Sols, Solides, Structures.
INPG-UJF-CNRS, RNVO - Alert Geomaterials - Grenoble, France.

Abstract Constitutive equations for geomaterials constitute a very intricate field. In the first part of this chapter, a synthetic view of constitutive formalism is presented. An intrinsic classification of all existing constitutive relation is deduced. Then examples of incrementally non-linear relations are given and some applications follow. A numerical study of the so-called "yield surfaces" is presented, and is followed by a discussion on the validity of the principle of superposition for incremental loading. Finally the question of the existence of a flow rule is discussed for both axisymmetric and 3D conditions. In agreement with discrete element computations, it is shown that a flow rule can exist in 2D and not in 3D.

1 Introduction

For many years the study of the mechanical behaviour of geomaterials and its description by constitutive relations has been developed in the framework characterized by isotropic linear elasticity (Hooke's law) and by solid friction (Coulomb's law). However, since the end of the 1960's the development of more powerful numerical methods such as finite element method and the use of high-performance computers has brought to the fore of the scene the question that is becoming a crucial one: what constitutive relation for geomaterials must be introduced into a computer code? Thirty years later, the choice is large and the state of affairs confused; we will try to classify the various existing constitutive relations into some general classes with respect to the structure of the constitutive relationships.

Any user of finite element codes must be able to characterize the capacities of any implemented constitutive model in order to interpret correctly the numerical results obtained and also in order to know if the physical phenomena, that the user considers to have an important influence on the behaviour of the engineering work being analyzed, can be effectively taken into account by the constitutive relation used.

The first part of this chapter is devoted to a presentation of the incremental formulation of constitutive relations. Two main reasons have made such an incremental presentation indispensable. The first one is physical and is the fact that, as soon as some plastic irreversibilities are mobilized inside the geomaterial, the global constitutive functional, which relates the stress state $\sigma(t)$ at a given time t to the strain state $\epsilon(t)$

history up to this time, is a priori very difficult to formulate explicitly since this functional is singular at all stress-strain states (or more precisely non-differentiable, as we shall show). An incremental formulation enables us to avoid this fundamental difficulty. The second reason is numerical and stems from the fact that geomaterial behaviour, and the modelling of engineering works generally, exhibit many non-linearity sources which imply that the associated boundary value problem must be solved by successive steps linked to increments of boundary loading. Such finite element codes need therefore, the constitutive relations to be expressed incrementally.

2 Principle of determinism

In all that follows, it is assumed generally small transformations, an approximation often justified in civil engineering by the fundamental uncertainty of the spatial variation of mechanical properties. Thermo-mechanical effects will also be disregarded by noting that the influence of temperature variations on geomaterials behaviour is generally linked with particular situations, which are often not without interest if we think of the freezing of soils to increase their strength or on the contrary of increasing the temperature to accelerate the primary consolidation of clays. Finally we assume that geomaterials are simple media in the sense of Truesdell and Noll (1965), a hypothesis very widely accepted for the description of the mechanical behaviour of media.

2.1 Principle of determinism in the large

The first expression of the principle of the determinism is obtained by stating that the stress state $\sigma(t)$ at a given time t is a functional of the history of the tangent linear transformation up to this time t. It implies that it is necessary to know all the loading path in order to deduce the associated response path.

From a mathematical point of view, this is stated by the existence of a stress functional $\mathcal{F}$:

$$\sigma(t) = \mathcal{F}\left[\mathbf{F}(\tau)\right], \qquad -\infty < \tau \leq t \qquad (2.1)$$

where $\mathbf{F}(\tau)$ is the tangent linear transformation at time τ; also called deformation gradient. The deformation gradient $\mathbf{F}$ is the jacobian matrix of the position $\phi(X, \tau)$ of the material point X at time τ.

The existence of such a functional, and not a function, is related to an essential physical characteristic: for irreversible behaviours the knowledge of the strain $\epsilon(t)$ at time t does not enable us to determine the stress and vice versa. For example, we can think of viscous or plastic materials where a given level of stress can be related to an infinite number of different strain states.

This functional $\mathcal{F}$ will be called linear if :

$$\forall \lambda \in \mathbb{R} : \mathcal{F}\left[\lambda \, \mathbf{F}(\tau)\right] \equiv \lambda \, \mathcal{F}\left[\mathbf{F}(\tau)\right]$$

$$\forall \mathbf{F}_1, \mathbf{F}_2 : \mathcal{F}\left[\mathbf{F}_1(\tau) + \mathbf{F}_2(\tau)\right] \equiv \mathcal{F}\left[\mathbf{F}_1(\tau)\right] + \mathcal{F}\left[\mathbf{F}_2(\tau)\right]$$

In such a case, the material response to a sum of histories will be simply equal to the sum of the responses to each history. This constitutes Boltzmann's principle and it is the basis of linear viscoelasticity theory, but it is not at all valid in elastoplasticity where, when one doubles for example the strain, the stress is obviously not doubled, due to the non-linear behaviour.

This stress-strain relationship, which must be studied in the framework of non-linear functionals, is moreover non-differentiable as soon as there exist some plastic irreversibilities. Owen and Williams (1969) showed in fact that the assumptions of non-viscosity and of differentiability of the stress functional $\mathcal{F}$, imply that there is no internal dissipation. In other words a non-viscous material whose constitutive functional is differentiable is necessarily elastic.

Basically, this is due to the fact that, in plasticity, the tangent loading modulus is not equal to the tangent unloading modulus.

Therefore, if one wants to describe the behaviour of anelastic materials (geomaterials are essentially of this kind) by using a stress-strain relationship, this relation must be formulated by a non-linear and non-differentiable functional.

There is clearly a need, therefore, to study constitutive relations using an incremental formulation and no longer a global one. Note that incremental form tends to rate form when the 'time' increment tends to zero. In the following, the terms increment or rate are employed in the same sense.

We are now going to introduce an incremental formulation using a second statement of the principle of determinism.

2.2 Principle of determinism in the small

Let us consider a deformed solid, and let us assume that only quasi-static loading will be applied to it at a continuously varying rate.

The second principle of determinism, which can be called 'in the small' to distinguish it from the first one 'in the large', is obtained by stating that a small load applied during time increment dt induces a small uniquely determined response.

We denote by $d\epsilon = \mathbf{D}\, dt$, the incremental strain tensor of order two equal to the product of strain rate tensor of order two, $\mathbf{D}$ (symmetric part of transformation rate $\mathbf{L}$: $\mathbf{L} = \dot{\mathbf{F}}\,\mathbf{F}^{-1}$) and time increment dt, and by $d\sigma = \hat{\sigma}\, dt$ the incremental stress tensor, equal to the product of an objective time derivative of Cauchy stress tensor σ and dt. Thus the second determinism principle implies, from a mathematical point of view, the existence of a tensorial function $\mathbf{H}$ relating the three quantities:

$$\mathbf{H}_\chi\,(d\epsilon, d\sigma,\ dt) = 0 \qquad (2.2)$$

What are the properties of this tensorial function $\mathbf{H}$? The first remark concerns the fact that $\mathbf{H}$ depends on the previous stress-strain history. This history is generally characterized by some scalar and tensorial variables which will appear as parameters χ in relation (2.2). These parameters describe, as far as possible, the actual deformed state of the solid. Following the various constitutive theories, they are sometimes called 'memory variables', 'hardening parameters', 'internal variables', etc.

Secondly, $\mathbf{H}$ must satisfy the objectivity principle; this means that $\mathbf{H}$ must be independent of any movement of the observer relative to the solid. Thus $\mathbf{H}$ is an isotropic function of all its arguments: $d\boldsymbol{\epsilon}$, $d\boldsymbol{\sigma}$ and also the state tensorial variables χ, which characterize its present deformed state. But it is an anisotropic function of $d\boldsymbol{\epsilon}$ and $d\boldsymbol{\sigma}$.

Finally, $\mathbf{H}$ is essentially a non-linear function as soon as there are some plastic irreversibilities, since, if $\mathbf{H}$ was linear, the functional $\mathcal{F}$ would be differentiable, a property which excludes the existence of non-viscous irreversibilities.

On the contrary, for linear viscoelasticity this function $\mathbf{H}$ is linear and takes the following form:

$$d\boldsymbol{\epsilon} = \mathbf{M}\,d\boldsymbol{\sigma} + \mathbf{C}\,dt$$

where $\mathbf{M}$ is the (constant) elastic tensor of order four and $\mathbf{C}$ the creep rate tensor of order two of the material.

We remark here also that this determinism principle, which has very wide applications in physics by connecting notions of cause and effect, is not always satisfied, particularly in cases such as bifurcation situations. In such conditions a continuous variation of state variables can induce, at the bifurcation point, a sudden change in the evolution of the system with loss of uniqueness as a result of local imperfections not taken into account by the analysis (see examples given by Darve (1984)). At such a bifurcation point an identical cause in terms of state variables can have different effects; such a situation is common in geomaterial behaviour when a shear band appears, by localisation of plastic strains, while the deformation, up to this state, is respecting mainly a diffuse mode (Vardoulakis et al., 1978).

3 Application to non-viscous materials

In formulating constitutive relations it is often more convenient to replace the stress tensor $\boldsymbol{\sigma}$ and strain tensor $\boldsymbol{\epsilon}$, of second order, by two vectors of $\mathbb{R}^6$ defined in a six-dimensional related space:

$$\underline{\sigma} = \begin{pmatrix} \sigma_{11} \\ \sigma_{22} \\ \sigma_{33} \\ \sqrt{2}\,\sigma_{23} \\ \sqrt{2}\,\sigma_{31} \\ \sqrt{2}\,\sigma_{12} \end{pmatrix} \quad \text{and} \quad \underline{\epsilon} = \begin{pmatrix} \epsilon_{11} \\ \epsilon_{22} \\ \epsilon_{33} \\ \sqrt{2}\,\epsilon_{23} \\ \sqrt{2}\,\epsilon_{31} \\ \sqrt{2}\,\epsilon_{12} \end{pmatrix}$$

We will utilize greek indices, ranging from 1 to 6, in such a case and reserve latin indices, ranging from 1 to 3, to characterize tensorial components in the original three-dimensional space. The $\sqrt{2}$ coefficients, which appear in the definition of these vectors, allow us to preserve the original metric unchanged (isometric mapping). For example,

$$\|\boldsymbol{\sigma}\|^2 = \sigma_{ij}\,\sigma_{ij} = (\sigma_{11})^2 + (\sigma_{22})^2 + (\sigma_{33})^2 + 2(\sigma_{23})^2 + 2(\sigma_{31})^2 + 2(\sigma_{12})^2 = \sigma_{\alpha}\sigma_{\alpha}$$

The constitutive relation (2.2) will be written in the new notation as

$$H_{\chi}(d\epsilon_{\alpha}, d\sigma_{\beta},\ dt) = 0$$

where now H is a vectorial function (of $\mathbb{R}^6$ space) of 13 variables.

Now let us particularize, in this section, the class of materials considered and restrict the study to the case of non-viscous materials or to cases of loading for which a given geomaterial exhibits negligible viscosity. This means that the loading rate (characterized by time gradation on loading path) has no influence on material constitutive behaviour: at whatever rate the given loading path is followed, the response path remains unchanged. In other words the class of behaviour considered is rate independent.

This restriction of the constitutive law implies that the constitutive function H, which relates dϵ and dσ, is independent of the time increment dt during which the incremental loading has been applied. Therefore H is independent of dt and, in an equivalent manner if **H** is regular, one can now study the vectorial function **G**:

$$d\epsilon_\alpha = \mathbf{G}_\alpha(d\sigma_\beta) \tag{3.1}$$

3.1 Homogeneity of G

We have just seen that the behaviour of a non-viscous material is independent of the loading rate. This means that, in order to verify the one-to-one relation on a given loading path and its related unique response path, there is a dependency condition between the two rate vectors D and $\dot{\sigma}$. For instance, if the loading path is followed twice as fast, the response path will also be followed at exactly twice the rate. **G** is invariant with respect to the rates.

From a mathematical point of view this independence of non-viscous behaviours on loading rates implies the following identity:

$$\forall \lambda \in \mathbb{R}^+ : \mathbf{G}_\alpha(\lambda \, d\sigma_\beta) = \lambda \, \mathbf{G}_\alpha(d\sigma_\beta) \tag{3.2}$$

which states that if the stress rate is multiplied by any positive scalar λ, the strain rate response is also multiplied by the same scalar λ.

This is the first property of **G**: **G** is a homogeneous function of degree 1 in dσ with respect to the positive values of the multiplying parameter. This homogeneity property must not be confused with that of 'positively homogeneous' functions, which is given by;

$$\forall \lambda \in \mathbb{R} : f(\lambda \, d\sigma) = |\lambda| \, f(d\sigma).$$

3.2 Non-linearity of G

Let us recall that we want to describe plastic irreversibilities, and consider the incremental strains, which are produced by a return incremental load -dσ applied after dσ. If the behaviour is purely elastic, this stress cycle would induce strain cycle equally closed (with no permanent strain). Therefore for elastic behaviour:

$$\forall \, d\sigma \in \mathbb{R}^6 : \mathbf{G}(-d\sigma) = -\mathbf{G}(d\sigma)$$

In a more general manner, **G** is linear for elasticity since a regular correspondence relationship between σ and ϵ implies the existence of a linear relation between dσ and dϵ. In other words we know, in fact, that the principle of superposition for incremental loading is valid in elasticity and implies

$$\forall \, d\sigma^{(1)}, d\sigma^{(2)} \in \mathbb{R}^6 \times \mathbb{R}^6 : \mathbf{G}\left(d\sigma^{(1)} + d\sigma^{(2)}\right) = \mathbf{G}\left(d\sigma^{(1)}\right) + \mathbf{G}\left(d\sigma^{(2)}\right)$$

This incremental linearity, which is a characteristic property of elasticity if we consider non-viscous materials, must obviously not be confused with an eventual global linearity. The existence of such a global linear relationship between σ and ϵ describes linear-elastic behaviour, while an incrementally linear relation is characteristic of all elastic constitutive relations.

Since we want to describe the behaviour of essentially anelastic media such as geomaterials, we need therefore to take into account non-linear function **G**.

3.3   Anisotropy of G

The last property of **G** that we will emphasize is its anisotropy. If we consider all the arguments of **G**, which means the six components of dσ as well as the memory parameters χ characterizing the previous strain history of the medium, the objectivity principle implies that **G** is an isotropic function of the whole of its tensorial arguments. In fact a constitutive relation, describing the behaviour of a material, must be independent of any frame of reference used to define or formulate it, this frame possibly being in motion with respect to the material. This expresses a very obvious physical reality: whatever the movement of an observer looking at the deformation of a homogeneous sample under loading, the relation between dσ, dϵ and χ, which describes the incremental mechanical properties of this sample, must remain invariant.

However, let us now consider a deformed anisotropic sample of a geomaterial and carry out an experiment which consists of rotating the principal axes of incremental loading with respect to the sample. We do not obtain the same incremental response merely rotated in the same manner.

So nature requires us to consider functions **G** anisotropic with respect to dσ and fixed parameters χ. This anisotropy is directly linked to the geometrical meso-structure of the material, which is gradually modified by the strain (particularly irreversible) history. We have seen that this history can be characterized by scalar and tensorial state parameters χ.

In simple cases this anisotropy is a priori directly imposed by the choice of these state parameters. If we consider, for example, only scalar memory parameters (such as void ratio) defined independently of any frame, owing to the objectivity principle the function **G** will be an isotropic function, which is not supported by experiments.

If we add to these scalar memory parameters one single tensor variable (such as the stress tensor), it is easy to see that **G** will be an orthotropic function of dσ, the axes of orthotropy being identified with the principal axes of stress. In this case, it means that **G** is invariant by symmetry with respect to any plane containing two principal stress directions.

In the more general case of state variables with at least two non-commutating tensorial variables of second order, anisotropy is a priori anything whatever. Orthotropy then becomes a constitutive assumption, which must be considered as an approximation to the real behaviour of the material for a classes of loading in which stress and strain principal axes rotate.

Having described the three main properties of **G**, we will now focus on the first one (the homogeneity of degree 1) to see the mathematical consequences of such property.

Let us recall for this purpose Euler's identity for homogeneous regular functions of degree 1 by writing it for a function of two variables, for example, as:

$$\forall\, x, y \in \mathbb{R} \times \mathbb{R} : f(x,y) = x\frac{\partial f}{\partial x} + y\frac{\partial f}{\partial y}$$

where partial derivatives $\partial f/\partial x$ and $\partial f/\partial y$ are homogeneous functions of degree 0. Therefore the six functions G_α, which are homogeneous functions of degree 1 of the six variables $d\sigma_\beta$, satisfy the identity,

$$G_\alpha(d\sigma_\beta) = \frac{\partial G_\alpha}{\partial(d\sigma_\beta)}d\sigma_\beta, \quad \alpha, \beta \in \{1, 2, \cdots, 6\}^2$$

with summation on the repeated index β. If we note,

$$M_{\alpha\beta}(d\sigma_\beta) = \frac{\partial G_\alpha}{\partial(d\sigma_\beta)}, \quad \alpha, \beta \in \{1, 2, \cdots, 6\}^2$$

we obtain the following constitutive relation,

$$d\epsilon_\alpha = M_{\alpha\beta}(d\sigma_\gamma)d\sigma_\gamma, \quad \alpha, \beta, \gamma \in \{1, 2, \cdots, 6\}^3$$

where the 6×6 functions $M_{\alpha\beta}$ are homogeneous of degree 0 of the six variables $d\sigma_\gamma$. Therefore they are in an equivalent manner functions only of the direction of $d\sigma$, characterized by unit vector $\underline{u}$:

$$u_\gamma = \frac{d\sigma_\gamma}{\|d\sigma\|}, \quad \gamma \in \{1, 2, \cdots, 6\}$$

with

$$\|d\sigma\| = \sqrt{d\sigma_{ij}d\sigma_{ij}} = \sqrt{d\sigma_\alpha d\sigma_\alpha}, \quad i, j \in \{1, 2, 3\}^2, \quad \alpha \in \{1, 2, \cdots, 6\}$$

finally we obtain $(\alpha, \beta, \gamma \in \{1, 2, \cdots, 6\}^3)$

$$d\epsilon_\alpha = M_{\alpha\beta}(u_\gamma)d\sigma_\gamma \tag{3.3}$$

Equation (3.3) is the general expression for all rate independent constitutive relations. The existence of an elasto-plastic constitutive tensor $\mathbf{M}$ has been proven, as well as its essential dependency of state variable χ and the direction of $d\sigma$

This constitutive matrix $\mathbf{M}$, which then can be called 'tangent constitutive matrix', should be utilized to solve bifurcation problems by strain localization into shear bands (Darve, 1984) and also preferably used in finite element computations. This matrix in fact describes, at best, the material behaviour round-about a given incremental loading direction.

It is clear that a given constitutive relation will be characterized by only one tangent constitutive tensor after its definition has been given. But from this tangent tensor it is possible to build an infinite number of other constitutive tensors, which can be called 'secant constitutive tensors' and which describe exactly the same constitutive behaviour.

The relation (3.3) will now, allow us to propose a classification of all the existing rate independent constitutive relations with respect to their intrinsic structure.

4 Main classes of rate independent constitutive relations

The incremental non-linearity, which is also that of plastic irreversibilities, results, as we have just seen, from the constitutive tensor $\mathbf{M}$ varying with the direction of the incremental loading. The mode of the chosen directional variation is very much influenced by the constitutive model used. It is precisely this mode of description of incremental non-linearity that we are going to take as a guide for this review of main classes of existing constitutive relations.

First of all we need to define the notion of 'tensorial zone' (Darve and Labanieh, 1982). We will call a tensorial zone any domain in the incremental loading space on which the restriction of $\mathbf{G}$ is a linear function. In other words the relation between $\mathbf{d\epsilon}$ and $\mathbf{d\sigma}$ in a given tensorial zone is incrementally linear.

If we denote Z the tensorial zone being considered, the definition implies:

$$\forall \boldsymbol{u} \in Z : \mathbf{M}(\boldsymbol{u}) \equiv \mathbf{M}^z$$

In zone Z, the constitutive relation is characterized by a unique tensor $\mathbf{M}^z$.

If $\boldsymbol{u}$ belongs to Z, $\lambda \boldsymbol{u}$ belongs also to Z for all real positive values λ. Therefore a zone is defined by a set of half-infinite straight lines, whose apex is the same and is at the origin of the incremental loading space. Tensorial zones then comprise adjacent hypercones, whose common apex is this origin.

What does the constitutive relation become on the common boundary of two (or several) adjacent tensorial zones?

If $\mathbf{M}^{z_1}$ and $\mathbf{M}^{z_2}$ are constitutive tensors attached respectively to tensorial zones Z_1 and Z_2, we must obviously satisfy the continuity requirement of the response to loading directions $\boldsymbol{u}$:

$$\forall \boldsymbol{u} \in Z_1 \cap Z_2 : \mathbf{M}^{z_1} \boldsymbol{u} \equiv \mathbf{M}^{z_2} \boldsymbol{u}$$

or equivalently

$$\forall \boldsymbol{u} \in Z_1 \cap Z_2 : (\mathbf{M}^{z_1} - \mathbf{M}^{z_2}) \boldsymbol{u} \equiv 0 \tag{4.1}$$

Relation (4.1) can be called a 'continuity condition' for zone changes. This condition forbids, in particular, an arbitrary choice of constitutive tensors in two adjacent tensorial zones.

We will see further that conventional elastoplastic relations satisfy this condition by means of a class of neutral loading paths, while other relations may not satisfy it. Gudehus (1979) studied in this way the continuity of several constitutive relations for axisymmetrical loading paths with fixed principal axes, as did Di Benedetto and Darve (1983) with rotating principal axes.

The "response-envelopes", as proposed by Gudehus (1979), constitute geometrical diagrams which characterize completely a constitutive relation at a given stress strain state after a given strain history. At this state, all the the incremental loading, having the same norm but oriented in all directions, are considered and all the incremental responses are plotted. the extremities of the response vectors form a surface which is called "response-envelope". Some examples will be given later in this chapter.

This criterion of the number of tensorial zones that we have chosen to classify the main classes of non-viscous constitutive relations, since this number characterizes the structure of the relation. In fact the directional dependency of the constitutive tensor **M** with respect to the incremental loading can be described either in a discontinuous manner, by a finite number of different tensors related to the same number of tensorial zones (we will call such relations 'incrementally piece wise linear'), or in a continuous manner by a continuous variation of the constitutive tensor with the direction of the incremental loading (relations called 'incrementally non-linear').

4.1 Constitutive relations with one tensorial zone

The first class of relations that we are going to look at is related to the simplest assumption that only one tensorial zone exists. Therefore

$$\forall \boldsymbol{u} : \mathbf{M}(\boldsymbol{u}) \equiv \mathbf{M}$$

We have here the class of elastic laws, since there is a unique linear relation between dϵ and dσ (incremental linearity):

$$\mathrm{d}\epsilon_\alpha = M_{\alpha\beta}\mathrm{d}\sigma_\beta \quad \alpha, \beta \in \{1, 2, \cdots, 6\}^2$$

These elastic laws may be isotropic or anisotropic, linear or non-linear (global non-linearity). The isotropic and linear case corresponds to the well known Hooke's law with its two constitutive constants: Young's modulus and Poisson's ratio.

4.2 Constitutive relations with two tensorial zones

We now find laws with two tensorial zones, which can then be called 'loading zones' and 'unloading zones'. These two tensorial zones are separated by a hyperplane in dσ space or in dϵ space.

These laws may be either of hypoelastic type with a unique loading-unloading criterion (for example, models proposed by Guélin (1980)) or of elastoplastic type with one regular plastic potential (for example, models proposed by Mroz (1967); Prevost (1978), or Dafalias and Hermann (1980)).

The model developed by Duncan and Chang (1970) is a non-linear isotropic hypoelastic model with a specific loading-unloading criterion. It is easy to verify that such a model cannot be continuous at the frontier between the two zones; we only mention it here as a reminder.

The classical elastoplastic relations utilize an elastic matrix $\mathbf{M}^e$, related to the unloading zone, and an elastoplastic matrix $\mathbf{M}^{ep}$, related to the loading zone.

Let us study the continuity condition for this last case. The assumption of the additive decomposition of the incremental strain into its elastic part (reversible) and its plastic part (irreversible) allows us to write:

$$\mathrm{d}\epsilon = \mathrm{d}\epsilon^e + \mathrm{d}\epsilon^p$$

The plastic incremental strain $\mathrm{d}\epsilon^p$ is given by a flow rule which is often specified in terms of a plastic potential g:

$$\mathrm{d}\epsilon^p = \frac{\partial g}{\partial \sigma}\,\mathrm{d}\lambda$$

where $\mathrm{d}\lambda$ is called the plastic multiplier. Let us recall the continuity condition (4.1) at the interface of the elastic zone and the elastoplastic zone, which may be written here as:

$$(\mathbf{M}^e - \mathbf{M}^{ep})\,\boldsymbol{u} = 0$$

or,

$$\mathbf{M}^e\mathrm{d}\sigma - \mathbf{M}^{ep}\mathrm{d}\sigma = \mathrm{d}\epsilon^e - \mathrm{d}\epsilon = \mathrm{d}\epsilon^p = 0$$

This identity must be satisfied for any $\mathrm{d}\sigma$ belonging to the hyperplane, which is the frontier between the loading and unloading zones.

The loading condition is obtained by writing that the incremental stress is directed outwards from the elastic limit $f(\sigma) = 0$, defined by a yield function $f(\sigma, \chi)$, where χ is the set of internal variables. This function governs the onset of plastic deformations. It follows that:

$$\mathrm{d}\sigma \cdot \frac{\partial f}{\partial \sigma} > 0$$

In the same manner, the unloading condition is given by,

$$\mathrm{d}\sigma \cdot \frac{\partial f}{\partial \sigma} < 0$$

The equation of the hyperplane, the frontier between the two zones, in $\mathrm{d}\sigma$ space, is then given by:

$$\mathrm{d}\sigma \cdot \frac{\partial f}{\partial \sigma} = 0$$

An incremental path that satisfies this condition is called 'neutral loading'. The continuity condition $\mathrm{d}\epsilon^p = 0$ must then be satisfied by all neutral loadings;

$$\mathrm{d}\sigma \cdot \frac{\partial f}{\partial \sigma} = 0$$

By developing the consistency condition $\mathrm{d}f=0$, which means that the elastic limit is displaced by the stress point for loading (or in other words that the stress state must remain on the yield surface), we can demonstrate that $\mathrm{d}\lambda$ is proportional to $\mathrm{d}\sigma \cdot (\partial f/\partial \sigma)$. So the continuity condition is always fulfilled owing to the existence of neutral loading.

This large class of elastoplastic constitutive relations with a unique plastic potential is itself divided into several subclasses following the assumptions which allow precise expressions of a plastic potential.

First of all, this plastic potential $g(\sigma)$ may or may not coincide with the elastic limit surface $f(\sigma)$. Where there is such a coincidence, the elastoplastic material is called 'associated', otherwise it is 'non-associated'. From experiments, geomaterials have to be considered as non-associated, except for saturated undrained clays studied under total stresses.

This difference between associated and non-associated flow rules has a major consequence in terms of stability and bifurcation analysis. This will be developed in chapter on "Continuous and discrete modelling of failure in geomechanics".

The plastic potential may also vary with the strain history and be dependent on state variables (called also 'hardening parameters') which characterize the memory of geomaterial (see chapter by R. Nova on "Developments of elasto-plastic strainhardening models for soil behavior").

In damage theory, equivalent notions are introduced. The "elastic limit" (or "yield surface" or "loading surface") is replaced by the surface where damage appears. This surface is used to define a loading-unloading criterion. The differential equation giving the damage evolution corresponds to the elasto-plastic "flow-rule". The "continuity condition" is fulfilled through a consistency relation (see the chapter by G. Pijaudier Cabot on "Continuum damage modelling in geomechanics").

4.3   Constitutive relations with four tensorial zones

To this class belong the hypoelastic constitutive relations with two loading-unloading criteria proposed by Davis and Mullenger (1978) and elastoplastic models with a double plastic potential (Lade (1977); Loret (1981); Nova and Wood (1979(@) and Vermeer (1978), for example). This means that, after a given strain history, the direction of the plastic incremental strain vector is no longer defined in a unique way by the normal to a unique plastic potential (eventually a function of the history) but may coincide with two normals to two different potentials following the direction of the present incremental stress vector.

With respect to each of these two plastic mechanisms, the incremental stress may be considered to be a loading or an unloading. Therefore we obtain the following four possibilities:

1. $(\partial f_1/\partial \sigma) \cdot \mathrm{d}\sigma > 0$: loading for criterion 1; p_1
2. $(\partial f_1/\partial \sigma) \cdot \mathrm{d}\sigma < 0$: unloading for criterion 1; e_1
3. $(\partial f_2/\partial \sigma) \cdot \mathrm{d}\sigma > 0$: loading for criterion 2; p_2
4. $(\partial f_2/\partial \sigma) \cdot \mathrm{d}\sigma < 0$: unloading for criterion 2; e_2

Finally, four different constitutive tensors can be obtained:

$$\mathbf{M}^{e_1 e_2}, \mathbf{M}^{p_1 e_2}, \mathbf{M}^{e_1 p_2}, \mathbf{M}^{p_1 p_2}$$

Each of them is associated with a certain tensorial zone in $\mathrm{d}\sigma$ space. The various continuity conditions can be satisfied by applying both of the consistency equations $\mathrm{d}f_1 = 0$ and $\mathrm{d}f_2 = 0$, which enable us to obtain two arbitrary plastic multipliers $\mathrm{d}\lambda_1$ and $\mathrm{d}\lambda_2$ proportional to $(\partial f_1/\partial \sigma) \cdot \mathrm{d}\sigma$ and to $(\partial f_2/\partial \sigma) \cdot \mathrm{d}\sigma$ respectively (directly for independent plastic mechanisms or by solving a system of 2 equations for dependent mechanisms). The neutral loading paths then belong to two hyperplanes whose equations are given by,

$$(\partial f_1/\partial \sigma) \cdot \mathrm{d}\sigma = 0$$

$$(\partial f_2/\partial \sigma) \cdot \mathrm{d}\sigma = 0$$

These two hyperplanes enable us to define four tensorial zones in dσ space. The reasoning and presentation will be the same for hypoelastic models with two loading-unloading criteria.

4.4 Constitutive relations with eight tensorial zones

By further increasing the number of tensorial zones we now find incrementally octo-linear relations with eight tensorial zones.

Two models belong to this class: the elastoplastic model with three plastic potentials proposed by Aubry et al. (1991) and the incremental octolinear relation developed by Darve and Labanieh (1982). The basis of the law with three plastic potentials is essentially the same as for the models with a double potential, which we have presented above. Concerning the incremental octolinear relation, the basis of this model derives from the assumption of eight tensorial zones in dσ space. It satisfies all the different continuity conditions for fixed principal stress and fixed principal strain axes (Gudehus, 1979) and has been generalized (Darve, 1980, 1990) to arrive at an 'incrementally non-linear of second order' constitutive relation, which will be presented in section 5.

4.5 Constitutive relations with an infinite number of tensorial zones

We could say that they have an infinite number of tensorial zones, since each direction of dσ space is linked to a given tangent constitutive tensor which varies in a continuous manner with this direction. They may be of three different types. The endochronic models (Valanis, 1971; Bazant, 1978) take into account non-linearity by introducing an intrinsic 'time' which is always increasing. The hypoplastic models by Kolymbas (1977) or Chambon et al. (1994) assume a priori a direct non-linear relation between dϵ and dσ with also a scalar variable always positive. Thirdly we find models based on a non-linear interpolation between given constitutive responses, the non-linearity being linked to the kind of interpolation rule used (Darve, 1980). Another model has been proposed by Dafalias (1986) in the framework of his 'bounding surface' theory of plasticity.

The interest in incrementally non-linear constitutive relations is based mainly on the fact that it is not necessary to postulate the existence of either an elastic limit surface (since any purely elastic domain has disappeared) and the linked loading-unloading condition or a plastic potential. In fact the non-linearity of the incremental relation allows a direct description of the different behaviours in 'loading' and in 'unloading'. For this reason the incrementally non-linear relations may be closer to the physics that governs deformation of geomaterials.

NTZ$^{(*)}$	Constitutive tensors	classes of constitutive relations
1	**M**	Elasticity Hyperelasticity Hypoelasticity in its strict sense
2	$\begin{vmatrix} \mathbf{M}^e \\ \mathbf{M}^{ep} \end{vmatrix}$ or $\begin{vmatrix} \mathbf{M}^+ \\ \mathbf{M}^- \end{vmatrix}$	Elastoplasticity with 1 plastic potential or Hypoelasticity with 1 load-unload crit.
4	$\begin{vmatrix} \mathbf{M}^{e_1 e_2} \\ \mathbf{M}^{p_1 e_2} \\ \mathbf{M}^{e_1 p_2} \\ \mathbf{M}^{p_1 p_2} \end{vmatrix}$	Elastoplasticity with 2 plastic potentials
4	$\begin{vmatrix} \mathbf{M}^{++} \\ \mathbf{M}^{-+} \\ \mathbf{M}^{+-} \\ \mathbf{M}^{--} \end{vmatrix}$	Hypoelasticity with 2 load-unload crit.
8		Elastoplasticity with 3 plastic potentials Octolinear incremental relation
∞	$\mathbf{M}(\mathrm{d}\sigma/\|\mathrm{d}\sigma\|)$	Endochronic models Hypoplasticity Incrementally non-linear relations

Table 1 : The main classes of non-viscous constitutive relations.   (*)NTZ : Number of Tensorial Zones.

5 Incrementally non-linear constitutive relations of second order

Let us recall relation (3.3):

$$\mathrm{d}\epsilon_\alpha = M_{\alpha\beta}(u_\gamma)\mathrm{d}\sigma_\gamma \tag{5.1}$$

with

$$u_\gamma = \frac{\mathrm{d}\sigma_\gamma}{\|\mathrm{d}\sigma\|} \quad \gamma \in \{1, 2, \cdots, 6\}$$

We consider polynomial series expansions for 36 functions $M_{\alpha\beta}$ of 6 variables u_γ:

$$M_{\alpha\beta}(u_\gamma) = M_{\alpha\beta}^1 + M_{\alpha\beta\gamma}^2 u_\gamma + M_{\alpha\beta\gamma\delta}^2 u_\gamma u_\delta + \cdots \tag{5.2}$$

From equation (5.1) and equation (5.2) it follows that:

$$\mathrm{d}\epsilon_\alpha = M_{\alpha\beta}^1 \mathrm{d}\sigma_\beta + \frac{1}{\|\mathrm{d}\sigma\|} M_{\alpha\beta\gamma}^2 \mathrm{d}\sigma_\beta \mathrm{d}\sigma_\gamma + \cdots \tag{5.3}$$

The first term of equation (5.3) describes the elastic behaviors and both the first terms are incrementally non-linear constitutive relations of second order.

In this paper we will consider only loading paths which are defined in fixed incremental stress-strain principal axes. In such a case the expression of the model is the following (for a more general presentation see Darve and Dendani (1989) or Darve (1990)) :

$$\begin{pmatrix} d\epsilon_1 \\ d\epsilon_2 \\ d\epsilon_3 \end{pmatrix} = \frac{1}{2} \left[\mathbf{N}^+ + \mathbf{N}^- \right] \cdot \begin{pmatrix} d\sigma_1 \\ d\sigma_2 \\ d\sigma_3 \end{pmatrix} + \frac{1}{2 \, \|d\sigma\|} \left[\mathbf{N}^+ - \mathbf{N}^- \right] \cdot \begin{pmatrix} d\sigma_1^2 \\ d\sigma_2^2 \\ d\sigma_3^2 \end{pmatrix} \qquad (5.4)$$

with

$$\|d\sigma\| = \sqrt{d\sigma_i \, d\sigma_i}, \quad i \in \{1, 2, 3\}$$

The two matrices $\mathbf{N}^+$ and $\mathbf{N}^-$ have the following form (for more details see Darve and Dendani (1989) or Darve (1990)):

$$\mathbf{N}^+ = \begin{bmatrix} \dfrac{1}{E_1^+} & -\dfrac{\nu_{12}^+}{E_2^+} & -\dfrac{\nu_{13}^+}{E_3^+} \\ -\dfrac{\nu_{21}^+}{E_1^+} & \dfrac{1}{E_2^+} & -\dfrac{\nu_{23}^+}{E_3^+} \\ -\dfrac{\nu_{31}^+}{E_1^+} & -\dfrac{\nu_{32}^+}{E_2^+} & \dfrac{1}{E_3^+} \end{bmatrix} \quad \mathbf{N}^- = \begin{bmatrix} \dfrac{1}{E_1^-} & -\dfrac{\nu_{12}^-}{E_2^-} & -\dfrac{\nu_{13}^-}{E_3^-} \\ -\dfrac{\nu_{21}^-}{E_1^-} & \dfrac{1}{E_2^-} & -\dfrac{\nu_{23}^-}{E_3^-} \\ -\dfrac{\nu_{31}^-}{E_1^-} & \dfrac{\nu_{32}^-}{E_2^-} & \dfrac{1}{E_3^-} \end{bmatrix} \qquad (5.5)$$

Where E_i and ν_{ij} are, respectively, tangent moduli and tangent Poisson's ratio on 'generalized triaxial paths'. These paths correspond to the conventional triaxial paths (the two constant lateral stresses are equal), but on 'generalized triaxial paths' the lateral stresses are fixed but independently. The superscript (+) means 'compression' in the axial direction for these paths and the superscript (-) means 'extension'.

The behavior of the studied material for these specific paths is assumed to be given by laboratory triaxial tests and described by analytical expressions. It is in these expressions that the constitutive constants appear as well as the state variables (stress tensor and void ratio) and the memory parameters (which are of two types of discontinuous and continuous nature). Thus $\mathbf{N}^+$ and $\mathbf{N}^-$ are depending on state variables and memory parameters.

It is clear from equation (5.4) that this relation is homogeneous of degree one with respect to $d\sigma$. Thus, it describes a rate independent behaviour. Equation (5.4) is also non linear in $d\sigma$, it means that it can describe plastic irreversible strains : for an elementary stress cycle $(d\sigma, -d\sigma)$ the irreversible strain is equal to:

$$\begin{pmatrix} d\epsilon_1 \\ d\epsilon_2 \\ d\epsilon_3 \end{pmatrix} = \frac{1}{\|d\sigma\|} \left[\mathbf{N}^+ - \mathbf{N}^- \right] \cdot \begin{pmatrix} d\sigma_1^2 \\ d\sigma_2^2 \\ d\sigma_3^2 \end{pmatrix}$$

The recoverable strain has not an elastic nature since it is equal to:

$$\begin{pmatrix} d\epsilon_1 \\ d\epsilon_2 \\ d\epsilon_3 \end{pmatrix} = \frac{1}{2} \left[\mathbf{N}^+ + \mathbf{N}^- \right] \cdot \begin{pmatrix} d\sigma_1 \\ d\sigma_2 \\ d\sigma_3 \end{pmatrix} - \frac{1}{2 \, \|d\sigma\|} \left[\mathbf{N}^+ - \mathbf{N}^- \right] \cdot \begin{pmatrix} d\sigma_1^2 \\ d\sigma_2^2 \\ d\sigma_3^2 \end{pmatrix}$$

It means that it is not possible to decompose in an additive manner the strain into an elastic part and a plastic one. Elasticity and plasticity are intrinsically mixed into the constitutive relation.

Now we will see how equation (5.4) is degenerating in the one dimensional case and in the cases of an elastic material and a perfectly plastic one.

For the one dimensional case (5.4) degenerates into the following scalar expression:

$$d\epsilon = \frac{1}{2}\left(\frac{1}{E^+} + \frac{1}{E^-}\right) d\sigma + \frac{1}{2}\left(\frac{1}{E^+} - \frac{1}{E^-}\right)|d\sigma| \tag{5.6}$$

It is easy to verify that relation (5.6) is able to describe any rate independent one dimensional behaviour with one single expression (5.6). If one wants to interpret (5.4) with conventional elasto-plastic concepts, it is necessary to introduce a loading-unloading criterion and we obtain:

$$\begin{cases} d\sigma \geq 0 \quad : \quad d\epsilon = \dfrac{1}{E^+}d\sigma \qquad \text{(loading)} \\[2ex] d\sigma \leq 0 \quad : \quad d\epsilon = \dfrac{1}{E^-}d\sigma \qquad \text{(unloading)} \end{cases}$$

If the behaviour of the material is elastic, we obtain the same behaviour for 'compressions' and 'extensions' (as defined previously) which implies:

$$\mathbf{N}^+ = \mathbf{N}^-$$

Relation (5.4) degenerates into the following :

$$\begin{pmatrix} d\epsilon_1 \\ d\epsilon_2 \\ d\epsilon_3 \end{pmatrix} = [\mathbf{N}] \cdot \begin{pmatrix} d\sigma_1 \\ d\sigma_2 \\ d\sigma_3 \end{pmatrix} \tag{5.7}$$

Equation (5.7) is exactly the general expression of anisotropic non-linear elasticity in identical incremental stress-strain principal axes.

Finally, it is also possible to exhibit the equations of perfect plasticity as a degenerating case. Let us write relation (5.4) in the following form:

$$\begin{pmatrix} d\epsilon_1 \\ d\epsilon_2 \\ d\epsilon_3 \end{pmatrix} = [\mathbf{N}(u)] \cdot \begin{pmatrix} d\sigma_1 \\ d\sigma_2 \\ d\sigma_3 \end{pmatrix} \tag{5.8}$$

By inverting (5.8) we obtain:

$$d\sigma = \mathbf{N}^{-1}(u)\, d\epsilon$$

Conditions for perfect plasticity imply:

$$\begin{cases} d\sigma \quad = \quad \quad \quad 0 \\ \|d\epsilon\| \quad : \quad \quad \text{undetermined} \end{cases}$$

It follows therefore that:

$$\det \mathbf{N}^{-1}(u) = 0 \tag{5.9}$$

which represents the plastic condition.

$$\mathbf{N}^{-1}(u)\, d\epsilon = 0 \tag{5.10}$$

is a generalized flow rule, since the solution of equation (5.10) while equation (5.9) is verified gives the direction of $d\epsilon$ but not its intensity. Condition (5.10) is directionally dependent with respect to $d\sigma$ which reflects the fact that the yield surface is locally deformed into a vertex. This question is discussed more in detail in section 8.

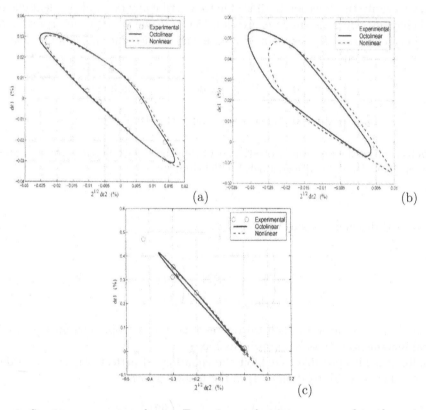

Figure 1. Strain response-envelopes. Experimental points compared to the octo-linear model (continuous lines) and to the non-linear model (discontinuous lines) at 3 different stress states: an isotropic stress state of 100 kPa (a), and 2 deviatoric stress states ($\sigma_1 = 200$ kPa, $\sigma_2 = 100$ kPa (b); $\sigma_1 = 400$ kPa, $\sigma_2 = 100$ kPa (c))

The incrementally 'octo-linear' model is given by the following relation:

$$\begin{pmatrix} d\epsilon_1 \\ d\epsilon_2 \\ d\epsilon_3 \end{pmatrix} = \frac{1}{2}\left[\mathbf{N}^+ + \mathbf{N}^-\right] \cdot \begin{pmatrix} d\sigma_1 \\ d\sigma_2 \\ d\sigma_3 \end{pmatrix} + \frac{1}{2}\left[\mathbf{N}^+ - \mathbf{N}^-\right] \cdot \begin{pmatrix} |d\sigma_1| \\ |d\sigma_2| \\ |d\sigma_3| \end{pmatrix} \tag{5.11}$$

The incremental structure of this model is the same as an elastoplastic relation with 3 plastic potentials. The constitutive constants for both the incremental non-linear model and octo-linear model are the same.

Both these constitutive relations (the non-linear one and the octo-linear model) can be viewed as interpolations between the generalized triaxial responses. The octo-linear model corresponds to a linear interpolation rule, while the incrementally non-linear model describes a quadratic non-linear interpolation. Indeed it is possible to define a family of incrementally non-linear relations depending on a positive scalar ρ as proposed by (Darve and Laouafa, 2000):

$$
\begin{pmatrix} d\epsilon_1 \\ d\epsilon_2 \\ d\epsilon_3 \end{pmatrix} = \frac{1}{2} \left[\mathbf{N}^+ + \mathbf{N}^- \right] \cdot \begin{pmatrix} d\sigma_1 \\ d\sigma_2 \\ d\sigma_3 \end{pmatrix} + \frac{\sqrt{1+\rho}}{2} \left[\mathbf{N}^+ - \mathbf{N}^- \right] \cdot \begin{pmatrix} \dfrac{d\sigma_1^2}{\sqrt{d\sigma_1^2 + \rho \|d\sigma\|^2}} \\ \dfrac{d\sigma_2^2}{\sqrt{d\sigma_2^2 + \rho \|d\sigma\|^2}} \\ \dfrac{d\sigma_3^2}{\sqrt{d\sigma_3^2 + \rho \|d\sigma\|^2}} \end{pmatrix}
$$

For $\rho = 0$, this relation coincides with the octo-linear model, while for $\rho \to +\infty$ it converges toward the incrementally non-linear relation of second order.

According to the construction of "response-envelops" as presented in section 4. (which characterize completely a constitutive relation at a given stress-strain state after a given strain history), octo-linear and non-linear models have been compared to experimental results (Royis and Doanh, 1998), for axisymmetric conditions, Figure 1 shows such a comparison between experimental points and theoretical curves.

It is interesting to note the degeneration of the response-envelopes to a straight line close to the plastic limit condition. This straight line corresponds to a flow rule in axisymmetric conditions (i.e. same incremental strain direction for any incremental stress loading direction).

6 Application to the analysis of the yield surfaces

In the elastoplastic theory yield surfaces are classically defined as an elastic limit surface in the six-dimensional physical stress space. All stress states inside the yield surface can be reached without any inelastic strain. Experimentally it has appeared that it was very difficult to detect the stress state, along a given stress path, where the first plastic strain is developing, because (particularly for geomaterials) the yield surface of materials before any loading is defined by a very small domain (Hicher, 1985) and the process of inducing plastic strains along a given loading path is gradual and continuous particularly for the first applied loading. For that many authors have chosen a strain intensity criterion in order to obtain an objective procedure to plot the graph of the yield surfaces and to compare them together upon rational bases.

The same difficulty is met with our model since plastic strains are always coexisting with elastic ones, even if for small strain $\mathbf{N}^+$ and $\mathbf{N}^-$ are very close and the behaviour thus rather elastic. The principle of the numerical procedure which is chosen in order to simulate the experimental procedure is depicted in Figures 2 and 3. After a given loading path (which is here in this example a stress isotropic loading) at a given stress state (here $\sigma_1 = \sigma_2 = \sigma_3 = 100$kPa), we consider the family of axisymmetrical stress paths ($\sigma_2 = \sigma_3$) which are directed in all the axisymmetrical stress directions and we stop the computation in the successive directions when the strain intensity criterion is reached : $\|\Delta\epsilon\| = \|\Delta\epsilon\|_{limit}$. Here,

$$\|\Delta\epsilon\|_{limit} = \sqrt{(\Delta\epsilon_1)^2 + 2(\Delta\epsilon_3)^2} = 3 \ 10^{-4}$$

The stress paths are radial and rectilinear, while the strain responses are obviously not necessarily rectilinear even if the strain levels are rather small. The integration of the constitutive model on these rectilinear stress paths is stopped when the strain intensity criterion is fulfilled. 360 radial stress paths are considered, each one differing from the previous one by one degree. The figure obtained in the associated strain bissector plane is a circle, because of this same limit value, while in the stress bissector plane we obtain numerically the trace of the 'yield surface'. The yield surface is clearly more elongated in the isotropic direction since the stiffness is the largest in this direction. The sign of angles is kept into the transformation but not their values as it is possible to notice on Figures 2 and 3. By using this tool it is interesting to investigate the evolution of yield

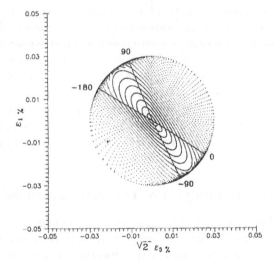

Figure 2. Principle of numerical simulation of the experimental procedure to obtain yield surfaces (in the strain space). The strain intensity criterion is equal to: $\|\Delta\epsilon\| = 3 \ 10^{-4}$, (Darve et al., 1995)

surfaces with the stress-strain history, particularly in order to exhibit the main hardening mechanism: kinematic, isotropic or rotational. Figures 4 and 5 depict examples of such results for the loose sand (Figure 4) and for the dense sand (Figure 5) in axisymmetrical cases. In these cases the numerical modeling involves an integration of the model along some proportional stress paths, which are represented by continuous black lines and correspond to constant lateral pressure paths and constant mean pressure paths, until some limit stress states noticed on the figures by the black points. From these stress states

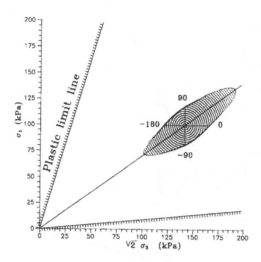

Figure 3. Principle of numerical simulation of the experimental procedure to obtain yield surfaces (in the stress space). The strain intensity criterion is equal to: $\|\Delta\epsilon\| = 3\ 10^{-4}$, (Darve et al., 1995)

reversals were simulated by the model until the stress states noticed by cross points. It was from these last stress states that the numerical procedure to plot the yield surface was applied.

The 'plastic limit lines', which are plotted on Figure 4 and Figure 5, correspond to the classical and usual Mohr-Coulomb plastic criterion. A plastic limit surface is introduced inside the constitutive model as a stress surface which bounds the stress states that can be reached by any loading path. When the stress state is approaching this limit surface, at least one of the tangent moduli E_i tends to zero. At this time no constitutive softening is introduced inside the model. It means that the experimental characterizing of the plastic limit surface necessitates tests maintaining homogeneous stress-strain field as long as possible in order to delay the strain localisation phenomenon.

Three main conclusions can be exhibited from both Figures 4 and 5:

1. the main mechanism which influences the evolution of the yield surfaces is a kinematic hardening,
2. there is a small isotropic hardening, but the other largely more important factor is the influence of the shape of the plastic limit surface which curves the yield surfaces when they are approaching this limit surface,
3. a loose sand has rounder yield surfaces than the dense one.

Figure 6 depicts yield surfaces in a deviatoric stress plane (the mean pressure is equal to 300 kPa) for the dense Hostun sand. The initial yield surface is rigorously a circle because of the initial overall isotropy. Each yield surface has been plotted after a stress loading path followed until the black points seen on the figure. The same method has been adopted as previously but without stress reversals. At the stress states (marked by the black points) 360 radial rectilinear stress paths are applied (each one differs from

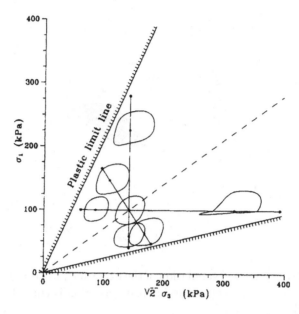

Figure 4. Numerical simulation of yield surfaces in a bissector stress plane for the loose Hostun sand. The material is loaded until the black points, then unloaded until the cross points before plotting the yield surface. (Darve et al., 1995)

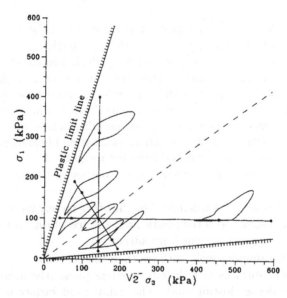

Figure 5. Numerical simulation of yield surfaces in a bissector stress plane for the dense Hostun sand. The material is loaded until the black points, then unloaded until the cross points before plotting the yield surface. (Darve et al., 1995)

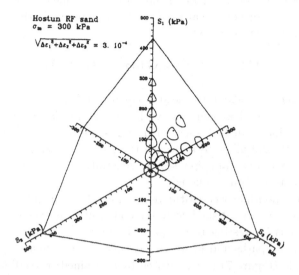

Figure 6. Numerical simulation of yield surfaces in a deviatoric stress plane for the dense Hostun sand. The material is loaded until the black points, before plotting the yield surface.(Darve et al., 1995)

the adjacent ones by one degree in direction). The loading is stopped when the strain intensity criterion is reached. As previously the main hardening is kinematic. A quite astonishing point is the fact that the form of the plastic limit surface influences very fast, from the isotropic state, the yield surfaces whose form reflects and anticipates in a quite faithful way the pattern of the limit surface.

7 Application to the principle of superposition for incremental loading

As recalled briefly in section 2., this 'principle' implies that the material response to a sum of incremental loading is equal to the sum of the associated incremental responses. This 'principle' can be considered as being formulated in a comparable way to Boltzmann's principle. While the last one is applied to a finite loading (that means a certain loading history), the first one considers only incremental loading. It can be said that the first one is a principle of superposition 'in the small', while Boltzmann's principle is a principle of superposition 'in the large'.

Boltzmann's principle implies that the behaviour is linear elastic or linear visco-elastic, since the constitutive functional must be linear. The principle of superposition for incremental loading implies that the constitutive incremental function is linear, which implies for the described behaviour to be non-linear elastic or non-linear visco-elastic.

If we consider only elasto-plastic strains and if we assume the general validity of the principle, the constitutive operator $\mathbf{G}$ (equation (3.1)) is linear and thus describes only elastic behaviour (linear or non-linear elasticity). More precisely the principle is verified strictly inside the same tensorial zone :

$$\mathbf{G}\left(d\sigma^1 + d\sigma^2\right) = \mathbf{G}\left(d\sigma^1\right) + \mathbf{G}\left(d\sigma^2\right),$$

if and only if $d\sigma^1$ and $d\sigma^2$ belong to the same tensorial zone.

Experimentally it is well known that with servo-controlled machines this principle is approximately verified. In fact such apparatuses are not able to follow exactly the required loading path and allow to approach the path considered by a sequence of approximations. The experimental responses are generally considered as satisfying as soon as the approximations are 'small' enough.

For incrementally non-linear constitutive relations the 'principle' of superposition for incremental loading is never satisfied since the constitutive incremental function is non-linear. Therefore it is interesting to compare the responses provided by the model for rectilinear proportional paths and for piece-wise linear approximated paths with bends.

The first example (Figure 7) is constituted by a 'drained triaxial' loading performed on the dense Hostun sand, which means path with a constant lateral pressure. This path is decomposed into several pieces divided into two parts: one is an incremental isotropic loading ($d\sigma_1 = d\sigma_3$) and the other is a constant mean pressure loading ($d\sigma_1 + 2d\sigma_3 = 0$). The length of the 'steps' varies and five cases have been considered : 1, 2, 4, 8 and 16 kPa. Two main conclusions can be exhibited from results depicted on Figure 7:

1. when the length of the 'steps' decreases, the response diagrams converge to certain asymptotical behaviour,

2. these converged response curves are not confounded with the responses to the proportional loading, but remain satisfactorily close.

It is also interesting to notice that the total length of the rectilinear proportional path and the decomposed paths (even for 'small' lengths of 'steps') are completely different (the ratio of both the lengths is equal to $\sqrt{2}$). The same conclusions can be make by analyzing the results depicted in Figure 8 in the case of the loose Hostun sand.   The same method is also considered in order to investigate the behaviour under an 'undrained' loading path (in fact isochoric loading) performed on loose Hostun sand. The following decomposition has been adopted in axisymmetric conditions:

$$\left\{ \begin{array}{lll} \dfrac{q}{p'} & = & \text{constant and} \quad d\epsilon_1 < 0 \\ p' & = & \text{constant and} \quad d\epsilon_1 > 0 \end{array} \right\}$$

with deviatoric stress $q = \sigma_1 - \sigma_3$ and the mean effective pressure $p' = (\sigma_1' + 2\,\sigma_3')/3$.

As the first path ($q/p' = cst$) is dilatant and the second one contractant, it is thus possible to fulfill the isochoric condition:

$$\Delta\epsilon_1 + 2\,\Delta\epsilon_3 = 0$$

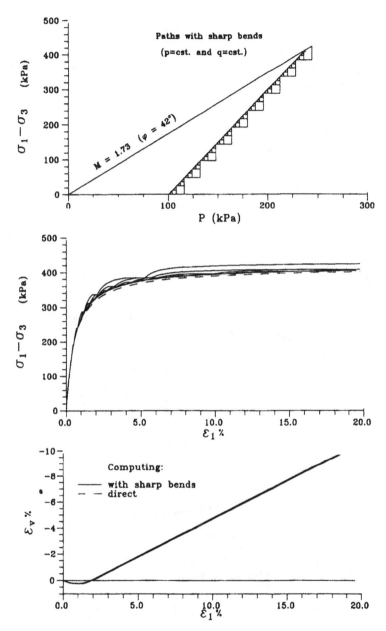

Figure 7. Comparison between numerical responses obtained from the incrementally non-linear model in the case of a rectilinear proportional loading (with a constant lateral pressure)- dash lines- and of 'staircase' paths - continuous lines. Dense Hostun sand (Darve et al., 1995).

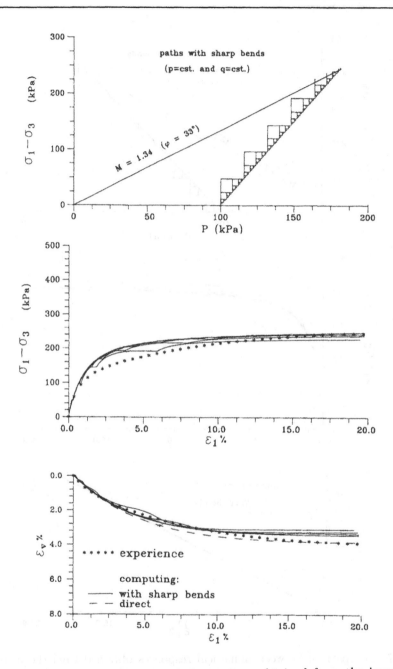

Figure 8. Comparison between numerical responses obtained from the incrementally non-linear model in the case of a rectilinear proportional loading (with a constant lateral pressure)- dash lines- and of 'staircase' paths - continuous lines. Loose Hostun sand (Darve et al., 1995).

The decomposition is depicted in Figure 9. The first path is applied until a given path length is reached:

$$\sqrt{(\Delta p)^2 + (\Delta q)^2} = 2\,\text{kPa}$$

The results are depicted in Figures 10 and 11 and show that there is not any significant influence of the kind of decomposition on the quantitative responses. Figure 11 allows us to verify that when the length of the sharp bends is decreasing the response curves converge and that the asymptotic diagram is rather close (but not coinciding) to the response issued from the rectilinear proportional isochoric strain path:

$$\begin{cases} d\epsilon_1 > 0 \\ d\epsilon_2 = d\epsilon_3 \\ d\epsilon_1 + 2\,d\epsilon_3 = 0 \end{cases}$$

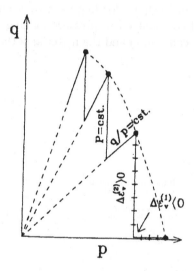

Figure 9. The considered decomposition of the loading path in the case of an isochoric loading

The four chosen lengths are equal to :

$$\sqrt{(\Delta p)^2 + (\Delta q)^2} = 1, 2, 4, 8\,\text{kPa}$$

Finally the computation is stopped only because of an artefact : for small levels of stress both the decomposition paths become contractant. Details can be found in Darve (1996).

8 Application to the analysis of flow rules

In section 6. the question of the existence of yield surfaces has been discussed through the use of the incrementally non-linear model and it has been shown that it was possible

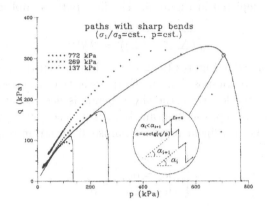

Figure 10. Simulation of an isochoric loading path by a path decomposed into radial unloading ($q/p' = cst$) and constant mean pressure loading ($p' = cst$). Three different initial isotropic stresses are considered and the material is the loose Hostun sand (Darve et al., 1995)

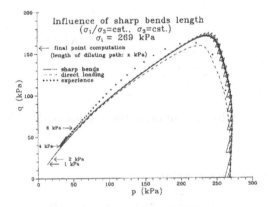

Figure 11. Study of the convergence of response curves when the decomposed paths length is decreasing from 8 to 1 kPa. The responses to the isochoric proportional strain path are plotted with dash lines (Darve et al., 1995)

to exhibit, as an approximation, some yield surfaces from a constitutive relation which does not consider them into its formulation. In the same spirit, the question of flow rule is discussed in this section.

Generally speaking, equation 3.3 can be inverted and it gives:

$$d\underline{\sigma} = \mathbf{M}^{-1}(\underline{u})d\underline{\epsilon}$$

with $\underline{u} = d\underline{\sigma}/\|d\underline{\sigma}\|$. The existence of limit stress states implies: $d\underline{\sigma} = \underline{0}$ with $\|d\underline{\epsilon}\|$ arbitrary large.

Thus:

$\det\mathbf{M}^{-1}(\underline{u}) = 0$ is a a plastic limit condition and $\det\mathbf{M}^{-1}(\underline{u})d\underline{\epsilon} = 0$ is a flow rule, which gives (when the limit stress state has been reached) the direction of $d\underline{\epsilon}$ but not its intensity.

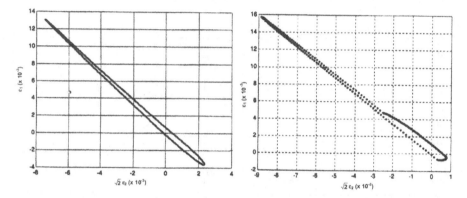

Figure 12. Flow rule in axisymmetric conditions for loose Hostun sand: degenerating response- envelops. Incrementally non-linear model on the left, octo-linear model on the right ($q = 1000$ kPa, $\sigma_3 = 500$ kPa, $\sqrt{\Delta\sigma_1^2 + 2\Delta\sigma_3^2} = 1$ kPa) (Fernandez, 2002)

It is important to note that this flow rule is singular since the direction of $d\underline{\epsilon}$ depends on $\underline{u}$ (i.e. the direction of $d\underline{\sigma}$). For the octo-linear model where the constitutive relation is incrementally piece-wise linear, the singularity is pyramidal (as pointed out by Hill (1967)). For the incrementally non-linear relation, not any particular shape can be expected: it is a general "vertex effect" (Hill, 1967).

Now a questions appears. From fig. 1, it seems that in axisymmetric conditions a flow rule is exhibited both by the octo-linear model and the non-linear relation.

Let us consider first the question from an analytical point of view with the octo-linear model. In axisymmetric conditions and fixed stress-strain principal axes, it comes with 5.5 notations:

$$\begin{pmatrix} d\epsilon_1 \\ \sqrt{2}d\epsilon_3 \end{pmatrix} = \begin{bmatrix} \frac{1}{E_1^{\pm}} & -\sqrt{2}\frac{V_3^{1\pm}}{E_3^{\pm}} \\ -\sqrt{2}\frac{V_1^{3\pm}}{E_1^{\pm}} & \frac{1-V_3^{3\pm}}{E_3^{\pm}} \end{bmatrix} \begin{pmatrix} d\sigma_1 \\ \sqrt{2}d\sigma_3 \end{pmatrix} \tag{8.1}$$

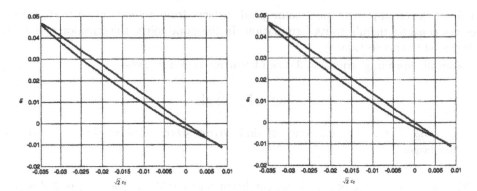

Figure 13. Flow rule in axisymmetric conditions for dense Hostun sand degenerating response- envelops. Incrementally non-linear model on the left, octo-linear model on the right ($q = 2000$ kPa, $\sigma_3 = 500$ kPa, $\sqrt{\Delta\sigma_1^2 + 2\Delta\sigma_3^2} = 1$ kPa) (Fernandez, 2002)

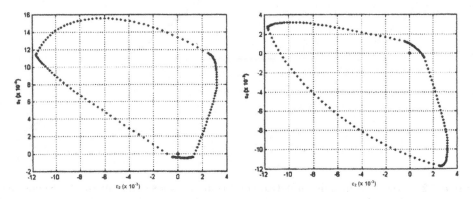

Figure 14. 3D response-envelops for loose Hostun sand. Octo-linear model ($q = 1000$ kPa, $\sigma_3 = 500$ kPa, $\sqrt{\Delta\sigma_1^2 + \Delta\sigma_2^2 + \Delta\sigma_3^2} = 1$ kPa) (Fernandez, 2002)

or by inverting relation 8.1:

$$\begin{pmatrix} d\sigma_1 \\ \sqrt{2}d\sigma_3 \end{pmatrix} = \frac{1}{1 - V_3^{3\pm} - 2V_1^{3\pm}V_3^{1\pm}} \begin{bmatrix} E_1^{\pm}\left(1 - V_3^{3\pm}\right) & \sqrt{2}E_1^{\pm}V_3^{1\pm} \\ \sqrt{2}E_3^{\pm}V_1^{3\pm} & E_3^{\pm} \end{bmatrix} \begin{pmatrix} d\epsilon_1 \\ \sqrt{2}d\epsilon_3 \end{pmatrix} \qquad (8.2)$$

The plastic limit condition is given by:

$$E_1^{\pm} E_3^{\pm} = 0 \qquad (8.3)$$

In the axisymmetric compression case, it comes:

$$E_1^+ = 0$$

and equation 8.2 implies at the limit stress state ($d\sigma_1 = d\sigma_3 = 0$):

$$V_1^{3+}d\epsilon_1 + d\epsilon_3 = 0 \qquad (8.4)$$

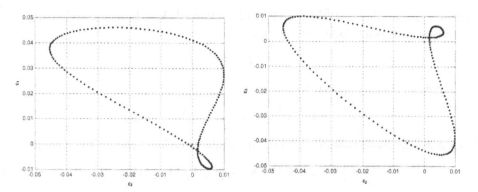

Figure 15. 3D response-envelops for dense Hostun sand. Incrementally non-linear model ($q = 2000$ kPa, $\sigma_3 = 500$ kPa, $\sqrt{\Delta\sigma_1^2 + \Delta\sigma_2^2 + \Delta\sigma_3^2} = 1$ kPa) (Fernandez, 2002)

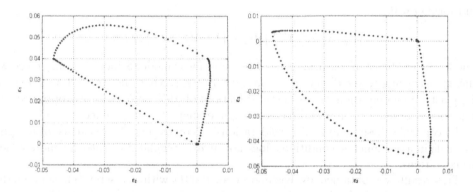

Figure 16. 3D response-envelops for dense Hostun sand. Octo-linear model ($q = 2000$ kPa, $\sigma_3 = 500$ kPa, $\sqrt{\Delta\sigma_1^2 + \Delta\sigma_2^2 + \Delta\sigma_3^2} = 1$ kPa)

which is precisely the equation of a flow rule. The same reasoning applies in axisymmetric extensions, where it comes:

$$E_3^- = 0 \qquad \text{(plastic limit condition)}$$

and:

$$(1 - V_3^{3-})d\epsilon_1 + 2V_3^{1-}d\epsilon_3 = 0 \qquad \text{(flow rule)}$$

It can be concluded that, in axisymmetric conditions, the octo-linear model implies the existence of a regular flow rule.

This conclusion is not true for plane strain condition ($d\epsilon_2 = 0$), where:

$$
\begin{pmatrix} d\sigma_1 \\ d\sigma_3 \end{pmatrix} = \frac{1}{d}
\begin{bmatrix}
E_1^\pm \left(1 - V_2^{3\pm}V_3^{2\pm}\right) & E_1^\pm \left(V_1^{3\pm} + V_1^{2\pm}V_2^{3\pm}\right) \\
E_3^\pm \left(V_3^{1\pm} + V_3^{2\pm}V_2^{1\pm}\right) & E_3^\pm \left(1 - V_1^{2\pm}V_2^{1\pm}\right)
\end{bmatrix}
\begin{pmatrix} d\epsilon_1 \\ d\epsilon_3 \end{pmatrix}
\tag{8.5}
$$

with: $d = 1 - V_1^{2\pm}V_2^{1\pm} - V_2^{3\pm}V_3^{2\pm} - V_1^{3\pm}V_3^{1\pm} - V_1^{2\pm}V_2^{3\pm}V_3^{1\pm}$ In plane strain compression, the plastic limit condition is still given by:

$$E_1^+ = 0,$$

but the flow rule corresponds to:

$$\left(V_3^{1\pm} + V_3^{2\pm}V_2^{1\pm}\right) d\epsilon_1 + \left(1 - V_1^{2\pm}V_2^{1\pm}\right) d\epsilon_3 = 0 \tag{8.6}$$

Even if it is assumed compression in plain strain direction 2 ($d\sigma_2 > 0$), it comes:

$$\left(V_3^{1\pm} + V_3^{2\pm}V_2^{1+}\right) d\epsilon_1 + \left(1 - V_1^{2+}V_2^{1+}\right) d\epsilon_3 = 0 \tag{8.7}$$

Equation 8.7 means that, according to the sign of $d\sigma_3$, 2 different flow rules are obtained. For $d\sigma_3 > 0$:

$$\left(V_3^{1+} + V_3^{2+}V_2^{1+}\right) d\epsilon_1 + \left(1 - V_1^{2+}V_2^{1+}\right) d\epsilon_3 = 0$$

and for $d\sigma_3 < 0$:

$$\left(V_3^{1-} + V_3^{2-}V_2^{1+}\right) d\epsilon_1 + \left(1 - V_1^{2+}V_2^{1+}\right) d\epsilon_3 = 0$$

Due to the existence of tensorial zones (the octo-linear model is incrementally piece-wise linear), according to the direction of ($d\sigma_1$, $d\sigma_3$) vector, when the stress state is reaching the plastic limit condition, two different directions of ($d\epsilon_1$, $d\epsilon_3$) vector are obtained. The flow rule is no more regular and has a triangular shape ("vertex" effect).

    To conclude these analytical considerations, it comes that the existence of a regular flow rule in axisymmetric conditions does not involve regular flow rules for more general 3D conditions.

    This result was obtained by Kishino et al. (2001) with a discrete element method applied to assemblies of spherical beads and also by Calvetti et al. (2002). The response-envelops (see their definition in section 4.) were computed and compared for axisymmetric conditions and for 3D conditions.

In axisymmetry, the loading circle (constituted by the extremities of the incremental stress vectors) was considered in the bisecting stress plane ($\sigma_2 = \sigma_3$), while in 3D the loading circle was on a plane perpendicular to the bisecting plane and to the stress vector corresponding to the center of the circle. Thus both the loading circles have the same center and the same radius, but they are situated in two orthogonal planes.

    The qualitative simulation of Kishino's discrete numerical results is presented on Figures 12, 13, 14, 15, and 16. Figures 12 and 13 illustrate the existence of a flow rule in axisymmetric conditions with both the non-linear model and the octo-linear one and with both the loose Hostun sand (Figure 12) and dense Hostun sand (Figure 13). The computation has been performed degree by degree on the stress loading circle, what means that there are 360 points on each figure. In the case of the octo-linear relation, the incremental relation between $d\underline{\sigma}$ and $d\underline{\epsilon}$ is linear in each tensorial zone. The response-envelopes are thus constituted by parts of ellipses. It appears clearly that, independently of any $d\underline{\sigma}$ direction, $d\underline{\epsilon}$ vector is directed approximately to the same direction. Let us recall here that, in these constitutive models, it is not possible to distinguish between elastic and plastic strains. Thus total incremental strains are considered for the plots.

For Figure 14, the stress loading circle is now perpendicular to the bisecting plane but the center of the circle is the same as before. The strain response-envelops are plotted in projection on (ϵ_1 versus ϵ_2) and (ϵ_3 versus ϵ_2) planes. It appears clearly that the notion of flow rule is no more valid to analyze these results in the considered case of loose Hostun sand.

Figure 15 and 16 present the same kind of results for dense Hostun sand. The same qualitative conclusion emerges: not any regular flow rule can be detected on these figures. The crossing point which appears on Figure 15 could be an artefact due to the projection of a 3D curve on two planes.

A general conclusion can be drawn by emphasizing the fact that these constitutive relations can describe, as a good approximation, a regular flow rule for axisymmetric conditions and not any flow rule for 3D conditions. More precisely, in these last conditions, the flow rule is singular and presents a vertex like structure.

Acknowledgements

This work has been performed in both frameworks of the European project DIGA and of the national French projects ROMICO and PIR. Their supports are gratefully acknowledged.

Bibliography

Aubry D., Hujeux J. C., Lassoudire F and Meimon Y., *A double memory model with multiple mechanisms for cyclic soil behaviour*, In Proc. of Int. Symp. on Numerical Models in Geomechanics, R. Dungar, G. N.Pande & J. Studer (eds), Balkema, Rotterdam, pages 3–13, 1991.

Bazant Z. P., *Endochronic inelasticity and incremental plasticity*, In Int. J. Solids Struct., 14 pages 691–714, 1978.

Calvetti F., Tamagnini C. and Viggiani V., *On the incremental behavior of granular soils*, In Numerical Models in Geomechanics, Pande and Pietruszczak (eds.), Zwets and Zeitlinger (publ.), pages 3–9, 2002.

Chambon R., Desrues J., Hammad W. and Charlier R., *CLoE, a new rate-type constitutive model for geomaterials. Theoretical basis and implementation*, In Int. J. Num. Anal. Meth. Geomech., 18 pages 253–278, 1994.

Dafalias Y. F., *Bounding surface plasticity. i. mathematical foundation and hypoelasticity*, In J. Eng. Mech., 112 (9) pages 966–987, 1986.

Dafalias Y. F. and Hermann L. R, *A bounding surface soil plasticity model*, In Proc. Symp. on Soils under Cyckic and Transient Loading, G. N. Pande & O. C. Zienkiewicz (eds), Balkema, 1 pages 335–345, 1980.

Darve F., *Une loi rhéologique incrémentale non-linéaire pour les solides*, In Mech. Res. Comm., 7 (4) pages 295–212, 1980.

Darve F., *An incrementally non-linear constitutive law of second order and its application to strain localization*, In Mechanics of Engineering Materials, C. S. Desai & R. H. Gallagher. John Wiley, London, pages 179–196, 1984.

Darve F., *Incrementally non-linear constitutive relationships*, In Geomaterials Constitutive Equations and Modelling, F. Darve ed., Elsevier Applied Science, pages 213–238, 1990.

Darve F., *Liquefaction phenomenon of granular materials and constitutive instability*, In Int. Journal of Engineering Computations, 7 pages 5–28, 1996.

Darve F. and Labanieh S., *Incremental constitutive law for sands and clays, simulation of monotonic and cyclic test*, In Int. J. Num. Anal. Meth. Geomech., 6 pages 243–273, 1982.

Darve F., Flavigny E. and Meghachou M., *Yield surfaces and principle of superposition revisited by incrementally non-linear constitutive relations*, In Int. Journal of Plasticity, 11(8) pages 927–948, 1995.

Darve F. and Laouafa F., *Instabilities in granular materials and application to landslides*, In Mech. Cohes. Frict. Mater., 5(8) pages 627–652, 2000.

Davis R. D. and Mullenger G., *A rate-type constitutive model for soils with critical state*, In Int. J. Num. Anal. Meth. Geomech., 2 pages 255–282, 1978.

Di Benedetto H. and Darve F., *Comparaison de lois rhéologique en cinématique rotationnelle*, In J. Mécan. Théor. Appl., 2(5) pages 769–798, 1983.

Duncan J. M. and Chang C. Y., *Non-linear analyses of stress and stress and strain in soils*, In J. Soil. Mech. and Found. Div., ASCE, 95(SM5) pages 1629–1653, 1970.

Darve F. and Dendani H., *An incrementally non-linear constitutive relation and its predictions*, In Constitutive Equations for granular non-cohesive soils, Cleveland ed. A.S. Saada & Bianchini, Balkema, Rotterdam, pages 237–254, 1989.

Fernandez E., *Internal Report*, 3S Laboratory, 2002.

Gudehus G., *A comparison of some constitutive laws for soils under radially loading symmetric loading and unloading*, In Proc. 3rd Int. Conf. Num. Meth. Geomech., W. Whittke (ed), Balkema, 4 pages 1309–1324, 1979.

Guélin P., *Note sur l'hystérisis mécanique*, In J. Mécan., 19(2) pages 217–247, 1980.

Hicher P. Y., *Comportement mécanique des argiles saturées sur divers chemins de sollicitations monotones et cycliques : application à une modélisation élasto-plastique et visco-élastique*, Ph D ,Ecole Centrale de Paris, 1985.

Hill R., *Eigenmodal deformations in elastic-plastic continua*, In J. Mech. Phys. Solids, 15 pages 371–386, 1967.

Kishino Y., Akaizawa H. and Kaneko K. *On the plastic flow of granular materials*, In Powders and Grains, Kishino (ed.), Zwets and Zeitlinger (publ.), pages 199–202, 2001.

Kolymbas D., *A rate dependent constitutive equation for soils*, In Mech. Res. Comm., 4(6) pages 367–372, 1977.

Lade P. V., *Elasto-plastic stress theory for cohesionless soils with curved yield surfaces*, In Int. J. Solids Struct., 13 pages 1019–1035, 1977.

Loret B., *Formulation d'une loi de comportement élasto-plastique des milieux granulaires*, Thèse de D.I., Ecole Polytechnique, 1981.

Mroz Z., *On the description of anisotropic work hardening*, In J. Mech. Phys. Sci., 15 pages 163–175, 1967.

Nova R. and Wood D. M., *A constitutive model for sand in triaxial compression*, In Int. J. Num. Anal. Meth. Geomech., 3 pages 255–278, 1979.

Owen D. R. and Williams W. O., *On the time derivatives of equilibrated response functions*, In ARMA, 33(4) pages 288–306, 1969.

Prevost J. H., *Plasticity theory for soil stress-strain behaviour*, In J. Eng. Mech. Div., ASCE, 104(EM5) pages 1177–1194, 1978.

Royis P. and Doanh T., *Theoretical analysis of strain response envelopes using incrementally non-linear constitutive equations*, In Int. J. Num. Anal. Meth. Geomechanics, 22 pages 97–132, 1998.

Truesdell C. and Noll W., *The non-linear field theories of mechanics*, In Handbuch Phys. II, Springer, 1965.

Valanis K. C., *A theory of viscoplasticity without a yield surface*, In Archives of Mechanics, 23 pages 517–551, 1971.

Vardoulakis I., Goldscheider M. and Gudehus G., *Formation of shear bands in sand bodies as a bifurcation problem*, In Int. J. Num. Anal. Meth. Geomech., 2 pages 99–128, 1978.

Vermeer P. , *A double hardening model for sand*, In Geotechnique, 28(4) pages 413–433, 1978.

Development of Elastoplastic Strain Hardening Models of Soil Behaviour

Roberto Nova[*][1]

[*]Department of Structural Engineering, Milan University of Technology (Politecnico), Milan, Italy

Abstract The paper presents the historical development of fundamental concepts related to the modelling of the non-linear and irreversible behaviour of soils. The concepts of failure, flow rule, successive yield surfaces and hardening rules are recalled. The original Granta Gravel and Cam Clay models are presented and discussed. The basic structure of the work-hardening model of Lade is sketched. The strain-hardening model of Nova and Wood, together with its successive modifications, is described in greater detail and numerous comparisons between experimental data and theoretical predictions are shown.

1 Introduction

Historically, Soil Mechanics problems have been tackled in different ways depending on the aim of the engineering analysis. If the major concern was to ensure safety against collapse, such as in the case of earth retaining walls, slope stability or the bearing capacity of foundations, soil was modelled as a rigid plastic continuum. If, on the contrary, the major concern were the settlements of a foundation or the soil pressures on a tunnel liner, soil was idealised as a linear elastic material, for which closed form solutions were available for the simplest, but also the commonest, problems.

In the former case, only the failure mechanism was investigated, whilst any deformation prior to failure was neglected. The first work conducted in this spirit is the famous "essai" by Augustin Coulomb (1776), presented at the French Academy of Sciences in 1773 and published three years later. In that paper the earth thrust against a retaining wall was determined by means of an equilibrium analysis of a wedge of soil at the verge of failure. The only additional hypothesis was that along the failure line, that is on the line delineating the wedge from the rest of the soil which remains undeformed, the limit strength of the soil is attained.

Such simple hypotheses lead to the correct solution in terms of the earth thrust. Also the correct distribution of the pressures on the wall could be determined, by assuming that the wedge did not remain rigid but that other failure lines were possible within it, so that the equilibrium analysis should be repeated for each such line. No information could be retrieved, however, on the wall displacements or on the effective pressure distribution when the wall was heavier than what strictly needed to ensure stability at the verge of failure.

[1] This paper is written within the framework of DIGA network (supported by the European Commission under contract n. HPRN-CT2002-00220) and ALERT Geomaterials.

In order to determine displacements or pressures far from the limit equilibrium condition, the latter approach was then followed. By imposing the equilibrium of each elementary volume of the soil mass, the following differential equations on stresses were derived:

$$\frac{\partial \sigma_{ij}}{\partial x_j} + \gamma \delta_{iz} = 0 \tag{1.1}$$

where σ_{ij} is the stress tensor, γ is the unit weight of the soil and δ_{iz} is a quantity which is equal to one if $i = z$ and zero otherwise, z being the vertical direction, taken positive upwards. In Equation 1.1 and henceforth, the Einstein summation convention on repeated indices is assumed to hold.

Since soil was idealised as a continuum, the fulfilment of the following compatibility equations was also imposed, in order to ensure that soil remains a continuum even after the deformation process:

$$\varepsilon_{ij} = -\frac{1}{2}\left(\frac{\partial U_i}{\partial x_j} + \frac{\partial U_j}{\partial x_i}\right) \tag{1.2}$$

where ε_{ij} is the strain tensor (compression taken as positive), whilst U_i is the displacement in the x_i direction.

The relation between strains and stresses was then specified via the constitutive equation:

$$\varepsilon_{ij} = C_{ijhk}\,\sigma_{hk} \tag{1.3}$$

where C_{ijhk} is a quadruple tensor.

In order to solve problems of practical interest, it was commonly assumed that C_{ijhk} was constant, isotropic and such that $C_{ijhk} = C_{hkij}$. Only two parameters were then necessary to completely specify soil behaviour: the Young modulus E and the Poisson ratio ν or, alternatively, the bulk modulus K and the shear modulus G. In that case the constitutive law (Equation 1.3) could be written in tensorial terms as:

$$\varepsilon_{ij} = \frac{1}{3}\frac{p}{K}\delta_{ij} + \frac{s_{ij}}{2G} \tag{1.4}$$

where δ_{ij} is the unit tensor, known also as Kronecker δ, p is the isotropic pressure, that is

$$p = \frac{1}{3}\sigma_{hk}\,\delta_{hk} \tag{1.5}$$

and s_{ij} is the stress deviator, defined as:

$$s_{ij} = \sigma_{ij} - p\delta_{ij} \tag{1.6}$$

Alternatively, Equation 1.3 could be rewritten in matrix form , by listing σ_{hk} and ε_{ij} in six dimensional vectors, as:

$$
\begin{Bmatrix} \varepsilon_x \\ \varepsilon_y \\ \varepsilon_z \\ \gamma_{xy} \\ \gamma_{xz} \\ \gamma_{yz} \end{Bmatrix} = \begin{bmatrix} 1/E & -v/E & -v/E & 0 & 0 & 0 \\ -v/E & 1/E & -v/E & 0 & 0 & 0 \\ -v/E & -v/E & 1/E & 0 & 0 & 0 \\ 0 & 0 & 0 & 1/G & 0 & 0 \\ 0 & 0 & 0 & 0 & 1/G & 0 \\ 0 & 0 & 0 & 0 & 0 & 1/G \end{bmatrix} \begin{Bmatrix} \sigma_x \\ \sigma_y \\ \sigma_z \\ \tau_{xy} \\ \tau_{xz} \\ \tau_{yz} \end{Bmatrix} \tag{1.7}
$$

where γ_{ij} are the engineering shear strains, i.e. twice the strains ε_{ij} , whilst τ_{ij} are the corresponding shear stresses.

Under properly defined boundary conditions on loads and displacements, the system composed of Equations 1.3, 1.4 and 1.6 admits a unique solution. The state of stress and strain and the corresponding displacements could be then determined in each point of the soil for given loading or imposed displacements.

In some simple cases, it was possible to found the solution even in closed form. Kelvin (1855), Boussinesq (1885) and Cerruti (1882) solved the three fundamental problems, that are, respectively: point load within an infinite mass, point vertical load on a semi-infinite mass, point horizontal load on a semi-infinite mass. By applying the principle of superposition, intrinsically linked to the linearity of the problem considered, it was then easy to determine the complete solution for a plethora of problems of practical interest , see for instance Poulos and Davis (1974).

This approach suffers from two fundamental shortcomings, however. First, it does not take the presence of water into account. Secondly, it assumes that the stress-strain behaviour of soil is linear, what is in obvious contradiction with experimental evidence.

Both points were tackled by Terzaghi and partially solved. As far as the problem of water in soil pores is concerned, Terzaghi (1925) postulated that deformability and strength of soil depend on effective stresses σ_{ij}' , defined as:

$$
\sigma_{ij}' \equiv \sigma_{ij} - u\delta_{ij} \tag{1.8}
$$

where u is the pore water pressure. According to that principle, whose validity was subsequently experimentally demonstrated by Rendulic(1937) , the constitutive law for the soil skeleton should be then rewritten as :

$$
\varepsilon_{ij} = C_{ijhk}\sigma_{hk}' \tag{1.9}
$$

where also C_{ijhk} may be a function of σ_{hj}' in case non-linear behaviour is considered.

In the following, the validity of the effective stress principle will be taken for granted and stresses will always mean effective stresses, unless otherwise specified. It is worth noting that if there is water in the pores another unknown, the pore water pressure, should be determined. In order to do that, other equations, the balance of mass and the Darcy law, should be added to the system of Equations 1.3, 1.4, 1.9. The interaction of soil skeleton and pore water in boundary value problems will not be considered henceforth, however.

As far as deformability is concerned, Terzaghi (1923) noted that the oedometric compressibility of a soil sample was different when it was subjected to a vertical stress that was less than or equal to the maximum ever experienced. In both cases soil compressibility was not a constant but depended on the state of stress. Soil behaviour appeared then to be at the same time non-linear and irreversible. Elasticity theory could not be therefore rigorously applied.

It was too difficult , however, to take such a non-linearity into account. This was done only in the very simple case of one dimensional loading, such as the case of a long and wide embankment insisting on a compressible layer of limited thickness. For general loading conditions, elasticity was still widely employed, taking care to choose appropriate values for the moduli, which were assumed to depend on the state of stress and the previous loading history of the soil deposit. Indeed, no other possibility existed until the advent of computers and computer oriented procedures, such as the finite element method.

2   Fundamentals of the theory of plasticity

Coulomb's essai concerns the calculation of the earth thrust against a retaining wall, the stability of a masonry column and the stability of a stone arch. These apparently distinct topics are in fact intimately connected by two important aspects. Firstly, in all cases the strength of the material can be seen as due to cohesion and friction between bricks or stones or between soil grains. Secondly, the conceptual path followed to solve those problems is the same, since in all cases the equilibrium condition at impending failure is studied. In one paper. Coulomb established the foundations of the theory of plasticity and the basic principles of the mechanics of geomaterials as applied to the theory of masonry structures and to geotechnical engineering.

The larger impetus to the development of the theory of plasticity came however from the study of metals. Tresca (1868) suggested that, for a metal specimen subjected to a general stress state, yield would occur when the shear stress on any plane reached a limiting value. In terms of principal stresses the Tresca yield criterion can be expressed as:

$$\max\left\{|\sigma_1 - \sigma_2|, |\sigma_1 - \sigma_3|, |\sigma_2 - \sigma_3|\right\} - \sigma_L = 0 \qquad (2.1)$$

where σ_L is the strength in pure tension (or compression).

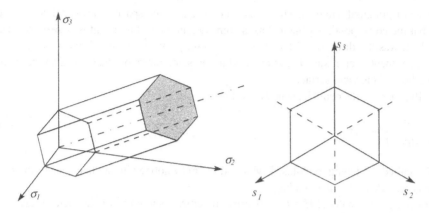

Figure 1. Tresca failure condition a) in the principal stress space b) section in the deviatoric plane.

From a geometrical standpoint, condition 2.1 represents a prism in the principal stress space, whose axis is parallel to the hydrostatic axis $(\sigma_1 = \sigma_2 = \sigma_3)$ and whose intersection with a plane orthogonal to it, the so called deviatoric plane, is a regular exagon (Figure 1). It was assumed that for a generic stress state represented by a point within the prism, it was possible to load the specimen by means of any stress increment $d\sigma_{hk}$ and that the corresponding strain increment $d\varepsilon_{ij}$ could be obtained via Equation 1.7. If the stress point lies on the surface given by Equation 2.1, however, the direction of the possible stress increment is restricted since the material is unable to sustain any stress state outside the surface given by condition 2.1. Either the increment is directed inward the prism or along the yield surface itself. In the former case unloading occurs and purely elastic strains take place. In the latter case, plastic as well as elastic strains occur. If the elementary volume of material is kinematically unconstrained, unlimited strains develop.

Purely for mathematical convenience, von Mises (1913) suggested to employ the following equation as a yield condition instead of Equation 2.1:

$$\sigma_1^2 + \sigma_2^2 + \sigma_3^2 - \sigma_1\sigma_2 - \sigma_1\sigma_3 - \sigma_2\sigma_3 - \sigma_L^2 = 0 \tag{2.2}$$

From a geometrical standpoint Equation 2.2 represents a circular cylinder in the principal stress space. It is easy to verify that such a cylinder is circumscribed to the Tresca exhagon. The two criteria give the same failure stress in pure tension or compression, whilst the latter predicts that the maximum shear stress in pure torsion is about 1.155 times larger than that predicted by the former. It was subsequently experimentally demonstrated by Lode (1926) and, more accurately, by Taylor and Quinney (1931) that the Mises yield criterion is much closer to experimental data for copper, aluminium and mild steel.

The experiments conducted by the latter authors on hollow tubes subjected to tension and torsion demonstrated also other important properties of the plastic strain increments. It was shown first that the modulus of the strain increment vector did not depend directly on the modulus of the stress increment vector, as in the elastic theory. It was instead linearly related to the projection of the stress increment vector onto the gradient of the yield surface in the stress space.

If the space of principal strain increments is superposed on the space of principal stresses, in such a way that the corresponding principal axes coincide, an interesting result was also obtained. No matter which was the direction of the stress increment, provided it caused plastic strains, the plastic strain increment vector was always directed in the same direction, that was also the direction of the gradient of the yield surface.

Analytically, such results could be stated as follows:

$$d\varepsilon_{ij}^{p} = \frac{1}{H}\frac{\partial f}{\partial \sigma_{ij}}\frac{\partial f}{\partial \sigma_{hk}}d\sigma_{hk} \tag{2.3}$$

where H is called hardening modulus. Taylor and Quinney noticed in fact that a small increase in strength was recorded after first yielding.

For a perfectly plastic material $H = 0$, and stress increments should lie on the yield surface.

Equation 2.3 gives the most general expression of plastic strain increments for a hardening material with an associated flow rule. That name comes from that the yield or loading function plays in fact two roles at the same time. It specifies when plastic strains occur and determines also which is the direction of the plastic strain increment vector. In principle, at least, one may think to separate those two aspects. We may then define a new function g, called plastic potential, such that the plastic strain increment vector is directed as its gradient. In that case, the loading function f only specifies whether plastic strains occur or not and the material is said to obey a non-associated flow rule. Plastic strain increments are then given by the following equation:

$$d\varepsilon_{ij}^{p} = \frac{1}{H}\frac{\partial g}{\partial \sigma_{ij}}\frac{\partial f}{\partial \sigma_{hk}}d\sigma_{hk} \tag{2.4}$$

If H is different from zero, when plastic strains occur the loading function is modified accordingly. Various possibilities exist, a priori. The yield surface can grow (or shrink) homothetically, that is it may change dimensions but not in shape, see Figure 2a. In that case hardening (or softening) is said to be isotropic. Another possibility is that the yield function translates and rotates in the stress space changing neither shape nor dimensions, see Figure 2b. In that case hardening is said to be kinematic. More complex possibilities exist, however. The model that seems to take all the relevant phenomena into account is that proposed by Mroz (1967), which combines isotropic and kinematic hardening, as shown in Figure 2c. For the sake of simplicity, however, we shall henceforth consider isotropic hardening(or softening), only.

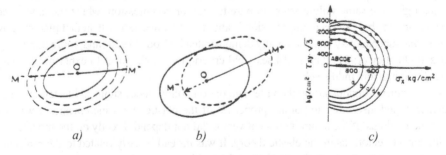

Figure 2. Types of hardening: a)isotropic, b) kinematic, c) mixed.

Note that the hardening modulus specifies the amount of plastic strain rates. It is, in a sense, conceptually similar to the elastic modulus E or G. H is not a constant, however. It is instead a function of the plastic strain history of the material. Many models assume for instance that H is a function of the plastic work undergone by the material element. Such models are usually termed workhardening models. Other models assume instead that H is a function of a scalar combination of plastic strains, without making explicit reference to the plastic work. Such models are usually termed strain hardening models.

It is worthmentioning that Equation 2.4 could have been derived also in a different way. Assume that a loading function $f = f\left(\sigma_{hk'}, k_l\left(\varepsilon_{rs}^p\right)\right)$ exists such that:

if $f < 0$

or $f = 0$ and $df < 0$

only elastic strains may occur.

If instead

$f = 0$ and $df = 0$

then plastic as well as elastic strains occur. In that case

$$\frac{\partial f}{\partial \sigma_{hk}} d\sigma_{hk} + \frac{\partial f}{\partial k_l} dk_l = 0 \tag{2.5}$$

where

$$dk_l = \frac{\partial k_l}{\partial \varepsilon_{rs}^p} d\varepsilon_{rs}^p \tag{2.6}$$

By definition of plastic potential:

$$d\varepsilon_{rs}^p = \Lambda \frac{\partial g}{\partial \sigma_{rs}} \tag{2.7}$$

where Λ is a non-negative scalar, known as plastic multiplier. By substituting Equation 2.7 in Equation 2.6 and this in turn in Equation 2.5 and then solving for Λ, one obtains again Equation 2.4, having defined the hardening modulus as:

$$H \equiv -\frac{\partial f}{\partial k_l} \frac{\partial k_l}{\partial \varepsilon_{rs}^p} \frac{\partial g}{\partial \sigma_{rs}} \tag{2.8}$$

Equation 2.5 is known as the Prager consistency rule.

When plasticity theory started to be applied to Soil Mechanics problems in modern terms, hardening was initially neglected so that first yielding and failure were assumed to coincide. The generally acknowledged criterion was the so-called Mohr-Coulomb condition, that could be expressed as:

$$\sigma_M - \sigma_m - (\sigma_M + \sigma_m) \sin\varphi' - \cos\varphi' = 0 \tag{2.9}$$

where σ_M and σ_m are the largest and smallest principal stress, respectively, c' is the cohesion whilst φ' is known as the friction angle. By comparing Equations 2.1 and 2.9, it turns out that the former can be obtained from the latter by putting friction to zero. Figure 3 shows the intersection of the pyramid given by Equation 2.9 with the deviatoric plane. The resulting irregular exhagon clearly shows the difference in strength in compression $(\sigma_1 \geq \sigma_2 = \sigma_3)$ and extension $(\sigma_1 \leq \sigma_2 = \sigma_3)$ in triaxial loading conditions.

That criterion was quite convenient when two-dimensional problems were considered, but was cumbersome when general loading conditions were expected. Drucker and Prager (1952) suggested to modify Equation 2.9 as follows:

$$\sigma_1^2 + \sigma_2^2 + \sigma_3^3 - \sigma_1\sigma_2 - \sigma_1\sigma_3 - \sigma_2\sigma_3 - (Mp + T)^2 = 0 \tag{2.10}$$

By comparing Equations 2.2 and 2.10 it turns out that the former is a special case of the latter. Equation 2.10 gives a cone in the principal stress space whose axis is the hydrostatic axis, p. Depending on the chosen values of M and T, the Drucker-Prager cone can be made circumscribed or inscribed to the Mohr-Coulomb pyramid.

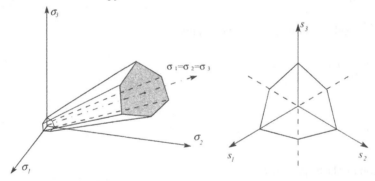

Figure 3. Coulomb failure condition a) in the principal stress space b) section.in the deviatoric plane

Both these criteria have been extensively used for the solution of practical boundary value problems, either with an associated flow rule or with a non-associated flow rule. The most important difference between the two criteria is that the Drucker-Prager condition is mathematically more convenient, but assumes the same strength in triaxial compression and extension, what is far from experimental evidence. Special attention should be paid, therefore, in choosing the constitutive parameters M and T for matching the experimental data.

Another criterion was more recently proposed by Matsuoka and Nakai(1974), which has the same mathematical advantages of the Drucker-Prager condition and describes the failure condition for non-cohesive soil in much closer agreement to actual experimental data. That criterion can be simply written as:

$$I_3 - \alpha I_1 I_2 = 0 \tag{2.11}$$

where:

$$I_1 = \sigma_1 + \sigma_2 + \sigma_3 = 3p \tag{2.12}$$

$$I_2 = \sigma_1\sigma_2 + \sigma_1\sigma_3 + \sigma_2\sigma_3 \tag{2.13}$$

$$I_3 = \sigma_1\sigma_2\sigma_3 \tag{2.14}$$

are the three invariants of the stress tensor. In Figure 4 a comparison between the Mohr-Coulomb and the Matsuoka-Nakai criteria is shown, together with experimental data obtained in true triaxial tests on Toyoura Sand, after Matsuoka and Nakai (1985). The strength difference in triaxial compression and extension is well matched by both criteria. Equation (19), however is much closer to experimental data in tests other than conventional triaxial. In addition, the curvilinear shape of the Matsuoka-Nakai criterion allows the failure condition to be described by a single equation instead of six, as in the case of the Mohr-Coulomb condition.

It is perhaps interesting to note that the Matsuoka-Nakai condition can be also written as:

$$\tan^2 \varphi_{12} + \tan^2 \varphi_{13} + \tan^2 \varphi_{23} = const \tag{2.15}$$

where

$$\varphi_{ij} = sin^{-1} \frac{\sigma_i - \sigma_j}{\sigma_i + \sigma_j} \tag{2.16}$$

are the mobilised friction angles in the planes x_i, x_j. Similarly the Mohr-Coulomb criterion can be written as:

$$\max \frac{\sigma_i - \sigma_j}{\sigma_i - \sigma_j} = sin\varphi \tag{2.17}$$

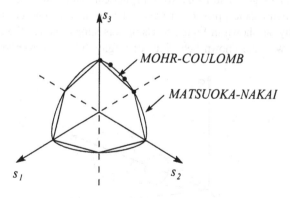

Figure 4. Comparison of Coulomb and Matsuoka-Nakai failure conditions with experimental data obtained in true triaxial tests on Toyoura sand specimens (after Matsuoka and Nakai, 1985).

The Matsuoka-Nakai condition appears then as a convenient modification of the Mohr-Coulomb criterion in the same spirit as the von Mises criterion is a modification of the Tresca equation. The von Mises condition could be
in fact equally well expressed as:

$$(\sigma_1 - \sigma_2)^2 + (\sigma_1 - \sigma_3)^2 + (\sigma_2 - \sigma_3)^2 = const \tag{2.18}$$

The formal correspondence between Equations 2.1 and 2.17 on one hand and Equations 2.18 and 2.15 on the other is evident.

If a granular material is also cohesive, the Matsuoka-Nakai condition can be easily generalised. It is in fact sufficient to define a modified stress tensor σ_{ij}^*, such that

$$\sigma_{ij}^* = \sigma_{ij} + c' \cot \varphi' \tag{2.19}$$

and to substitute σ_{ij} with σ_{ij}^* in Equations 2.11 or 2.15 to get the desired condition for a material with friction and cohesion.

3   The change of paradigm

Although it was evident that a sample of virgin soil could undergo irrecoverable strains even under hydrostatic loading conditions, that is very far from failure, the theory of plasticity was used in Soil Mechanics to describe collapse conditions, only, until the last years of the fifties. At that time two papers appeared that opened a large field of research directed towards the mathematical formulation of a model for soil as an elastic plastic strain hardening material.

Drucker, Gibson and Henkel (1957) put forward the idea of considering soil as a material of this type by analogy with metal plasticity. They suggested that a convenient way to represent the yield locus for a soil element was a cone with a circular cross section centred on the hydrostatic axis, as for the Drucker-Prager yield locus, with a spherical cap on the open end of the Drucker-Prager cone. As the hydrostatic pressure increases during an isotropic consolidation, the cone expands homothetically, as shown in Figure 5. Using this simple model, the authors found good qualitative agreement with the known behaviour of virgin soils in conventional drained triaxial tests.

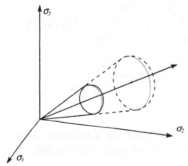

Figure 5. The Drucker, Gibson and Henkel conceptual model

At the end of their paper they wrote a sentence that appears to-day as prophetic:' It is possible that soil behaviour can be described by a theory of plasticity. Metals are complicated, inhomogeneous and anisotropic materials as are soils. Nevertheless, the behaviour of metals can be described by work hardening, stress-strain relations for homogeneous material with initial isotropy'.

One year later Roscoe, Schofield and Wroth (1958) published an extensive study of the progressive yielding of samples of both cohesive and granular materials. Although the elastic plastic nature was not in evidence and the hardening relations were not expressed in a way usual in the hardening theory of plasticity, it is apparent that the authors perfectly appreciated that the behaviour of soil could be described by strain hardening stress strain relations. Following their theory, once some fundamental parameters had been obtained experimentally, it was possible to predict the loading path and the ultimate state of any soil sample in drained and undrained compression tests. It was then possible to compare quantitatively experimental and calculated results.

Since the publication of those two papers a tremendous amount of research was devoted to finding reliable experimental data together with a rational interpretation of test results, particularly in Cambridge under the direction of Ken Roscoe. This effort produced several models for soil, some of which will be reviewed henceforth.

3.1 Granta Gravel

This model was developed by Schofield and Wroth (1968) at the University of Cambridge as a simple mean to introduce the slightly more complex model known as Cam Clay , Granta and Cam being the names of the upper and lower reaches, respectively, of the river that gives Cambridge its name.

Only triaxial compression tests are considered, so that only two independent parameters fully characterize the state of stress within the sample. Two convenient quantities are the mean effective pressure p' and the deviatoric stress q, which are defined as:

$$p' = \frac{1}{3}\left(\sigma_a' + 2\sigma_r'\right) \tag{3.1}$$

$$q = \sigma_a' - \sigma_r' \tag{3.2}$$

where σ_a' and σ_r' are the effective axial and radial stress, respectively. The ratio between q and p' will be called stress ratio, η. Correspondingly, the state of strain is also characterized by two parameters, the volumetric strain v and the deviatoric strain ε, defined as :

$$v = \varepsilon_a + 2\varepsilon_r \tag{3.3}$$

$$\varepsilon = \frac{2}{3}\left(\varepsilon_a - \varepsilon_r\right) \tag{3.4}$$

where ε_a and ε_r are the axial and radial strain, respectively. Stresses and strains are taken positive in compression. The strain increments have been chosen so that the work per unit volume:

$$dW = \sigma_a d\varepsilon_a + 2\sigma_r d\varepsilon_r \qquad\qquad (3.5)$$

could also be expressed as:

$$dW = p'dv + q\, d\varepsilon \qquad\qquad (3.6)$$

This condition is necessary in order to have normality of the plastic strain increment vector to the yield surface.

In Granta Gravel the material behaviour is assumed to be rigid plastic but allowing for hardening or softening. As a consequence, the total strains coincide with the plastic strains and the loading power is then equal to the power dissipated during plastic deformation. Assuming purely frictional behaviour, by analogy with Amonton's law (the so called Coulomb's law of friction), the dissipated work per unit volume can be expressed as:

$$dW = M\, p'\, d\varepsilon \qquad\qquad (3.7)$$

where M is a simple frictional constant. The normality rule is assumed to hold. From the definition of plastic potential. Equation 2.7 and recalling that elastic strains are assumed to be nil, it is then possible to derive that:

$$\frac{dv}{d\varepsilon} = \frac{\partial f / \partial p'}{\partial f / \partial q} \qquad\qquad (3.8)$$

where f is the loading function, that can be made explicit in terms of q. From Equations 3.6, 3.7 we have that:

$$\frac{dv}{d\varepsilon} = M - q/p' \qquad\qquad (3.9)$$

Equation 3.8 becomes therefore:

$$\frac{\partial f / \partial p'}{\partial f / \partial q} = M - q/p' \qquad\qquad (3.10)$$

which can be integrated to give:

$$f(p,q) = q + Mp'\ln\frac{p'}{p_c} = 0 \qquad\qquad (3.11)$$

Equation 3.11 gives a family of yield loci in the plane p', q. The parameter of this family is p_c, which is by definition the value of p' for $q = 0$. This parameter is a function of the loading history of the sample and is called preconsolidation pressure. The yield locus is given by the curve shown in Figure 6. Note that, since the shape of the yield locus is the same no matter which is the stress level, in the terminology of plasticity theory, this implies that hardening is assumed to be isotropic.

For a given stress rate, the direction of the strain rate is uniquely determined by the state of stress before the increment via the plastic potential, which coincides with the yield locus in this case, since normality has been assumed to hold. The amount of strain is not yet defined, however.

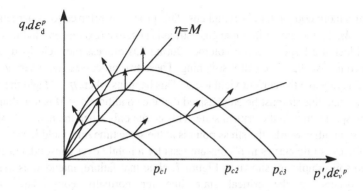

Figure 6. Granta Gravel yield function.

In order to do that, Schofield and Wroth assumed that it was possible to define a curve in the p', q, V space towards which every state of the soil tends for large deformations, V being the specific volume of the soil sample, that is :

$$V = 1 + e \tag{3.12}$$

where e is the void ratio. This line is called critical state line and is formally given by the following equations:

$$q = M \ p' \tag{3.13}$$

$$V = \Gamma - \lambda \ln p' \tag{3.14}$$

where Γ and λ are two material constants. By definition of p_c, from Equation 3.11 we get:

$$V = \Gamma - \lambda \left(\ln p_c - 1 \right) \tag{3.15}$$

Although formulated in a way different from usual theory of plasticity, Equation 3.15 plays the role of a hardening rule. In fact, eliminating p_c from Equations 3.11, 3.15 we get:

$$q = Mp' \left(1 + \frac{\Gamma - V}{\lambda} - \ln p' \right) \tag{3.16}$$

It is possible therefore to determine the volumetric strain associated with any increment in the state of stress, and from it to determine also the shear strain increment by means of the normality rule. The incremental problem is then completely solved.

Equation 3.16 represents a closed surface that must contain all the permissible states of the specimen, and for this reason is called the state boundary surface (SBS).

From Equation 3.15 we see that every yield locus is associated with a definite specific volume. From Equation 3.9 we see that when $\eta < M$, the increment in volumetric strain is positive (compression): it is then possible from a certain stress point on a yield locus at specific volume V_1 to reach a yield locus associated with a smaller volume. In this way the yield locus expands and the specimen becomes more compact, i.e. strengthens. This process is referred to as hardening. On the other hand if $\eta > M$, the increment in volumetric strain is negative (expansion) and it is possible to reach only yield loci at a larger specific volume. During this process the yield locus shrinks and the specimen becomes weaker -We call it softening. These two processes end when $dv = 0$ and it is no more possible to leave the current yield locus. This happens when $\eta = M$ and the state of soil lies on the critical state line. It must be emphasized that the condition $\eta = M$ is not alone sufficient to ensure that the specimen is in the critical state. This is true only when the material exhibits plastic deformation or, in other words, the stress point is at the same time on a yield locus.

Examples of stress strain curves in p' constant tests for a lightly over consolidated and a highly overconsolidated soil sample are shown in Figure 7. Note that failure, intended as maximum deviatoric stress occurs on the critical state line for normally consolidated or slightly overconsolidated soils, whilst it occurs on a curved line well above the critical state line for highly overconsolidated soils. Such a result is in good agreement with experimental evidence. Note also that for the former type of soil, volumetric strains are contractive, whilst for the latter they are dilative, as experimentally observed.

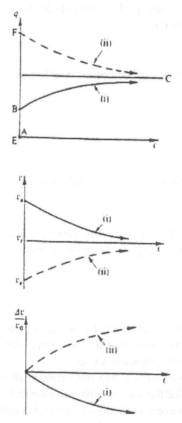

Figure 7. Granta Gravel predictions in p' constant tests for slightly and heavily overconsolidated specimens (after Schofield and Wroth, 1968).

3.2 Cam Clay

The predictions of Granta Gravel are very crude, however. Indeed, Granta Gravel was not really intended for modelling the behaviour of sand or other granular materials, but simply to introduce another model, sligthly more complex, to interpret test results for clay. This model was called Cam Clay. The only difference between the two is that, in the latter, part of the volumetric strain is recoverable, whilst the shear strain remains purely plastic. Cam Clay is therefore an elastic plastic model.

The isotropic unloading-reloading curve is assumed to reduce to a straight line in the V, $\ln p'$ plane, so neglecting the observed hysteresis loop. The slope of this line, k, is smaller than the slope of the isotropic consolidation line, λ. Schofield and Wroth postulated that this relation was valid for any unloading-reloading stress path. Namy (1970) showed that this hypothesis has experimental support.

The expression of the yield locus is the same as in Granta Gravel. The state boundary surface becomes instead:

$$q = \frac{Mp'}{\lambda - k}\left(\Gamma + \lambda - k - V - \lambda \ln p'\right) \tag{3.17}$$

The yield locus is no longer associated with a definite specific volume and within it, it is now possible to have recoverable change in volume. Nevertheless, any yield locus is still associated with a definite plastic volumetric strain. A better interpretation of test data is obtained and , in particular, in an undrained test, yielding is possible before the critical state line is attained. Figure 8 shows calculated results for a resedimented clay in drained and undrained triaxial tests. It is apparent than Cam Clay describes clay behaviour better than Granta Gravel.

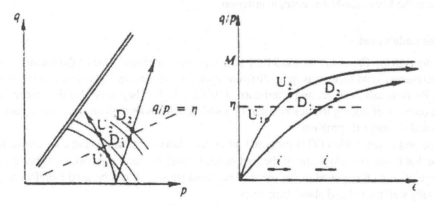

Figure 8. Cam Clay predictions in drained and undrained tests for a normally consolidated clay (after Schofield and Wroth, 1968).

3.3 Modified Cam Clay

Although the theoretical predictions were in good qualitative agreement with the actual behaviour of clay, quantitatively some refinements were necessary. Burland (1967) proposed to
change the expression of the yield locus and to use an ellipse passing through the origin of the axes
p' and q:

$$f = q^2 + M^2(p'-p_c)p' = 0 \qquad (3.18)$$

In this way an inconsistency of the original Cam Clay model was in fact eliminated. Under
purely isotropic conditions only volumetric strains occur for modified Cam Clay, whilst in the
previous model also shear strains could take place, as it is evident from Equation 3.9. Also, the
quantitative predictions for normally consolidated and sligthly overconsolidated clay samples were
indeed much better than with the original model, especially in undrained tests. Wroth (1976) employed such a model, modified to plane strain conditions, to predict the failure of a trial
embankment with a remarkable success. Since then, many other attempts to predict soil behaviour were performed , generally with satisfacory agreement between predictions and the observed
behaviour. Examples are for instance given in Britto and Gunn(1987).

The behaviour of highly overconsolidated clay was still badly predicted. In particular, the
stress level at which failure occurs was highly overestimated. For this reason for some practical
applications, the elliptic yield locus was used only in the hardening regime, as a cap in the original
Drucker, Gibson and Henkel model, see for instance Zienkiewicz, Humpheson and Lewis (1975).
Since in the deviatoric plane the yield locus is assumed to be circular, as in the Drucker-Prager
model, that is obviously in contrast with the experimental evidence. Gens and Potts (1988) further
modified the model by assuming as yield locus a curvilinear surface, whose cross section is similar
in shape to the Matsuoka-Nakai strength criterion.

3.4 The Lade's model

Attempts to use Cam Clay for modelling sand were not successful. One of the reasons was that
the experimental yield locus was quite different from the experimental plastic potential, as noted
first by Poorooshasb, Holubec and Sherbourne (1966), (1967). They were also the first to formulate the problem of finding the deformation of a soil sample under a given stress increment as an
incremental elastoplastic problem.

Some years later, Lade (1977) proposed an elastic plastic work hardening constitutive model
for sand that is still nowadays one of the best suited to interpret sand behaviour and has the great
advantage of being formulated in the space of principal stresses. It can be used therefore to interpret equally well triaxial and plane strain tests.

The strain increment tensor is assumed to be the sum of three components: the elastic strain in-
crement $d\varepsilon_{ij}^e$, the plastic collapse component $d\varepsilon_{ij}^c$ and the plastic expansive component $d\varepsilon_{ij}^p$,
so that

$$d\varepsilon_{ij} = d\varepsilon_{ij}^e + d\varepsilon_{ij}^c + d\varepsilon_{ij}^p \qquad (3.19)$$

The elastic strain increments are determined by means of a stress-strain relation which is formally identical to the Hooke law but for the fact that the Young modulus depends on the least principal stress in a non-linear way:

$$E_{ur} = K_{ur} \left(\frac{\sigma_3}{p_a} \right)^n \qquad (3.20)$$

K_{ur} and n are constitutive parameters that can be easily determined from the results of a drained triaxial compression test, by plotting in a log-log plot the average slope of the stress-strain curve in an unloading-reloading cycle and the corresponding confining pressure, whilst p_a is simply the atmospheric pressure.

Together with the Poisson ratio v, that can be determined from the corresponding lateral strains, they completely determine the unloading-reloading behaviour. Note, however, that such a relationship is hypoelastic , what means that, by performing a convenient stress cycle, work can be extracted from the sample. This has no dramatic consequence if only monotonic loading is involved, but uncontrollable errors can arise if the sample undergoes cyclic loading.

The plastic collapse strains are determined by means of an associated flow rule, assuming that the loading function is a spherical cap centred in the origin of axes. The equation of the cap is therefore:

$$f = f_c - I_1^2 - 2 I_2 = 0 \qquad (3.21)$$

Performing all the necessary derivations, plastic collapse strain increments are given by:

$$d\varepsilon_{ij}^c = \frac{dW_c}{f_c} \sigma_{ij} \qquad (3.22)$$

where the plastic work W_c is assumed to be linked to f_c by the following relationship:

$$W_c = C \, p_a \left(\frac{f_c}{p_a^2} \right)^p \qquad (3.23)$$

where C and p are experimental parameters that can be derived from the results of an isotropic compression test.

Finally, to derive plastic expansive strains it is assumed that the yield function is given by :

$$f = f_p - \left(I_1^3 / I_3 - 27 \right) \left(I_1 / p_a \right)^m = 0 \qquad (3.24)$$

where f_p is equal to η_1 at failure. The parameters η_1 and m are material constants that can be determined from the failure values of the deviator stress in triaxial compression tests. If $m > 0$, the failure locus is curved, as often experimentally observed. Figure 9 shows the yield function for expansive strains.

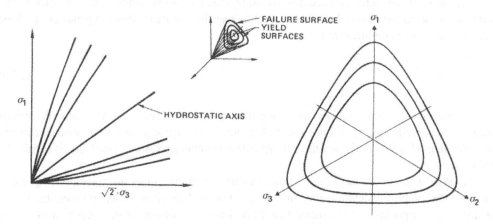

Figure 9. Yield function for expansive strains, according to Lade (1977).

For expansive plastic strains the flow rule is assumed to be non associated, the expression of the plastic potential is similar to that of the yield function:

$$g = g_p - I_1^3 + \left\{ 27 + \eta_2 \left(\frac{p_a}{I_1} \right)^m \right\} I_3 \qquad (3.25)$$

where η_2 , is given by the following equation:

$$\eta_2 = S\, f_p + R \sqrt{\frac{\sigma_3}{p_a}} + t \qquad (3.26)$$

S, R and t being experimental constants. In order to evaluate such constants it is necessary to determine v^p in triaxial compression tests, by determining the incremental ratio of plastic expansive strains directly from experimental results.

The amount of plastic strains can be finally calculated from the expression of the plastic work W_p . This is assumed to be related to f_p by means of the following equation:

$$f_p = a\, e^{-bW} P \left(\frac{W_p}{p_a} \right)^{1/q} \qquad (3.27)$$

where:

$$q = \alpha + \beta\, \sigma_3 / p_a \qquad (3.28)$$

$$a = \eta_1 \left(\frac{e\, p_a}{W_{p\,peak}} \right)^{1/q} \qquad (3.29)$$

$$b = \frac{1}{q\, W_{p\ peak}} \qquad\qquad (3.20)$$

$$W_{p\ peak} = P\ p_a \left(\frac{\sigma_3}{p_a}\right)^l \qquad\qquad (3.31)$$

and α, β, P and l are other constitutive parameters.

The way in which all these parameters could be derived from the results of triaxial tests is well illustrated in Lade (1989).

The description of triaxial test results is usually exceptionally good, Figure 10, but unfortunately such predictive capacity is somewhat lost when more complicated stress paths are followed, see for instance the results of the Cleveland Workshop, Figure 11. To overcome these difficulties, Lade and Kim (1989) proposed a revised version of this model.

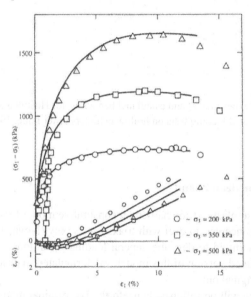

Figure 10. Comparison of experimental and calculated behaviour for Hostun sand in triaxial compression (after Lade, 1989).

Immediately after Lade, Vermeer (1978) and Nova and Wood (1979) published two other models for sand. Many other laws, based on the theory of plasticity with hardening, have been proposed since then. As far as virgin soils under monotonic loadings are concerned, however, the proposed modifications were not relevant for our scope, the model structure being the same. Different were instead the constitutive functions proposed. A comprehensive review is presented in Saada and Bianchini (1989). Important contributions were instead given by many authors for what concerns the behaviour within the yield surface (overconsolidated soils and cyclic loadings), but this goes beyond the scope of this presentation.

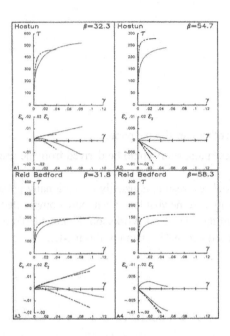

Figure 11. Comparison of experimental and calculated behaviour for Hostun and Reid Bedford sand in proportional loading tests on hollow cylinders (after Lade, 1989).

4   A strain hardening model for sand

In this section another elastoplastic strain hardening constitutive model for sand will be presented. This model was initially conceived to deal with triaxial tests only (Nova, 1977, Nova and Wood, 1979) and successively modified to describe general loading conditions (Nova, 1982). More recently (Nova, 1988), a refined version of this model was formulated in order to eliminate some of the shortcomings of the original one.

The historical sequence will be followed henceforth. The original model is in fact mathematically simpler and the physical meaning of the constitutive parameters is more evident. Once the basic concepts on which this model is founded are grasped, the passage to the more refined version is straightforward.

Before starting the presentation of the constitutive functions employed and the rationale for their choice, it is important to clarify which are the goals we intend to achieve when formulating a constitutive law for a soil.

Primarily we intend to find a simple 'explanation' for the observed test results. For instance we could be interested in understanding why the earth pressure coefficient at rest, K_0 , is related to the strength of the soil or why specimens of loose virgin sand show a peak in the stress-strain relation in undrained tests, whilst they do not in drained ones, or, again, why loose and dense sand show such a different behaviour in undrained tests, whilst they do not when tested in drained conditions.

The term 'explanation' is used clearly in a phenomenological sense, as common in every Civil Engineering problem. No recourse is made to the internal structure of the sample, but only the behaviour on the whole is considered. The soil specimen is then treated as a 'black box' subjected to a given input. The constitutive model is simply seen as a mean to predict the expected output, for any kind of possible input.

The second motivation for formulating a constitutive model is the need for a realistic description to solve boundary value problems of practical interest by means of a numerical technique, such as the finite element method, for instance . The predicted displacement pattern of a foundation will be obviously dependent on the constitutive law assumed to describe soil behaviour. Settlements and rotation under the combined action of vertical loading and overturning moment will be independent if the soil is considered to be linearly elastic and isotropic, whilst they will be coupled if the soil is considered to be elastic plastic strain hardening. The closer is the constitutive law to actual soil behaviour, the better will be the predictions of the foundation displacements.

This latter motivation has a considerable influence on the strategy followed to formulate the model. Simplicity of the model structure, low number of constitutive parameters, their easy and economic determination, become the governing criteria for the model formulation.

This is obviously paid with loss of accuracy in predictions. A good compromise between simplicity and accuracy is the goal to be achieved.

Is also important to note that in engineering practice, when sand is involved, in situ test results are very often the only data available to calibrate the model. Although theoretically not necessary, it is then of the greatest practical importance that constitutive parameters of the model could be easily linked to traditional soil constants, which can be determined from such tests via empirical, well established, relationships.

In the following, triaxial conditions will be considered first. The same variables as in Cam Clay will be employed, therefore.

4.1 Plastic potential

Before the basic concepts of strain hardening plasticity became familiar to researchers in Soil Mechanics, another one was very popular: the concept of stress-dilatancy relationship . From the analysis of experimental results obtained in triaxial tests on sand, Rowe (1962) noted in fact that, no matter which was the stress increment, the dilatancy i.e. the ratio between the increment of the volumetric strain and the increment of the axial strain, was always the same, for a given value of the stress ratio, i.e. the ratio between the axial stress and the confining pressure. In particular, Rowe found that a linear relationship between the stress ratio and the dilatancy fitted with reasonable accuracy the experimental observations. That relation is known since then as stress-dilatancy relationship.

Although with slightly different parameters, it is evident that Equation 3.9, which gives the plastic potential for Granta Gravel, expresses the same concept. The stress-dilatancy concept is therefore essentially the same as the concept of plastic potential for a rigid plastic material. At variance with elasticity, in fact, the ratio of the incremental strains depends on the state of stress only and not on the stress increment, provided that the latter is such to produce plastic strains.

The fact that the stress-dilatancy relationship depends on the ratio of the stresses and not on their absolute value is also very important. This means, in fact, that the plastic potential has the same shape no matter which is the stress level. For a model with associated flow rule as Granta Gravel or Cam Clay this is the same to say that hardening is isotropic.

To determine the plastic potential we may then conceive the following ideal test. Take a set of theoretically identical samples of virgin sand. Load then one sample by increasing axial stress and confining pressure in such a way that the ratio between the two remains constant. At a given value of the axial stress, for instance, perform a small cycle of loading-unloading and record the net strains after the completion of the cycle. Repeat that for different values of the axial stress and do the same for the other samples, each with a different value of the stress ratio. Collecting all the measured data it is possible to establish a stress-dilatancy relationship for the sand at hand.

Many authors, for instance Holubec (1966) in triaxial tests and Stroud (1971) in plane strain, actually performed such a test, or a similar one, in practice. Their results confirm the essence of the findings of Rowe and show that, for values of η not too small, the following linear relationship fits well the experimental data:

$$d \equiv \frac{dv^p}{d\varepsilon^p} = \frac{M - \eta}{\mu} \tag{4.1}$$

where d is defined as dilatancy, μ is an experimental parameter and M is the stress ratio for which dilation is zero as in Granta Gravel or Cam Clay. As we did for such models, we may then substitute Equation 4.1 in Equation 3.8 and integrate to obtain the expression of the plastic potential, that is:

$$g = q - \frac{Mp'}{1 - \mu}\left[1 - \mu\left(\frac{p'}{p_g}\right)^{\frac{1-\mu}{\mu}}\right] \tag{4.2}$$

where p_g is the abscissa of the intersection point between the plastic potential curve and the line $\eta = M$. Clearly, when $\mu = 1$ Equation 4.1 reduces to Equation 3.9 and the plastic potential takes consequently the same expression as that of Cam Clay. In this case, however, the yield function will be given a different expression.

Although Equation 4.1 fits the experimental data better than Cam Clay it has, however, the same shortcoming when η is small. It is evident, in fact, that for $\eta = 0$, the dilatancy is finite, what implies the possibility of deviatoric strains under, purely hydrostatic loading conditions. This contradicts the assumed hypothesis that the virgin material is isotropic. It is therefore necessary to modify such expression, at least for moderate values of η. A convenient way is to assume that:

$$\eta = \frac{c}{d} \tag{4.3}$$

Indeed Namy (1970) found experimental evidence for that by testing a silty clay whose behaviour is in many respects similar to that of a medium dense sand.

In order to avoid corners in the plastic potential, it is assumed that the two expressions merge continuously. This is possible if:

$$c = \frac{M^2}{4\mu} \tag{4.4}$$

For $\eta < M / 2$, the expression of the plastic potential is given then by the following equation:

$$g = q^2 + \frac{M^2}{4\mu}\left(p'^2 - p_c^2\right) = 0 \tag{4.5}$$

The abscissa of the transition point p_t is related to p_c by the following equation:

$$p_t = p_c / \sqrt{1+\mu} \tag{4.6}$$

while

$$p_g = p_t \left(\frac{1-\mu}{2\mu}\right)^{\frac{-\mu}{1-\mu}} \tag{4.7}$$

The assumed shape of the plastic potential is given in Figure 12.

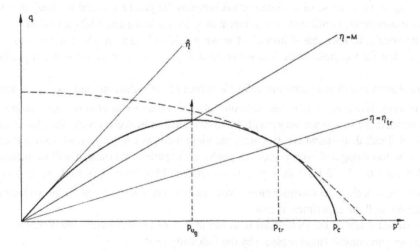

Figure 12. Plastic potential for the model of Nova and Wood (1979)

4.2 Yield function

The determination of the yield function is a difficult step in the formulation of an elastoplastic strain hardening model. In fact the transition from elastic to elastoplastic behaviour is not as sharp as the theory predicts, but a smooth transition occurs. This is complicated by the fact that the initial yield locus for a virgin soil vanishes into a point, the origin of axes, and by that the elastic behaviour is non linear, as we shall see better later on. From an experimental point of view, it is therefore quite difficult to discriminate between elastic and elastoplastic strains and to define the yield point. The experimental results obtained by Holubec (1966) and Cole (1967), although not precise in determining the yield locus for sand for the aforementioned reasons, showed , however, that its shape was indeed quite different from that of the plastic potential discussed above.

To determine more precisely the shape of the family of yield loci, Poorooshasb (1971) performed a complicated stress path. The specimen was initially isotropically consolidated and then subjected to an axial stress increase at constant confining pressure in a triaxial apparatus. It was then unloaded at constant deviator stress by decreasing the confining pressure. The specimen was then in an overconsolidated state. Upon further reloading at constant cell pressure, the yield point was recorded. The process was repeated with a new unloading and reloading so that several yield points were obtained. Unfortunately, when yielding occurs the yield locus expands, so that only two yield points belong in fact to the same yield locus. Although qualitatively, it was however possible to obtain in that way the shape of the yield locus, that was apparently independent of the stress level, as the plastic potential. Poorooshasb proposed then the following equations for the family of yield loci:

$$f = \eta - M + m \ln \frac{p'}{p_y} \tag{4.8}$$

where p_y is the abscissa of the intersection between the yield locus and the line $\eta = M$. This result was subsequently confirmed on a different sand by Tatsuoka and Ishihara (1974).

It is interesting to note that if instead of m we put M in Equation 4.8, it reduces to the Cam Clay expression for the yield locus. The experimental value of m, however, is much smaller than M.

Poorooshasb's equation is certainly valid for values of η which are not too small. When η is close to zero, however it is rather difficult either to reject or to confirm the validity of this equation because the strains are very small and discrimination between elastic and plastic strains is even more difficult than elsewhere. Certainly the yield locus as can be derived from that equation appears to be too elongated with respect to published experimental data. It will be therefore assumed that for $\eta \le M/2$ the flow rule is associated. This hypothesis finds some theoretical support on micromechanical considerations (Nova and Wood, 1979). It will be shown later on that this choice fits well the experimental data.

The assumed shape of the yield locus is shown in Figure 13. It is easy to show that p_y is linked to the isotropic preconsolidation pressure by the following relation:

$$p_y = \frac{p_c}{\sqrt{1 + \mu}} e^{-\frac{M}{2m}} \tag{4.9}$$

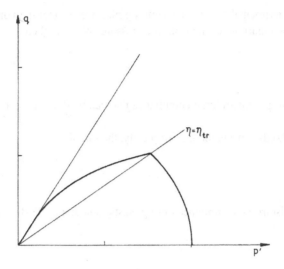

Figure 13. Yield function for the model of Nova and Wood (1979).

4.3 Hardening law

We can now specialize Equation 2.4 to evaluate the plastic multiplier. With the assumed expressions for plastic potential and yield function we note that the set of hidden variables that depend on the previous history of the material is in fact reduced to the single variable p_c. We still do not know how p_c depends on the plastic strains, but we may state that in general p_c will be a function of v^p and ε^p. By comparing Equations 2.4, 2.7 and 2.8, we see that the plastic multiplier can be written as:

$$\Lambda = -\frac{\dfrac{\partial f}{\partial p'}dp' + \dfrac{\partial f}{\partial q}dq}{\dfrac{\partial f}{\partial p_c}\left(\dfrac{\partial p_c}{\partial v^p}\dfrac{\partial g}{\partial p'} + \dfrac{\partial p_c}{\partial \varepsilon^p}\dfrac{\partial g}{\partial q}\right)} \qquad (4.10)$$

where the derivatives of the plastic potential and the yield function can be easily calculated explicitly. The only further step to determine Λ from Equation 4.10 is the experimental evaluation of the two partial derivatives $\partial p_c / \partial v^p$ and $\partial p_c / \partial \varepsilon^p$. For this we may note that, as a general rule, a partial derivative can be found by observing the variation of the value of the function when we give an increment to one independent variable and keep the other constant. In this case to calculate $\partial p_c / \partial v^p$ we must evaluate the variation of p_c with respect to v^p during a process in which ε^p is kept constant. We can see from the expression of the plastic potential that this occurs during isotropic consolidation.

By plotting the experimental results in a semi-log plot, we see that the volumetric strain varies almost linearly with the logarithm of the isotropic pressure. We have then:

$$dv = B\frac{dp_c}{p_c}$$

(4.11)

where B is the isotropic logarithmic compliance that can be determined by performing an isotropic consolidation test.

A similar relation holds if we unload, isotropically, the sample:

$$dv^e = B_e\frac{dp'}{p'}$$

(4.12)

where B_e is the isotropic compliance during elastic unloading. Plastic strain increments will therefore be given by:

$$dv^p = B_p\frac{dp_c}{p_c}$$

(4.13)

B_p being the difference between B and B_e; the partial derivative is finally obtained:

$$\frac{\partial p_c}{\partial v^p} = \frac{p_c}{B_p}$$

(4.14)

An analogous derivation can be performed for $\partial p_c / \partial \varepsilon^p$. From the plastic potential, we may note that constant plastic volumetric strains occur when a $\eta = M$ test is conducted. From tests performed by El Sohby (1964) we may argue that an expression similar to Equation 4.14 holds for $\partial p_c / \partial \varepsilon^p$. We have in fact:

$$\frac{\partial p_c}{\partial \varepsilon^p} = \frac{Dp_c}{B_p}$$

(4.15)

where D is another experimental parameter.

The plastic multiplier, and consequently plastic strain increments, can be therefore determined for every assigned stress increment.

Equations 4.14 and 4.15 can be integrated so that:

$$p_c = p_0 \exp\left\{\frac{v^p + D\varepsilon^p}{B_p}\right\}$$

(4.16)

The hardening law is then specified but for a reference pressure p_0. It will be assumed that if $p' < p_0$ the soil does not deform. In practice, p_0 is quite small and its value does not affect very much the calculated behaviour of a specimen.

Equation 4.16 shows also that a given p_c, and then a given yield locus, is associated to a linear combination of plastic strains. This will be used in the following section to propose an alternative way to determine the yield locus.

4.4 Elastic behaviour

The experimental behaviour of a sand in unloading-reloading is neither linear nor elastic. Hysteresis loops and ratcheting are commonly observed phenomena. A complicate constitutive law is then necessary if the behaviour in unloading reloading need to be modelled with precision, as for instance when cyclic loading is expected. If a sample is subject to monotonic loading only or to a very limited number of cycles of unloading-reloading, such phenomena as cyclic mobility, ratcheting or hysteresis are not relevant from an engineering viewpoint. We can therefore formulate a simple model that is not able to describe in detail the observed behaviour but that anyway is able to reproduce the features which are essential for an engineering analysis.

The elastic model that will be employed is therefore the following:

$$dv^e = B_e \frac{dp'}{p'} \tag{4.17}$$

$$d\varepsilon^e = \frac{2}{3} L_0 \, d\eta \tag{4.18}$$

Equation 4.17 is simply a generalisation of Equation 4.12 which was found to hold true for isotropic loading conditions. Experimental results obtained by Namy (1970) seem to give support to this hypothesis. Equation 4.18 is also a generalisation of what experimentally observed in triaxial tests on sand. The shear compliance in unloading is in fact approximately proportional to the inverse of the isotropic pressure.

It is worth noting that the constitutive law given by Equations 4.17 and 4.18 is hypoelastic. what means that the so called elastic strains are fully reversible if a closed load cycle is performed, but no elastic potential exists. To perform such a cycle it is necessary to add energy to the sample or , conversely, to extract work from the sample, depending on the order in which the different loading sequences are followed, see for instance Zytynski, Randolph, Nova and Wroth (1978). This is obviously in contrast with the second principle of Thermodynamics. The rationale for such a choice is therefore linked to the limited scope we want to achieve.

To avoid such problems, without abandoning the framework of elasticity, a more complicate law should be adopted for the elastic part of the deformation. Recently, Lade and Nelson (1987) proposed a new model which admits an elastic potential and seems to be also quite effective for practical purposes. Clearly even such a law can not model hysteresis or ratcheting, which are non-elastic phenomena.

We may finally note that, once a model for elastic strains has been chosen, plastic strains can be determined from total strains even when monotonic stress paths are followed. If we now consider Equation 4.16, we see that to each point of the p', q plane we can associate the value of a constant which is obtained as a linear combination of the plastic strains which were necessary to reach that stress point starting from the natural, stress free, state. The contours at equal value of such a constant give the yield loci, which can be determined experimentally without making recourse to cycles of unloading and reloading. This procedure was suggested by Nova and Wood (1978). An example is given in Figure 14. It is apparent that the shape of the family of yield loci determined by means of such a procedure is similar to the assumed shape of the yield loci postulated by Nova and Wood (1979).

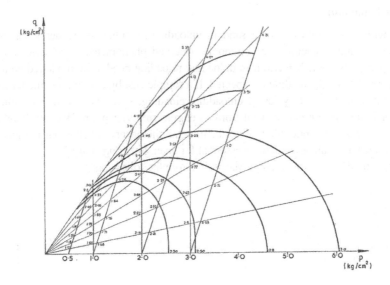

Figure 14. Experimental determination of the yield locus(after Nova and Wood, 1978).

4.5 Model performance

The simplicity of the model presented allows the stress-strain relation to be given in closed form for many stress paths of practical importance (Nova (1982)). This helps in the determination of constitutive parameters even when ad hoc test results are not available.

It is possible to derive analytically, for instance, that in drained test the failure value of the stress ratio, η_f is linked to the constitutive parameters by means of a simple equation:

$$\eta_f = M + \mu D \tag{4.19}$$

At failure the dilation is also simply given by D. If the drained test is a p' constant test, the value at which the sample starts to dilate is nothing else than M, while the initial tangent to the curve stress ratio-deviatoric strain is simply $\frac{2}{3}L_0$. Four parameters out of seven can be then retrieved directly from a p' constant test without performing complicated and laborious tests. Other two B_p and B_e can be derived from the isotropic consolidation necessary to bring the sample at the desired p' value. Only m can not be determined directly, but it is easy to choose its most convenient value by performing a parametric analysis and taking the value for which calculated results are closer to experimental data.

The model predictions, even if satisfactory, are clearly far from being perfect, however. It can be advisable then to check the derived values against the results of other tests and to see whether a more appropriate set of parameters exists or not. It is then convenient to seek other relations between the constitutive parameters. The theory predicts for instance that a zero volumetric strain test by proportional deformation is a test in which the stress ratio is constant and equal to:

$$\eta_u = M + \frac{B_e}{B}\mu D \tag{4.20}$$

This is also the asymptotic value to which tends the stress ratio in an undrained test. the stress ratio at failure in a drained test is then larger than in undrained conditions.

The stress ratio corresponding to an oedometric test is instead given by:

$$\eta_0 = \eta_u - 3/2\left(1 - B_e/B\right) \tag{4.21}$$

From the test data given for the Grenoble Workshop (Gudehus, Darve and Vardoulakis (1984)) it was possible to derive the constitutive parameters for Karlsruhe sand and to describe reasonably well the observed behaviour in triaxial constant cell pressure and oedometric test, Figures 15 and 16. Also, the constant volume value of the stress ratio was 1.51 which was within the interval experimentally determined.

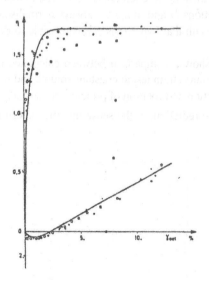

Figure 15. Karlsruhe sand - constant cell pressure compression test. After Nova (1982).

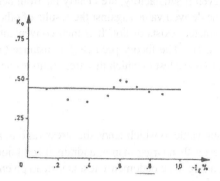

Figure 16. Karlsruhe sand - variation of Ko during an oedometric test. After Nova (1982).

To predict the behaviour in extension it was necessary to extend the model to cover situations other than triaxial compression and to take account of the Lode angle. It was decided to change the value of M in order that the locus for which volumetric strain rates are zero is a Mohr-Coulomb pyramid. The parameters D and m were scaled so that the ratio to the current value of M was constant whilst all the other parameters were left unchanged. The rationale for this choice comes from that in the closed form solutions D and m appear always normalised with respect to M as if they had the same 'dimensions'. With that hypothesis it was possible to describe very well triaxial tests on Fuji River sand.

Figures 17, 18 and 19 show a comparison between calculated and experimental results in a drained extension test, in a plane strain test at constant volume and in a plane strain test at constant volume after K_0 consolidation and rotation of principal stress axes. The last prediction is notable since it is an actual class A prediction, in the sense that the experimental data were not known a priori.

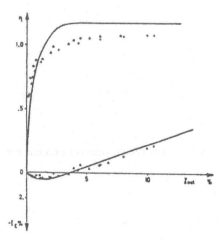

Figure 17. Karlsruhe sand – constant cell pressure extension test. After Nova (1982).

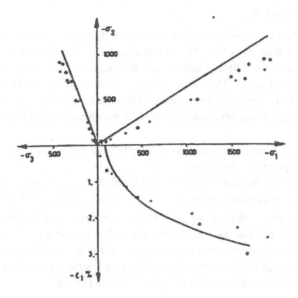

Figure 18. Karlsruhe sand - plane strain test at constant volume. After Nova (1982).

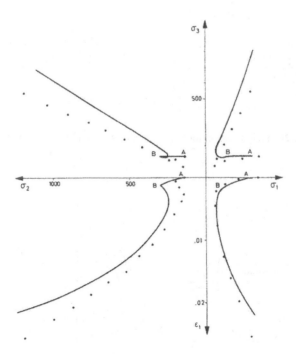

Figure 19. Karlsruhe sand – prediction of behaviour in constant volume plane strain test after Ko consolidation and rotation of principal stresses- after Nova (1982).

4.6 A further improvement

The model presented suffered from a fundamental shortcoming in that the extension to general loading conditions was done by scaling M, without altering the formal structure of the constitutive law. This is not correct, since if M depends on the Lode angle then it also depends, in a very complex way, on the state of stress. Therefore, all the derivatives of the yield function and of the plastic potential should be modified accordingly. Although conceptually simple, the explicit calculation of such derivatives was very cumbersome and the aforementioned shortcut was preferred instead.

This may give rise to erroneous predictions, especially in strain controlled tests where the derivatives of the plastic potential play a relevant role.

There was also another shortcoming related to the sharp change of the slope of the yield locus at $\eta = M / 2$. That causes an unrealistic kink in the calculated stress strain law. It was then decided to formulate another model leaving the structure unchanged but choosing more convenient expressions for the yield function and the plastic potential, whose shapes are shown in Figures 20 and 21. Also the hardening law was sligthly modified since it was assumed that the preconsolidation pressure depends on all the three invariants of the plastic strain tensor. The complete tensorial formulation of the modified constitutive law is given in Nova (1988).

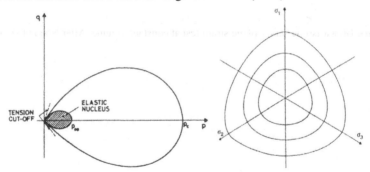

Figure 20. Yield function for 'sinfonietta classica'- after Nova (1988)

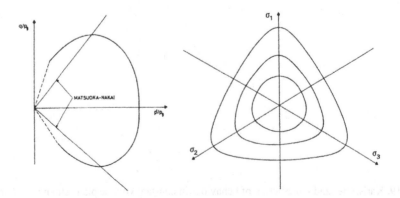

Figure 21. Plastic potential for 'sinfonietta classica'- after Nova (1988)

The performance of the new model was tested in the Cleveland Workshop. The constitutive parameters were calibrated starting from the results of triaxial compression and extension tests. Typical predictions for Hostun and Reid Bedford sand in hollow cylinder and cubic triaxial apparatus are shown in Figures 22, 23, 24. When the load path is monotonic, predictions are very good, whilst for complicated stress path involving unloading, like the circular test shown in Figure 24, predictions are very poor. This did not come at surprise since the hypoelastic modelling of the unloading reloading behaviour was clearly inappropriate for that type of test.

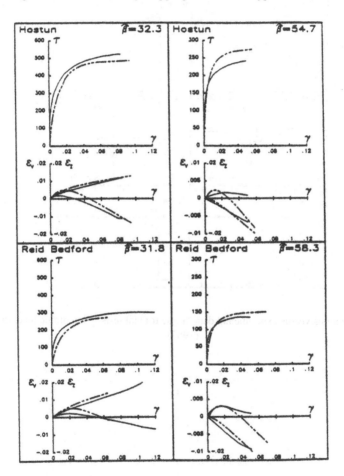

Figure 22. Predictions versus experimental data in hollow cylinder tests – monotonic loading - after Nova (1988).

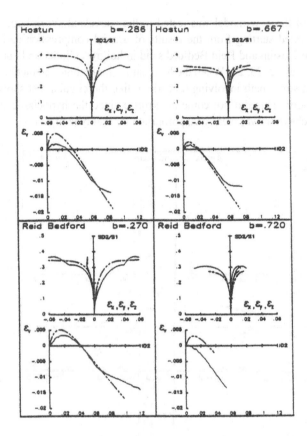

Figure 23. Predictions versus experimental data in true triaxial tests – monotonic loading - after Nova (1988).

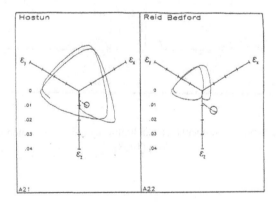

Figure 24. Predictions versus experimental data for circular stress tests – true triaxial apparatus – after Nova (1988).

4.7 Use of the model

The model presented is able to reproduce the observed behaviour with reasonable accuracy. It can be used therefore as a mean to understand better some phenomena, which may be difficult to grasp or to quantify on the basis of the experimental values alone.

Consider for instance the phenomenon of sand liquefaction in undrained conditions. Sand may liquefy either under monotonically increasing loads or under cycles of loading-unloading at constant stress amplitude. In the former case, liquefaction is possible only for very loose samples, whilst in the latter, even dense sand may reach that state, in the laboratory, provided the number of cycles is high enough.

The model discussed so far gives a justification for the occurrence of liquefaction under monotonic, or 'static', loading without any additional *ad hoc* hypothesis. We see in fact on Figure 25 that the predicted shape of the effective stress path in undrained conditions very much depends on the dilatancy parameter D. If D is very small or zero, what happens for very loose sands, the deviator stress reaches a peak and then decreases to very small values, even close to zero. Correspondingly, the pore water pressure increases up to the value of the confining pressure. A comparison between calculated and experimental results (after Castro, 1969) is shown on Figure 26, (Nova and Hueckel, 1981). If D is larger than zero, after the peak the deviator stress decreases but then increases again. Finally if D is larger than a critical value no peak exists. The line in the p', q plane which is the locus of the points for which the isotropic effective pressure starts to increase, known as phase transition line after Tatsuoka and Ishihara (1974), is the line for which $\eta = M$.

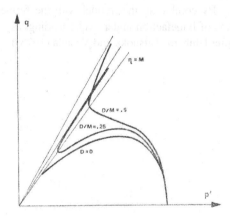

Figure 25. Effect of dilatancy parameter on undrained effective stress path after Nova and Hueckel (1981).

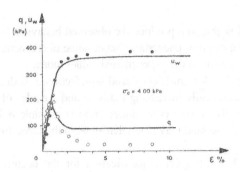

Figure 26. Comparison between calculated and experimental results for a sand undergoing static liquefaction after Nova and Hueckel (1981).

The reason for liquefaction under cyclic loading is different. After each cycle, if it is free to drain water out, the sample densifies, no matter which is its initial density. Correspondingly, the pore water pressure increases if drainage is not allowed for. This leads to a continuous decrease of the effective isotropic pressure and eventually to the failure of the sample.

In the framework of elastoplasticity, this stress path is within the yield locus. Elasticity can not clearly model such a ratcheting effect and a different model should be conceived. Nova and Hueckel (1980), for instance, formulated a model which is able to describe hysteresis and ratcheting within the yield surface. By combining that model with the former one, it was possible to describe also the phenomenon of liquefaction under cyclic loading, as shown in Figure 27 (Nova and Hueckel (1981) - data after Ishihara, Tatsuoka and Yasuda (1975)).

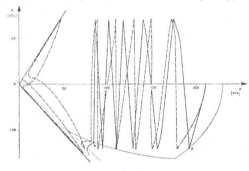

Figure 27. Modelling of cyclic liquefaction - after Nova and Hueckel (1981).

A constitutive model can also be used to analyse the effect of some systematic errors in experimental tests. The penetration of the lateral membrane into a triaxial sample, for instance, may considerably affect the measured volumetric strains in a test in which the confining pressure does not remain constant. By assuming a suitable equation to represent this penetration effect, Baldi and Nova (1983) studied its influence on the determination of the liquefaction potential in monotonic tests. Figure 28 shows that the calculated stress path is very much influenced by the compliance of the membrane. In fact if this is reduced, for instance by increasing the sample diameter, the susceptibility to liquefaction largely increases, as experimental results shown in Figure 29 clearly demonstrate.

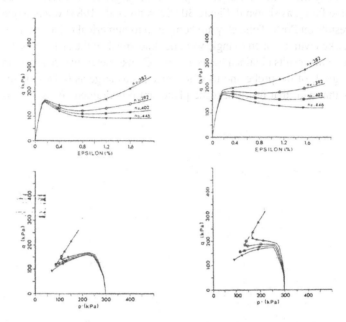

Figure 28. Effect of membrane compliance on calculated stress path in undrained tests after Baldi and Nova (1983).

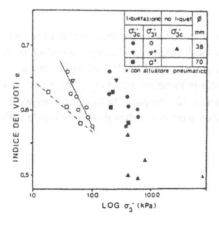

Figure 29. Modification of liquefaction potential due to reduced effect of membrane. After Baldi and Nova (1983).

In situ tests can also be modelled. It is generally accepted, for instance, that with a pressuremeter test the undrained cohesion is overestimated (Jamiolkowski et al., 1979). One of the reasons
is that plane strain conditions, which are assumed to apply in any theoretical interpretation of the
test, are not actually met. The shorter is the probe, the larger is the overestimation of the correct
plane strain value for c_u, as shown in Figure 30 (Borsetto et al., 1983) where a number of different
experimental results on Porto Tolle clay are compared to numerical calculations. It is worth noting
that the data marked with an open triangle were not known when the analysis was performed. They
make reference to the results obtained by means of a Camkometer, whose aspect ratio, i.e. the ratio
between the length L of the probe and its diameter D, is as large as 6. Despite that, the numerical
analysis shows that with that instrument the plane strain undrained cohesion is overestimated by
24%.

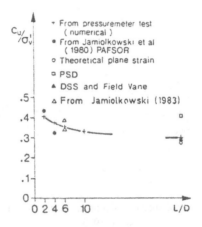

Figure 30. Effect of aspect ratio on back calculated value of C_u in a pressuremeter test.

Finally, a constitutive model can be used for the solution of boundary value problems of practical interest. In Figure 31 the mesh employed to analyse a section of the Underground Railway of
Milan is shown. The soil is a gravelly sand reinforced with grout in a region close to the excavation. The available geotechnical information was very poor, since only SPT tests were available.
Despite that, as shown in Figure 32, the predictions (full line) were not far from the observed
behaviour (dots) and indeed they were better than those obtainable with an elastic perfectly plastic
model (dotted lines).

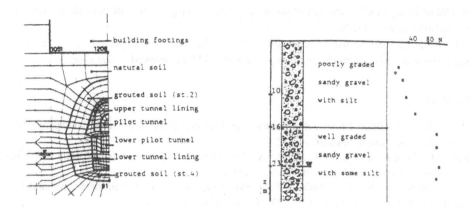

Figure 31. Analysis of a tunnel in grouted sand; mesh used and row geotechnical data. After Botti et al. (1988).

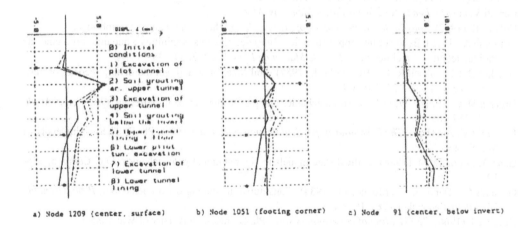

Figure 32. Comparison between calculated and observed settlements for the tunnel of Figure 31. After Botti et al. (1988).

4.8 Conclusions

An elastoplastic strain hardening model with a non-associate flow rule can describe with reasonable accuracy the observed behaviour of virgin sand. Such a model may be employed to understand better the experimental behaviour of sand or to interpret test results or eventually to solve boundary value problems.

The model has the advantage of having a simple constitutive structure, of being characterized by a limited number of constitutive parameters, easily determinable with laboratory or even in situ tests, and of being flexible enough to model any type of virgin soil. Not only sands were in fact tested but also clays and silts.

The model can not be used successfully when cyclic loading is expected, however. A model which takes hysteresis and ratcheting into account should be employed in that case.

References

Baldi G. , Nova R. (1983) Effetti della penetrazione della membrana in prove di liquefazione $\overline{R}$. *15° Cong. Naz. Geotecnica*, Spoleto, 1, 3-12.

Borsetto M. , Imperato L. , Nova R. , Peano A. (1983) Effects of pressuremeter of finite length in soft clay. *Proc. Int. conf. 'In Situ Testing'* Paris, 2, 211-215.

Botti E. , Canetta G. , Nova R. , Peduzzi R. (1988) An application of a strain hardening model to the design of tunnels in sand. *Proc. 6th ICONMIG*, Innsbruck, 3, 1641-1646.

Boussinesq J. (1885) *Applications des potentielles a l'etude de l'equilibre et du mouvement des solides elastiques*. Gauthier-Villars, Paris.

Britto A. M. , Gunn M. J. (1987) *Critical State Soil Mechanics via Finite elements*. Ellis Horwood.

Burland J. B. (1967) *Deformation of soft clay*. Ph. D., Thesis, University of Cambridge.

Castro G. (1969) *Liquefaction of sand*. Harvard SM Series N. 81

Cerruti V. (1882) *Rend. Accademia Lincei Mem. Fis. Mat.*

Cole E. R. (1967) *Soils in the simple shear apparatus*. Ph. D. Thesis University of Cambridge.

Coulomb A. (1976) Essai sur une application des règles des maximis et minimis a quelques problèmes de statique relatifs a l'architecture. *Mem. de Mat. et de Phys.* 7, 1773, 343-382, Paris.

Drucker D. C. , Gibson R. E. , Henkel D. J. (1957) Soil Mechanics and workhardening theories of plasticity. *Trans. ASCE*, 122, 338-346.

Drucker D. C. , Prager W. (1952) Soil mechanics and plastic analysis or limit design. *Quart. App. Math.* 10, 2, 157-165.

El-Sohby M. A. (1964) *The behaviour of particulate materials under stress*. Ph. D. Thesis, University of Manchester.

Gens A. , Potts D. M. (1988) Critical state models in computational geomechanics. *Eng. Comp.* 5, 178-197.

Gudehus G. , Darve F. , Vardoulakis I. (1984) Constitutive modelling of soil behaviour. *Proc. Grenoble Workshop*, September 1982, Balkema.

Holubec I. (1966) *The yielding of cohesionless soils*. Ph. D. thesis, University of Waterloo.

Ishihara K. , Tatsuoka F. , Yasuda S. (1975) Undrained deformation and liquefaction of sand under cyclic stresses. *Soils and Foundations* 15, 1, 29-44.

Jamiolkowski M. , Lancellotta R. , Marchetti S. , Nova R. , Pasqualini E., (1979) General report design parameters for soft clay , *7 ECSMFE*, Brighton, State of the art volume 1-39.

Kelvin Lord, (Thomson W.) (1855) *Quart J. of Math.*

Kim M. K. , Lade P. V. (1988) Single hardening constitutive model for frictional materials I. Plastic potential function. *Computers and Geotechnics*, 5, 4, 397-324.

Lade P. V. (1977) Elastoplastic stress-strain theory for cohesionless soil with curved yield surfaces. *Int. J. Solids abd Structures*, 13, 1019-1035.

Lade P. V. (1988) Double hardening constitutive model for soils parameter determination and predictions for two sands. *Proc. Cleveland Workshop Constitutive Equations for Granular Non-Cohesive Soils*, 367-382.

Lade P. V. , Nelson R. B. (1987) Modelling the elastic behaviour of granular materials. *Int. J. Num. An. Meth. Geom.* , 11, 5, 521-542.

Lade P. V. , Kim M. K. (1988) Single hardening constitutive model for frictional materials II. Yield criterion and plastic work contours. *Computers and Geotechnics*, 6, 1, 13-30.

Lade P. V. , Kim M. K. (1988) Single hardening constitutive model for frictional materials III. Comparison with experimental data. *Computers and Geotechnics*, 6, 1, 31-48.

Lode W. (1926) Versuche uber den Einfluss der mittleren Hauptspannung auf das Fliessen der Metalle Eisen, Kupfer und Nickel, 2, *Physik* 36, 913-939.

Matsuoka H. , Nakai T. (1974) Stress deformation and strength characteristics under three different principal stresses. *Proc. JSCE* 232, 59-70.

Matsuoka H. , Nakai T. (1982) A new failure condition for soils in three-dimensional stresses. *Proc. IUTAM conf. Deformation and Failure of Granular Materials*, Delft, 253-263.

Mises R. von (1913) Mechanik der festen Korper in plastisch deformablen Zustand. Gottinger Nachrichten, *Math. Phys. Kl.* 582-592.

Mroz W. (1967) On the description of anisotropic work hardening. *I. J. Mech. Phys. Solids* 15, 163-175.

Namy D. (1970) *An investigation of certain aspects of stress strain for clay soils.* Ph. D. thesis, Cornell University.

Nova R. (1977) On the hardening of soils. *Archiv. Mech. Stos.* 29, 3, 445-458.

Nova R. (1982) A model of soil behaviour in plastic and hysteretic ranges. Part I - monotonic loading. *Proc. Int. workshop 'Constitutive modelling of soils behaviour'* Villard de Lans, published 1984, 289-309.

Nova R. (1988) Sinfonietta classica: an exercise on classical soil modelling. *Proc. Symp. Constitutive Equations for Granular non-cohesive soils*, Cleveland.

Nova R. , Hueckel T. (1980) A geotechnical stress variables approach to cyclic behaviour of soils. *Proc. Behaviour of soils under cyclic and transient loading.* Swansea, 1, 301-314.

Nova R. , Hueckel T. (1981) A unified approach to the modelling of liquefaction and cyclic mobility. *Soils and Foundations* 21, 13-28.

Nova R. , Wood D. M. (1978) An experimental programme to define the yield function for sand. *Soils and Foundations* 18, 4, 77-86.

Nova R. , Wood D. M. (1979) A constitutive model for sand in triaxial compression. *I. J. Num: Anal. Meth. Geomech.* 3, 3, 255-278.

Poorooshasb H. B. (1971) Deformation of sand in triaxial compression. *4th Asian Reg. Conf. on Soil Mech.* , Bangkok, 1, 63-66.

Poorooshasb H. B. , Holubec I. , Sherbourne A. N. (1966) Yielding and flow of sand in triaxial compression. (Part I), *Can Geotech. J.* 3, 4, 179-190.

Poorooshasb H. B. , Holubec I. , Sherbourne A. N. (1967) Yielding and flow of sand in triaxial compression. (Parts II and III), *Can Geotech. J.* 4, 4, 376-397.

Poulos H. G. , Davis E. H (1974) *Elastic solutions for soil and rock mechanics.* Wiley.

Rendulic L. (1937) Ein Grundgesetz der Tonmechanik und sein experimenteller Beweis. *Bauingenieur*, 18, 459-467.

Roscoe K. H. , Schofield N. A. , Wroth C. P (1958) On the yielding of soils. *Geotechnique*, 8, 22, 53.

Rowe P. W. (1962) The stress dilatancy relation for static equilibrium of an assembly of particles in contact. *Proc. Roy. Soc.* , A269, 500-527.

Saada A. , Bianchini G. (1988) Constitutive equations for granular non cohesive soils. *Proc. Int. Workshop*, Cleveland 22-24 July 1987, Balkema.

Schofield N. A. , Wroth C. P (1968) *Critical State Soil Mechanics.* McGraw-Hill, London.

Stroud M. A. (1971) The behaviour of sand at low stress levels in the simple shear apparatus. *Ph. D. , Thesis.* University of Cambridge.

Vesic A. S. , Clough G. V. (1968) Behaviour of granular materials under high stresses. *J. Soil Mech. Found. Div.* , *Proc. ASCE*, 94, SM3, 661-688.

Taylor G. I. , Quinney H. (1931) The plastic distortion of metals. *Phil. Trans. Roy. Soc. A*, 230, 323-362.

Terzaghi K. (1923) Die berechnung der Durchlassigkeitsziffer des Tones aus dem Verlauf der Hydrody-namischen Spannungsuscheinungen. *Akad. Der Wissens. Wien, Math. -naturwiss. kl.* Part. IIa, 132, 3/4 125-138.

Terzaghi K. (1925) Erdbaumechanik auf Bodenphysicalischer Grundlage. Vienna, Denticke.

Tresca H. (1868) Memoire sur l'ecoulement des corps solides. *Mem. Pres par div. Savants* 18, 733-799.

Vermeer P. A. (1978) A double hardening model for sand. *Géotechnique* 28, 4, 413-433.

Wroth C. P. (1977) The predicted performance of a soft clay under a trial embankment loading based on the Cam Clay model. in *Finite Elements in Geomechanics*. G. Gudehus Editor, CH 6, Wiley.

Zienkiewicz O. C. , Humpheson C. , Lewis R. W. (1975) Associated and non-associated viscoplasticity and plasticity in Soil Mechanics. *Géotechnique* 25, 4, 671-689.

Zytynsky M. , Randolph M. F. , Nova R. , Wroth C. P. (1978) On modelling the unloading-reloading behaviour of soils. *I. J. Num. Anal. Meth. Geomech.* 2, 87-94.

Continuum damage modelling in geomechanics

Gilles Pijaudier-Cabot[1], Ludovic Jason[1]*, Antonio Huerta[2],and Jean-François Dubé[3]

[1]R&DO – GeM, Ecole Centrale de Nantes, France
[2]Laboratori de Càlcul Numèric, Universitat Politècnica de Catalunya, Barcelona, Spain
[3]LMGC/UMR5508, Université de Montpellier II, France
*also at EDF R&D, Clamart, France

Abstract: Continuum damage mechanics is a framework for describing the variations of the elastic properties of a material due to microstructural degradations. This chapter presents the application of this theory to the modelling of concrete. Several constitutive relations are devised, including incremental, explicit, and non local damage models. A general framework for damage induced anisotropy is also presented. Coupled damage and plasticity modelling is discussed. Finally, the issue of the experimental determination of the internal length in non local models is tackled.

3.1 Introduction

This chapter is concerned with the presentation of several continuum damage models and their implementation in non linear finite element analyses. Continuum damage means that the mechanical effects of progressive microcracking, void nucleation and growth are represented by a set of state variables which act on the elastic and/or plastic behaviour of the material at the macroscopic level. First, we will deal with the scalar damage model and present different versions of this type of constitutive relations. After having recalled how the local integration of the constitutive relation can be performed in the same spirit as for plasticity-based models, we will deal with more specific applications: cases where the evolution of damage is specified in an integrated form, where damage induced inelastic strains are introduced, where crack closure effects are modelled, and finally where damage is coupled to plasticity. The following section deals with damage induced anisotropy. A general framework for capturing directionality of damage is recalled. Coupling with plasticity and crack closure effects are also discussed. Non local damage is developed in the fourth section. We present integral and gradient models. In particular, a general scheme for the implementation of the gradient model, implemented into the finite element code "Code_Aster", is described. The fifth section deals with several computational issues related to damage such as the implementation of the model within secant or tangent algorithms. The chapter concludes on the problem of the calibration of the model parameters in damage models including the internal length. Inverse analysis of size effect tests on geometrically similar specimens is used for the experimental determination of these parameters.

3.2 Isotropic Damage Model

When damage is assumed to be isotropic, it is considered that it produces a degradation of the elastic stiffness of the material through a variation of the Young's modulus:

$$\varepsilon_{ij} = \frac{1+v_0}{E_0(1-d)}\sigma_{ij} - \frac{v_0}{E_0(1-d)}[\sigma_{kk}\delta_{ij}] \tag{1}$$

where v_0, E_0 and δ_{ij} are the Poisson's ratio and Young's modulus of the undamaged isotropic material and the Kronecker symbol respectively. The elastic (i.e. free) energy per unit mass of material is:

$$\rho\psi = \frac{1}{2}(1-d)\varepsilon_{ij}C^0_{ijkl}\varepsilon_{kl} \tag{2}$$

where C^0 is the stiffness tensor of the undamaged material. This energy is assumed to be the state potential. The damage energy release rate is defined as the variable associated to the damage state variable in the state potential:

$$Y = -\rho\frac{\partial\psi}{\partial d} = \frac{1}{2}\varepsilon_{ij}C^0_{ijkl}\varepsilon_{kl} \tag{3}$$

with the rate of dissipated energy:

$$\dot{\phi} = -\frac{\partial\rho\psi}{\partial d}\dot{d} \tag{4}$$

For an isotropic damage model, this equation reduces to:

$$\dot{\phi} = Y\dot{d} \tag{5}$$

The damage energy release rate is always positive and thus the rate of damage must be positive in order to comply with the second principle of thermodynamics.

The rules governing the evolution of damage can be defined following the same principles as those in elasto-plasticity. In fact, the general formalism is that of generalised standard materials where the loading function is assumed to be a pseudo potential of dissipation (Lemaitre, 1992). The evolution of damage requires the definition of a loading function:

$$f(Y,d) = Y - Y_0 - Z \tag{6}$$

where Y_0 is a parameter which defines the threshold of damage, and Z is the hardening-softening controlling variable. The evolution law is prescribed according to the normality rule:

$$\dot{d} = \dot{\lambda} \frac{\partial f}{\partial Y} \qquad (7)$$

with the Kuhn-Tucker conditions $\dot{\lambda} \geq 0, f \leq 0,$ and $\dot{\lambda} f = 0$. The evolutionary equations are completed by the definition of a hardening function:

$$\dot{\lambda} = H \dot{Z} \qquad (8)$$

where H can be regarded as the equivalent of a hardening-softening modulus in generalised plasticity. In most applications, this modulus is not constant. The major difference with plasticity is that the loading function and the evolution equation for damage are provided in an explicit form because these quantities are function of the total strain.

3.2.1 Damage Model with Crack Closure

Crack closure effects are of importance when the material is subjected to alternated loads, for instance reinforced concrete structures subjected to earthquakes. The fundamental property of the scalar damage model is that the time derivative of damage is constrained to be positive or zero (second principle of thermodynamics). During load reversals, however, micro cracks are closing progressively and the tangent stiffness of the material should increase while damage is constant (Fig. 1).

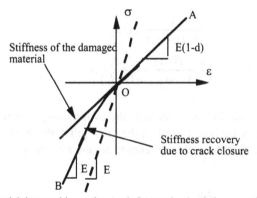

Material damaged in tension (path OA) and unloaded up to point B

Figure 1. Schematic response of a material subjected to uniaxial compression after being damaged in tension.

Within isotropic damage modelling, one solution is to introduce two damage scalars (d_t, d_c), instead of one, in order to separate the mechanical effect of micro cracking as a function of the type of loading. La Borderie (1991) developed the corresponding constitutive relations and applied it to concrete. The elastic energy is:

$$\rho\psi = \frac{1}{2}[\frac{1}{E_0(1-d_t)}<\sigma>_{ij}^+<\sigma>_{ij}^+ + \frac{1}{E_0(1-d_c)}<\sigma>_{ij}^-<\sigma>_{ij}^-$$
$$+\frac{v_0}{E_0}(\sigma_{ij}\sigma_{ij}-\sigma_{kk}^2)]+\frac{\beta_1 d_t}{E_0(1-d_t)}f(\sigma)+\frac{\beta_2 d_c}{E_0(1-d_c)}\sigma_{kk}$$

(9)

where $\langle\sigma\rangle^+$ is the positive part of the stress, i.e. the stress tensor expressed in its principal directions with the positive principal stresses only, and $\langle\sigma\rangle^- = \sigma - \langle\sigma\rangle^+$.

Inelastic strains due to the growth of damage are also taken into account in this model with a function $f(\sigma)$ introduced in order to represent the progressive vanishing of inelastic strains due to damage in tension when the material is subjected to compression:

$$f(\sigma) = \sigma_{kk} \text{ when } \sigma_{kk} \in]0,\infty] f(\sigma)$$
$$f(\sigma) = (1+\frac{\sigma_{kk}}{2\sigma_c})\sigma_{kk} \text{ when } \sigma_{kk} \in]-\sigma_c, 0]$$
$$f(\sigma) = -\frac{\sigma_c}{2}\sigma_{kk} \text{ when } \sigma_{kk} \in[-\infty, -\sigma_c]$$

(10)

where σ_c is the crack closure stress (material parameter). The stress-strain relation reads:

$$\varepsilon_{ij} = \frac{1}{E_0(1-d_t)}<\sigma>_{ij}^+ + \frac{1}{E_0(1-d_c)}<\sigma>_{ij}^-$$
$$+\frac{v_0}{E_0}[\sigma_{ij}-\sigma_{kk}\delta_{ij}]+\frac{\beta_1 d_t}{(1-d_t)}\frac{\partial f(\sigma)}{\partial \sigma_{ij}}+\frac{\beta_2 d_c}{(1-d_c)}\delta_{ij}$$

(11)

Since the model has two damage variables, the evolutionary equations include two yield functions. These functions are constructed on the same principles as the function used for the one scalar model, except that there are two energy release rates associated to the two damage variables respectively:

$$Y_t = -\rho\frac{\partial\psi}{\partial d_t}, \quad Y_c = -\rho\frac{\partial\psi}{\partial d_c}$$

(12)

The two loading functions are:

$$f_t(Y_t, d_t) = Y_t - Y_{0t} - Z_t, \quad f_c(Y_c, d_c) = Y_c - Y_{0c} - Z_c$$

(13)

and an associated model can be constructed with two independent hardening functions similar to Eq. (8).

Note that this constitutive relation cannot be integrated explicitly. The stress decomposition into positive and negative parts is not known a priori. Therefore, the conditions of evolution of damage are not known for each integration step. Same as in classical plasticity, the integration of the constitutive relations is performed with an iterative scheme: the equations of the system are the state laws and the evolution laws recast as follows in order to obtain a well-conditioned algebraic system:

$$
\begin{cases}
\varepsilon_{ij} = \dfrac{1}{E_0(1-d_t)} <\sigma>^+_{ij} + \dfrac{1}{E_0(1-d_c)} <\sigma>^-_{ij} \\[2mm]
+ \dfrac{v_0}{E_0}[\sigma_{ij} - \sigma_{kk}\delta_{ij}] + \dfrac{\beta_1 d_t}{(1-d_t)}\dfrac{\partial f(\sigma)}{\partial \sigma_{ij}} + \dfrac{\beta_2 d_c}{(1-d_c)}\delta_{ij} \\[2mm]
Y_t - Z_t = 0 \\[2mm]
Y_c - Z_c = 0 \\[2mm]
\dfrac{1}{(1-d_t)} = F_t(Y_t, Z_t) \\[2mm]
\dfrac{1}{(1-d_c)} = F_c(Y_c, Z_c)
\end{cases}
\qquad (14)
$$

Here, F_t, F_c are calculated from the integrated equation of evolution of damage instead of an incremental one. It does not change the general picture of the integration process, however. The system is written in a matrix form:

$$
[J]\bar{x} = 0 \quad \text{with } \bar{x}^T = \left\{ \sigma \quad Y_t \quad Y_c \quad \dfrac{1}{1-d_t} \quad \dfrac{1}{1-d_c} \right\}
\qquad (15)
$$

where matrix $[J]$ can be computed and programmed using formal calculus. A full Newton algorithm is implemented in order to solve this non-linear system. Note that there are two sorts of non-linearities: the first one is due to the evolution of damage (i.e. whether damage grows or not), the second is due to crack closure effects. For constant values of the damage variables, the material is orthotropic in the coordinate system of the principal stresses.

3.2.2 Integrated Damage Model

As we saw above, the evolution of damage is very often related to the state of strain. Moreover, it is defined in an explicit, integrated way, which is easier to handle. The damage loading function in Eq. (6) can be redefined as:

$$
f(\tilde{\varepsilon},\kappa) = \tilde{\varepsilon} - \kappa
\qquad (16)
$$

where $\tilde{\varepsilon}$ is a positive equivalent measure of strain and κ is a threshold value. The equation $f = 0$ represents a loading surface in strain space. For the uniaxial tensile case, the equivalent uniaxial strain in Eq. (16) is straightforward. It is the axial strain. However, for general states of stress, damage evolution should be related to some scalar quantity, function of the state of strain. There are, to this regard, several proposals. For example, an appropriate definition for metals is rooted in the elastic stored energy (Peerlings et al. 1998):

$$\tilde{\varepsilon} = \sqrt{\frac{1}{E} \, \varepsilon_{ij} C_{ijkl} \varepsilon_{kl}} \tag{17}$$

which is depicted in Figure 2. It is equivalent to the integrated version of the constitutive relation presented in section 2.1. For concrete, Mazars (1984) proposed the following form:

$$\tilde{\varepsilon} = \sqrt{\sum \left(<\varepsilon_i>_+ \right)^2} \tag{18}$$

where ε_i are the principal strains. A third possibility, which is also mentioned by Peerlings, is the modified von Mises definition:

$$\tilde{\varepsilon} = \frac{k-1}{2k(1-2v)} I_1 + \frac{1}{2k} \sqrt{\frac{(k-1)^2}{(1-2v)^2} I_1^2 - \frac{12k}{(1+v)^2} J_2} \tag{19}$$

I_1 and J_2 are the first invariant of the strain tensor and the second invariant of the deviatoric strain tensor respectively. Only the modified von Mises criterion leads to a new material parameter, namely the factor k. The parameter k is the ratio between uniaxial compressive and uniaxial tensile strengths. These criteria are plotted in figure 2 (with $k = 10$). The loading surfaces (Eqs. 17-19) are closed contours around the origin. The dashed lines represent the constant uniaxial compression and uniaxial tension stress paths.

The evolution of damage has the same form as in the previous sections:

$$\text{if } f(\tilde{\varepsilon}, \kappa) = 0 \text{ and } \dot{f}(\tilde{\varepsilon}, \kappa) = 0 \text{ then } \begin{cases} \dot{d} = h(\kappa) \\ \kappa = \tilde{\varepsilon} \end{cases} \text{ where } \dot{d} \geq 0$$

$$\text{otherwise } \begin{cases} \dot{d} = 0 \\ \dot{\kappa} = 0 \end{cases} \tag{20}$$

The function $h(\kappa)$ is specific, depending on different models in the literature. For tension only, exponential softening can be used:

$$h(\kappa) = 1 - \frac{\kappa_0}{\kappa} (1 - \alpha + \alpha e^{-\eta(\kappa - \kappa_0)}) \tag{21}$$

where κ_0, α, η are model parameters.

In order to capture the differences of mechanical responses of the material in tension and in compression, Mazars proposed to split the damage variable into two parts and used the equivalent strain defined in Eq. (18):

$$d = \alpha_t d_t + \alpha_c d_c \tag{22}$$

where d_t and d_c are the damage variables in tension and compression, respectively.

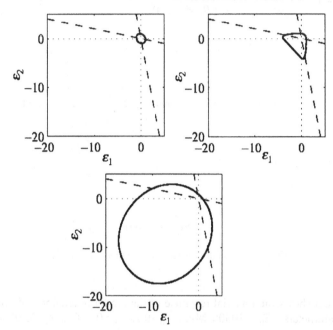

Figure 2. Contour plots for $\tilde{\varepsilon}$ for the elastic stored energy (top left), Mazars definition (top right), and the modified von Mises expression (bottom), after Peerlings et al. (1998).

They are combined with the weight coefficients α_t and α_c defined as functions of the principal values of the strains ε_{ij}^t and ε_{ij}^c, due to positive and negative stresses (see Mazars, 1984).

$$\varepsilon_{ij}^t = (1-d)C_{ijkl}^{-1}\sigma_{kl}^t, \quad \varepsilon_{ij}^c = (1-d)C_{ijkl}^{-1}\sigma_{kl}^c \tag{23}$$

$$\alpha_t = \sum_{i=1}^{3}\left(\frac{\langle \varepsilon_i^t \rangle\langle \varepsilon_i \rangle_+}{\tilde{\varepsilon}^2}\right)^{\beta}, \quad \alpha_c = \sum_{i=1}^{3}\left(\frac{\langle \varepsilon_i^c \rangle\langle \varepsilon_i \rangle_+}{\tilde{\varepsilon}^2}\right)^{\beta} \tag{24}$$

In uniaxial tension $\alpha_t = 1$ and $\alpha_c = 0$. In uniaxial compression $\alpha_c = 1$ and $\alpha_t = 0$. Hence, d_t and d_c can be obtained separately from uniaxial tests. The purpose of exponent β is to reduce the effect of damage on the response of the material under shear compared to tension (Pijaudier-Cabot et al. 1991). The evolution of damage is provided in an integrated form, as a function of the variable κ (Mazars, 1984):

$$d_t = 1 - \frac{\kappa_0(1 - A_t)}{\kappa} - \frac{A_t}{\exp(B_t(\kappa - \kappa_0))}$$

$$d_c = 1 - \frac{\kappa_0(1 - A_c)}{\kappa} - \frac{A_c}{\exp(B_c(\kappa - \kappa_0))}$$

(25)

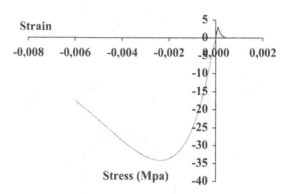

Figure 3. Uniaxial response of the model by Mazars (1984).

Figure 3 shows the uniaxial response of the model in tension and compression with the following parameters: $E_0 = 30000$ MPa, $v_0 = 0.2$, $\kappa_0 = 0.0001$, $A_t = 1$, $B_t = 15000$, $A_c = 1.2$, $B_c = 1500$, $\beta = 1$.

3.2.3 Damage coupled to plasticity

Elastic damage models or elastic plastic laws are not totally sufficient to correctly capture the constitutive behavior of concrete. In some cases, the calculation of the damage variable is a key point. For instance, it can become essential when coupled effects are considered (coupling between damage and permeability, damage and porosity ...). Picandet et al. (2001) observed in experiments that there is a relationship between damage and the gas permeability of the material. Damage, measured according to the unloading slope during cyclic compressive loading is related to the variation of permeability, independently from the type of concrete tested (normal concrete, high strength concrete, fiber-reinforced concrete). Thus, the capability of the constitutive model to capture the unloading behavior is essential if a proper evaluation of the permeability is required.

An elastic damage model is not appropriate as irreversible strains cannot be captured: a zero stress corresponds to a zero strain and the value of the damage is thus overestimated (figure 4a). An elastic plastic relation is not adapted (even with softening, see for example Grassl et al, 2002) as the unloading curve always follows the same elastic slope (figure 4b). An appealing alternative consists in combining these two approaches into an elastic-plastic-damage law. The softening behavior and the decrease in the elastic modulus are so reproduced by the damage part while the plasticity effect accounts for the irreversible strains. With this formulation, experimental unloading can be correctly simulated (figure 4c).

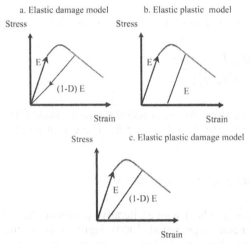

Figure 4. Unloading response of elastic damage, elastic plastic and elastic plastic damage models.

The plastic model is governed by the following set of equations:

$$d\varepsilon_{ij} = d\varepsilon_{ij}^e + d\varepsilon_{ij}^p$$
$$\sigma_{ij}^t = C_{ijkl}^0 \varepsilon_{kl}^e$$
$$d\varepsilon_{ij}^p = d\lambda.m_{ij}\left(\sigma', k_h\right) \tag{26}$$
$$dk_h = d\lambda.h\left(\sigma', k_h\right)$$

with the effective stress defined as

$$\sigma_{ij}^t = \frac{\sigma_{ij}}{(1-d)} \tag{27}$$

where ε, ε^e and ε^p are respectively the total, elastic and plastic strains, λ the plastic multiplier and k_h the hardening parameter ($0 \leq k_h \leq 1$). m and h are the flow vectors defined by :

$$
m_{ij} = \frac{\partial f}{\partial \sigma'_{ij}}
$$

$$
h = \frac{\sqrt{\dfrac{2}{3} \dfrac{\partial f}{\partial \sigma'_{ij}} \cdot \dfrac{\partial f}{\partial \sigma'_{ij}}}}{\zeta(\sigma')}
\tag{28}
$$

f is the yield surface and ζ a function of the stress (Etse and Willam, 1994).

$$
f = \bar{\rho}^2 (J'_2) - \frac{k(k_h, I'_1)\bar{\rho}_c^2(I'_1)}{r^2(\theta, I'_1)}
\tag{29}
$$

with I'_1 and J'_2 the effective stress invariant and θ the Lode angle function of the second and third stress invariant and ranging from $[-\pi/6 ; \pi/6]$. The evolution of the plastic multiplier is finally given by:

$$
f(\sigma', k_h) \leq 0, \quad d\lambda \geq 0, \quad f(\sigma', k_h).d\lambda = 0
\tag{30}
$$

This local problem is solved with an iterative procedure associated to a closest point projection algorithm (see Perez-Foguet et al, 2000). Figure 5 shows the evolution of the yield surface with an increasing hardening parameter k_h for simple compression $\bar{\rho} = K_1 \sqrt{J'_2}, \bar{\xi} = K_2 I'_1$, where K_1 and K_2 are two positive constants).

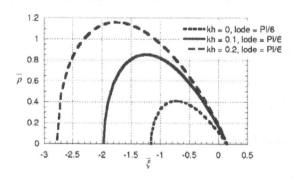

Figure 5. Plastic yield surface. Evolution with the hardening parameter for simple compression (Lode = $\pi/6$).

As the hardening parameter has a limited value of 1, once it has reached this critical level, hardening is not allowed any more and the yield surface becomes a failure one (constant effective stress).

Figure 6 gives the axial stress – strain curve for a loading in tension. As the development of damage is predominant during simple tension tests, the damage and plastic-damage models are almost equivalent.

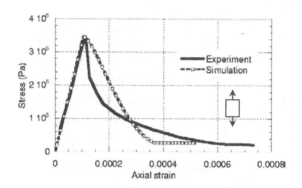

Figure 6. Stress strain curve for simple tension test. Elastic plastic damage formulation.

Cyclic compression is a second elementary test. Experimental results are taken from (Sinha et al, 1964). Figure 7 illustrates the numerical response for simple damage law (without any plasticity). With this type of relation, a zero stress corresponds to a zero strain.

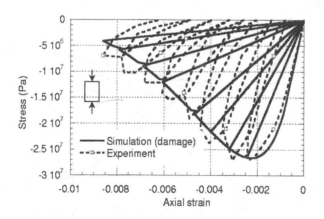

Figure 7. Cyclic compression test. Elastic damage simulation.

The numerical response of the elastic plastic damage model is given in figure 8. Damage simulates the global softening behaviour of concrete while the plastic part reproduces quantitatively the evolution of the irreversible strains. Experimental and numerical unloading slopes are thus similar, contrary to the simple damage formulation response. If this difference could seem negligible, it is in fact essential if a correct value of the damage needs to be captured. The elastic damage model overestimates d whereas the full plastic damage constitutive law provides more acceptable results.

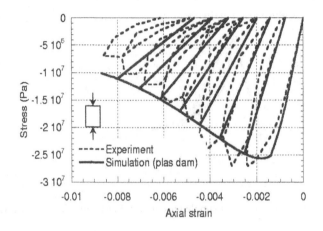

Figure 8. Cyclic compression test. Elastic plastic damage formulation.

Figure 9 and 10 illustrates the differences between the two approaches in term of volumetric behaviors.

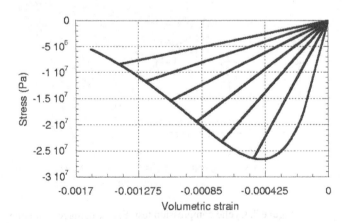

Figure 9. Cyclic compression test. Volumetric behaviour for elastic damage model.

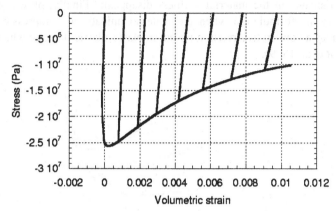

Figure 10. Cyclic compression test. Volumetric behaviour for the plastic damage model.

While, with the simple damage law, the volumetric strains keep negative values, the introduction of plasticity exhibits a change in the volumetric response from contractant (negative volumetric strains) to dilatant, a phenomenon which is experimentally observed (see Sfer et al, 2002 for example).

The introduction of plasticity associated with the development of damage plays thus a key role in the numerical simulation of a cyclic compression test. Moreover, the volumetric response, that was totally misevaluated by the elastic damage model, is correctly simulated with the full formulation. In figure 11, numerical results are compared with experiment for different levels of confinement pressures in triaxial experiments (P = 0, 1.5, 4.5, 9, 30 and 60 MPa). Simulations and experiment give similar results. The peak position is quantitatively well reproduced (except for 1.5 MPa).

The global evolution is also correct: the maximum of the axial stress increases with the pressure and the softening part is less and less significant. When the initial hydrostatic pressure takes higher values, the damage part of the model plays a minor role and plasticity effect becomes predominant.

3.3 Damage Induced Anisotropy

Microcracking is usually geometrically oriented as a result of the loading history on the material. In tension, microcracks are perpendicular to the tensile stress direction, in compression microcracks open parallel to the compressive stress direction. Although a scalar damage model, which does not account for directionality of damage, might be a sufficient approximation in usual applications, i.e. when tensile failure is expected with a quasi-radial loading path, damage induced anisotropy is required for more complex loading histories. The influence of crack closure is needed in the case of alternated loads: microcracks may close and the effect of damage on the material stiffness disappears. Finally, plastic strains need to be incorporated in the formulation when the material unloads in compression. The following section describes a constitutive relation based on elasto-plastic damage which addresses these issues (Fichant et al. 1999).

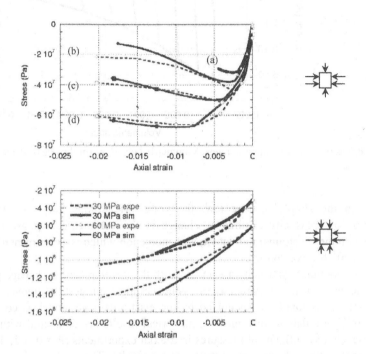

Figure 11. Triaxial test with increasing confinement. Axial stress - strain curves for low hydrostatic pressures. Straight lines (black markers) correspond to simulation, dotted lines (white markers) to experiment (on top (a) 0 MPa, (b) 1.5 MPa, (c) 4.5 MPa, (d) 9 MPa).

3.3.1 Principle

The model is based on the approximation of the relationship between the overall stress (latter simply denoted as stress) and the effective stress in the material defined here by the equation:

$$\sigma^t_{ij} = C^0_{ijkl}\varepsilon^e_{kl} \quad \text{or} \quad \sigma^t_{ij} = C^0_{ijkl}\left(C^{damaged}\right)^{-1}_{klmn}\sigma_{mn} \tag{31}$$

where σ^t_{ij} is the effective stress component, ε^e_{kl} is the elastic strain, and $C^{damaged}_{ijkl}$ is the stiffness of the damaged material. We define the relationship between the stress and the effective stress along a finite set of directions of unit vectors $\vec{n}$ at each material point:

$$\sigma = \left(1-d(n)\right)n_i\sigma_{ij}'n_j, \quad \tau = \left(1-d(n)\right)\sqrt{\sum_{i=1}^{3}\left(\sigma_{ij}'n_j-\left(n_k\sigma_{kl}n_l\right)n_i\right)^2} \qquad (32)$$

σ and τ are the normal and tangential components of the stress vector respectively. $d(n)$ is a scalar valued quantity which introduces the effect of damage in each direction $\bar{n}$.

The basis of the model is the numerical interpolation of $d(n)$ (called damage surface) which is approximated by its knowledge *over a finite set of directions*. The stress is solution of the virtual work equation:

find σ_{ij} such that $\forall \varepsilon_{ij}^{*}$

$$\frac{4\pi}{3}\sigma_{ij}\varepsilon_{ij}^{*} = \int_{S}\left(\left[(1-d(n))n_k\sigma_{kl}'n_ln_i+(1-d(n))\left(\sigma_{ij}'n_j-n_k\sigma_{kl}'n_ln_i\right)\right]\varepsilon_{ij}^{*}n_j\right)d\Omega \qquad (33)$$

Depending on the interpolation of the damage variable $d(n)$, several forms of damage induced anisotropy can be obtained. Here, it is defined by three scalars in three mutually orthogonal directions. It is the simplest approximation that yields anisotropy of the damaged stiffness of the material. The material is orthotropic with a possibility of rotation of the principal axes of orthotropy.

The stiffness degradation occurs mainly for tensile loads. Hence, the evolution of damage will be indexed on tensile strains. The evolution of damage is controlled by a loading surface f, which is similar to Eq. (6):

$$f(n) = n_i\varepsilon_{ij}^{e}n_j - \varepsilon_d - \chi(n) \qquad (34)$$

χ is an hardening softening variable which is interpolated in the same fashion as the damage surface. The initial threshold of damage is ε_d. The evolution of the damage surface is defined by an evolution equation inspired from that of an isotropic model:

if $f(n^{*}) = 0$ and $n_i^{*}d\varepsilon_{ij}n_j^{*} > 0$

$$\text{then}\begin{cases}dd(n^{*}) = \left[\dfrac{\varepsilon_d\left(1+a\left(n_i^{*}\varepsilon_{ij}^{e}n_j^{*}\right)\right)}{\left(n_i^{*}\varepsilon_{ij}^{e}n_j^{*}\right)^2}\exp\left(-a.\left(n_i^{*}\varepsilon_{ij}^{e}n_j^{*}-\varepsilon_d\right)\right)\right]n_i^{*}d\varepsilon_{ij}^{e}n_j^{*}\\[6pt] d\chi(n^{*}) = n_i^{*}d\varepsilon_{ij}^{e}n_j^{*}\end{cases} \qquad (35)$$

else $dd(\bar{n}^{*}) = 0, d\chi(n^{*}) = 0$

The model parameters are ε_d and a. Note that the vectors $\bar{n}^{*}$ are the three principal directions of the *incremental* strains whenever damage grows. After an incremental growth of damage, the new damage surface is the sum of two ellipsoidal surfaces: the one corresponding to the initial damage surface, and the ellipsoid corresponding to the incremental growth of damage.

the new damage surface is the sum of two ellipsoidal surfaces: the one corresponding to the initial damage surface, and the ellipsoid corresponding to the incremental growth of damage.

3.3.2 Coupling with plasticity

Same as for the isotropic model, we decompose the strain increment in an elastic and plastic one. The evolution of the plastic strain is controlled by a yield function which is expressed in term of the effective stress in the undamaged material. We have implemented in this anisotropic damage model the yield function due to Nadai (1950). It is the combination of two Drucker-Prager functions F_1 and F_2 with the same hardening evolution:

$$F_i = \sqrt{\frac{2}{3} J_2'} + A_i \frac{I_1'}{3} - B_i w \tag{36}$$

where J_2' and I_1' are the second invariant of the deviatoric effective stress and the first invariant of the effective stress respectively. w is the hardening variable and (A_i, B_i) are four parameters ($i = 1, 2$) which were originally related to the ratios of the tensile strength to the compressive strength denoted γ and of the biaxial compressive strength to the uniaxial strength denoted β:

$$A_1 = \sqrt{2} \frac{1-\gamma}{1+\gamma}, \quad A_2 = \sqrt{2} \frac{\beta-1}{2\beta-1}, \quad B_1 = 2\sqrt{2} \frac{\gamma}{1+\gamma}, \quad B_2 = \sqrt{2} \frac{\beta}{2\beta-1} \tag{37}$$

These two ratios are constant in the model: $\beta = 1.16$ and $\gamma = 0.4$. The evolution of the plastic strains is associated to these surfaces. The hardening rule is given by:

$$w = qp^r + w_0 \tag{38}$$

where q and r are model parameters, w_0 defines the initial reversible domain in the stress space, and p is the effective plastic strain.

3.3.3 Crack closure effects

A decomposition of the stress tensor into a positive and negative part is introduced again: $\sigma = \langle \sigma \rangle_+ + \langle \sigma \rangle_-$, where $< \sigma >_+$, and $< \sigma >_-$ are the positive and negative parts of the stress tensor. The relationship between the stress and the effective stress defined in Eq. (31) of the model is modified:

$$\sigma_{ij} n_j = (1 - d(n)) < \sigma >_{+ij}^t n_j + (1 - d_c(n)) < \sigma >_{-ij}^t n_j. \tag{39}$$

$d_c(n)$ is a new damage surface which describes the influence of damage on the response of the material in compression. Since this new variable refers to the same physical state of degradation as in tension, $d_c(n)$ is directly deduced from $d(n)$. It is defined by the same interpolation as $d(n)$ and along each principal direction i, we have the relation:

$$d_c^i = \left(\frac{d_i(1 - \delta_{ij})}{2} \right)^\alpha , \ i \in [1,3] \tag{38}$$

α is a model parameter. The constitutive relations contain 6 parameters in addition to the Young's modulus of the material and the Poisson's ratio. Their determination benefits from the fact that in tension, plasticity is negligible. Once the evolution of damage in tension has been fitted, the remaining parameters are fitted from a compression test. Figure 12 shows a typical uniaxial compression-tension response of the model corresponding to a concrete with a tensile strength of 3 MPa and a compressive strength of 40 MPa.

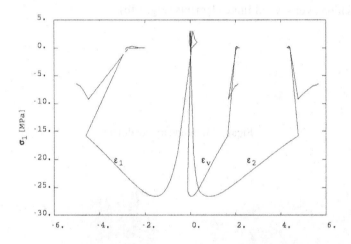

Figure 12. Uniaxial tension-compression response of the anisotropic model (longitudinal (1), transverse (2) and volumetric (v) strains as functions of the compressive stress)

3.4 Non Local Damage

We turn now attention to structural computations with the isotropic damage model. It is well known that when strain softening is encountered (see. Fig. 3 for instance), some difficulties arise.

Physically, after microcracking has taken place in the first part of the inelastic behavior of concrete, a macroscopic mechanism develops in the form of a damage band. The strains tend to localize in a specific area of finite dimension (macrocrack) but the model predicts this dimension to be zero, with a vanishing energy dissipation at failure. The stress response of a material point cannot be described locally but has also to consider its neighbourhood in order to take into account microcracks interaction (Askes, 2000). Mathematically (Benallal *et al*, 1993, Peerlings *et al*, 1996), when a material point enters the softening regime, the type of the governing equations changes from elliptic to hyperbolic. The mathematical description becomes ill-posed since the initial and boundary conditions that are meaningful for the elliptic problem are usually not relevant for the hyperbolic part. Numerically, the local formulation of the problem yields a response which is dependant on the smoothness and the orientation of the mesh. Let us consider for example a bar with a variable section under tensile loading whose boundary conditions are given in figure 13. In the damage distribution in figure 14 (a), damage localises only in the most loaded element. If we consider the extreme case of an infinitely fine mesh, this geometrical localisation in a single element yields a fracture without energy dissipation. The local formulation thus exhibits a mesh dependency problem and a regularisation technique has to be added in order to achieve a mesh independent result and damage localisation over several finite elements (Fig. 14b).

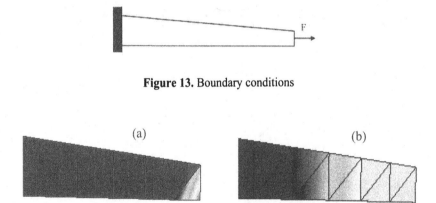

Figure 13. Boundary conditions

Figure 14. Damage distribution (a) local simulation (b) non local simulation. Clear zones correspond to damaged one.

It is now established that non locality, in a gradient or integral format, is mandatory for a proper, consistent, modelling of fracture (see e.g. de Borst et al. 1993). It avoids the difficulties encountered upon material softening and strain localisation. Within a single approach, it encompasses both crack initiation (for which continuum models are very well fitted) and crack propagation (for which discrete fracture approaches have been developed). We shall describe in the following integral and gradient non local damage models.

3.4.1 Integral Model

Consider for instance the scalar damage model in which evolution of damage is controlled by the equivalent strain $\tilde{\varepsilon}$ introduced by Mazars. The principle of nonlocal continuum models with local strains is to replace the equivalent strain $\tilde{\varepsilon}$ with its average (Pijaudier-Cabot and Bazant, 1987):

$$\bar{\varepsilon}(x) = \frac{1}{V_r(x)} \int_\Omega \psi(s)\tilde{\varepsilon}(s+x)\,ds \quad \text{with} \quad V_r(x) = \int_\Omega \psi(s)\,ds \tag{39}$$

where Ω is the volume of the structure, $V_r(x)$ is the representative volume at point x, and $\psi(s)$ is the weight function, for instance:

$$\psi(s) = \exp\left(-\frac{4|s|^2}{l_c^2}\right) \tag{40}$$

l_c is the internal length of the non local continuum. $\bar{\varepsilon}$ replaces the equivalent strain in the evolution of damage. In particular, the loading function becomes $f(\bar{\varepsilon}, \kappa) = \bar{\varepsilon} - \kappa$. As we will see in section 3.5, it should be noticed that this model is easy to implement in the context of explicit, total strain models. Its extension to plasticity and to implicit incremental relations is awkward. The local tangent stiffness operator relating incremental strains to incremental stresses becomes non symmetric, and more importantly its bandwidth can be very large due to non local interactions. This is one of the reasons why gradient damage models have become popular over the past few years.

3.4.2 Gradient damage model

A simple method to transform the above non local model to a gradient model is to expand the equivalent strain into Taylor series truncated for instance to the second order:

$$\tilde{\varepsilon}(x+s) = \tilde{\varepsilon}(x) + \frac{\partial \tilde{\varepsilon}(x)}{\partial x}s + \frac{\partial^2 \tilde{\varepsilon}(x)}{\partial x^2}\frac{s^2}{2!} + \dots \tag{41}$$

Substitution in Eq. (39) and integration with respect to variable s yields:

$$\bar{\varepsilon}(x) = \tilde{\varepsilon}(x) + c^2 \nabla^2 \tilde{\varepsilon}(x) \tag{42}$$

where c is a parameter which depends on the type of weight function in Eq. (39). Its dimension is m^2 and it can be regarded as the square of an internal length. Substitution of the new expression of the non local equivalent strain in the non local damage model presented above yields a gradient damage model. Computationally, this model is still delicate to implement because it requires higher continuity in the interpolation of the displacement field.

This difficulty can be solved if an implicit format of the gradient damage model is used. Eq. (42) is replaced with

$$\bar{\varepsilon}(x) - c^2 \nabla^2 \bar{\varepsilon}(x) = \tilde{\varepsilon}(x) \tag{43}$$

Here, the definition of the non local equivalent strain is implicit. It is the solution of a Fredholm equation. As shown by Peerlings et al. (1996), the implicit form is in fact an exact representation of the integral relation devised by Pijaudier-Cabot and Bazant (1987), provided a specific exponential weight function is used.

3.4.3 Extension of the Gradient Approach

In the anisotropic damage models described in section 3.3, the evolution of damage is directional. The evolution of damage is also directional in the microplane-based models and in the smeared crack models. The extension of the gradient damage model to anisotropy, or to any strain-based damage model, can be performed in a very systematic way (Godard, 2001, 2003). The Fredholm equation is written for each strain component ε_{nn} :

$$\bar{\varepsilon}_{nn} - c^2 \nabla^2 (\bar{\varepsilon}_{nn}) = \varepsilon_{nn} \tag{44}$$

Therefore, a non local (gradient type) strain tensor is computed within each iteration. It is from this tensor that an equivalent strain can be computed, or that directional damage growth can be devised depending on the type of damage model that is implemented. This extended form of a gradient model has been implemented in the general purpose finite element code *"Code_Aster"* at EDF.

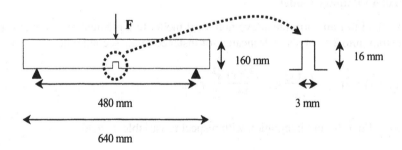

Figure 15. Bending beam.

3.5 Computer implementation of damage models

Since the major effect of damage is a secant stiffness reduction when microcracking progresses, a non-linear computational procedure based on the secant material stiffness seems quite appropriate. This type of algorithm is also widely used when an integral non local damage model is implemented. The major reason is that the tangent stiffness operator (in the case of loading) is far from being straightforward because of non locality.

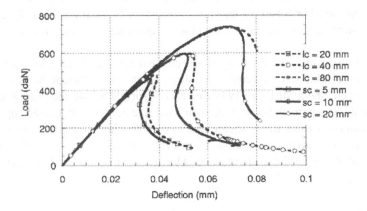

Figure 16. Comparison between two non local approaches for different values of the internal length parameter.

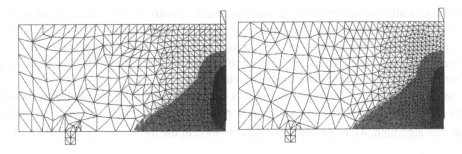

Figure 17. Distribution of damage. Dark zones correspond to damaged one (D = 1 for black zones, 0<D<1 for grey zones and D = 0 for white zones)

The integral relation due to the non local term is discretised according to the finite element mesh used for the analysis and an usual quadrature rule is employed for its evaluation. In finite element calculations, the weight function is chopped off: the weights that are less than 0.001

are set to zero. The actual volume of integration does not span over the entire volume of the solid and the calculation of the integrals requires less computer time and memory as the number of neighbouring integration points is reduced. It should be kept in mind, however, that the secant stiffness algorithm presents serious difficulties, mainly there is no proof of convergence of such an algorithm. In large scale finite element computations, a standard Newton-Raphson algorithm is highly preferable, as convergence is more robust. Such an algorithm relies on the derivation of the consistent tangent stiffness of the material response.

The implementation of the scalar damage model, coupled to the gradient approach is different. The equilibrium equations and the Fredholm equations are solved as a coupled problem. The non local strain tensor can be discretised according to the same finite element grid as the displacements. For a better robustness, a standard Newton-Raphson algorithm is used with a consistent tangent operator that ensures quadratic convergence, if properly implemented.

In order to illustrate results with the integral and gradient damage models, we consider the classical case of a three point bending beam containing a notch. Two different meshes are used (a COARse 890 element and a MEDium 1938 element). Figure 15 shows the beam geometry. Figure 16 presents the comparison between the non local integral and regularised approaches for the medium mesh (s_c is equal to the square root of c). Choosing the appropriate non local parameters, the two techniques are almost equivalent. Figure 17 presents the damage distribution in the half beam for the gradient model. One can see that it is mesh independent.

3.6 Model calibration

The calibration of the model parameters in the non local (integral or gradient) damage model is facing the difficulty of getting at the same time the parameters involved in the evolution of damage κ_0, A_t, B_t, and the internal length l_c. The first set is a point-wise piece of information while the influence of the internal length appears in a boundary value problem with non homogeneous strains (by definition, $\tilde{\varepsilon} = \bar{\varepsilon}$). It follows that it is not possible to calibrate the model parameters in the damage evolution law, without considering the computation of a complete boundary value problem up to failure. Hence the model parameter calibration has to rely on an inverse analysis technique.

The model parameters are optimised so that computational results are as close as possible to the experimental results. The optimisation process used here is based on a Levenberg-Marquardt algorithm, which performs the minimisation of the following functionnal $\Im(\vec{P})$:

$$\Im(\vec{P}) = \frac{1}{2} \sum_{i=1}^{3} \left(\frac{R^i(\vec{P}) - R^i_{exp}}{R^i_{exp}} \right)^2 \tag{45}$$

where $R_{exp}{}^i$ is the experimental response for test i ($i \in [1,3]$), and $R^i(\vec{P})$ is the numerical response corresponding to the vector of model parameters $\vec{P}$. As we will see next, the calibration cannot be performed on a single set of experimental data, e.g. three point bend experiments. One possibility is to perform the calibration from several experiments, and for instance size effect test data.

3.6.1 Size effect tests

Three point bend experiments have been carried out by Le Bellégo et al. (2000) on notched geometrically similar specimens of various height D = 80, 160, and 320 mm, of length L = 4D, and of thickness b = 40 mm kept constant. The length-to-height ratio is L/D = 4, the span-to-height ratio is $l/D = 3$, and the notch length is D/10 (Fig. 18). These are mortar specimens with a water-cement ratio of 0.4 and a cement-sand ratio 0.46 (all by weight). The maximal sand grain size is d_a = 3mm. Experimental and computational data will be also interpreted with the help of Bazant's size effect law, in one of its simplest form (Bazant and Planas, 1998). The nominal strength is obtained with the formula:

$$\sigma = \frac{3}{2} \frac{Fl}{b(0.9D)^2} \tag{46}$$

where F is the peak load. The size effect law reads:

$$\sigma = \frac{Bf_t'}{\sqrt{1 + D/D_0}} \tag{47}$$

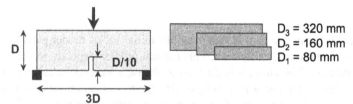

Figure 18. Geometry of the three point bend size effect tests

D_0 is a characteristic size, f_t' is the tensile strength of the material, and B is a geometry – related parameter.

Figure 19 shows the size effect law in a log-log plot. It can be seen that the nominal value of the material strength decreases as the size of the beam increases. This is typical of a damage process in which the process zone is size independent (controlled by the internal length) compared to the structure. For very small sizes, the process zone covers all the specimen and a material strength criterion is expected, while for infinitely large specimens the process zone size is negligible and Linear Elastic Fracture Mechanics should apply.

The calibration of the model parameters from inverse analysis relies first on an accurate finite element representation of each size of specimen. Experience shows that the accuracy of the finite element discretisation in the fracture process zone depends on the ratio of the size of the finite element to the internal length. Therefore, geometrically similar finite element meshes should not be used for the calibration. In order to avoid some possible mesh bias, the finite element size in the process zone should be kept constant and small enough compared to the internal length (at least one third of the internal length). Obviously, the internal length is not known a priori and this requirement is checked at the end of the calibration.

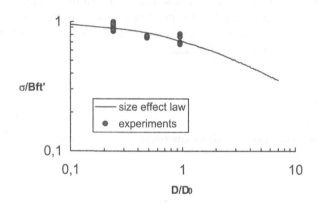

Figure 19. Experimental results of size effect in three point bend tests.

The optimisation technique has to start from an initial set of model parameters obtained from a manual (trial and error) calibration. This manual process depends on the exact form of the constitutive relation. Therefore, it is difficult to devise a general technique. Still it can be carried out in a rather systematic way as follow: first the model parameters are fitted such that the load deflection curve of one size of specimen is correctly described (usually the medium size). As an initial guess, the internal length lies between $3d_a$ and $5d_a$, where d_a is the maximum aggregate size of concrete according to Bazant and Pijaudier-Cabot (1989). The initial damage threshold and the parameters in the equation controlling the damage growth are varied only (in fact, the Young's modulus of the material could be obtained as well from the initial stiffness of the beam). Then, the internal length is adjusted, the other parameters being kept constant, so that the numerical responses for the three specimen sizes are as close as possible to the experimental results. This process requires many calculations, each trial iteration involves three finite element computations. Figure 20 shows such a fit obtained manually.

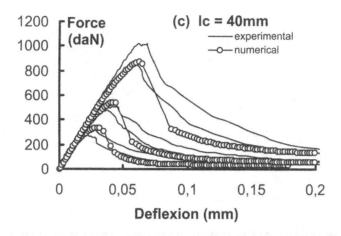

Figure 20. Manual fit of the damage parameters from size effect tests.

Because there are experimental errors and because the above calibration process is based on a variation of the internal length which is disconnected to the variation of the other model parameters, there are several sets of model parameters which may provide an acceptable fit of the tests data. By acceptable, we mean that the numerical responses are as much as possible within the extreme curves due to experimental dispersion (or within an arbitrary distance from them). The difference between these sets ought to be magnified. For this purpose, the size effect method is quite useful (see Le Bellégo et al. 2003).

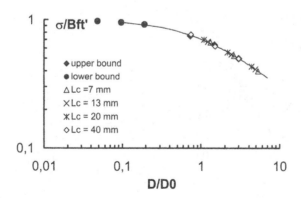

Figure 21. Interpretation of quasi-identical fits on the size effect plot.

Figure 21 shows how several fits obtained manually with different values of the internal length can be distinguished on the size effect plot.

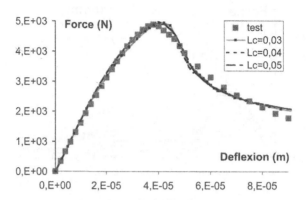

Figure 22. Automatic calibration of the model parameters on the middle size experiment for different - fixed - internal lengths.

For each set of model parameters, the constants in the size effect law have been fitted and the plot on Fig. 21 has been constructed. It is clear on this figure that none of them are consistent with the experimental data. While the quality of the fit is quite hard to decide upon on load – deflexion curves, it is much easier to examine it on the size effect plot.

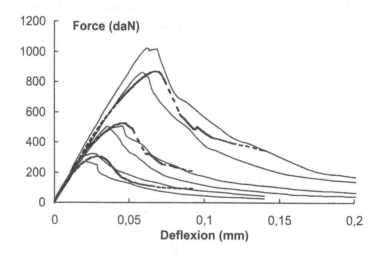

Figure 22. Automatic calibration of the model parameters. The dotted lines are the model responses and the plain lines are, for each size, the minimum and maximum curves obtained from the experimental data.

In order to exemplify the lack of objectivity of the model calibration (i.e. the lack of uniqueness of the result) when a single size of beam is used as the experimental data set, Fig. 22 shows the results of the automatic calibration on the middle size specimen. Three fixed different values of the internal length have been considered. For each one, the optimisation has been performed yielding different values of the damage parameters. The computed responses are the same, in good agreement with the experimental data. Bending of notched beams is quite often used for parameter identification because it is relatively easy to perform compared to tensile tests that are very sensitive to boundary conditions (Carmeliet, 1999). This figure shows that the model cannot be calibrated from inverse analysis of a single load deflexion curve.

Figure 23 shows the result of the optimisation from the three sizes of specimens simultaneously. Compared to the manual fit in Fig. 20, the present set of model parameter provides computational results that are closer to the experiments. We tried to change the initial set of model parameter in order to investigate its stability. Overall, the final sets did not differ very much. The largest difference was found on the internal length, which lied between 40 mm and 50 mm.

3.7 Closure

Continuous damage mechanics is a theory which aims at describing the mechanical effect of cracking and void growth in an elastic material. Isotropic and anisotropic damage models have been presented in this chapter. There is a rather generic technique which permits a rational derivation of damage models, given a level of anisotropy which is defined a priori. In order to be more representative of the concrete mechanical response, damage must be coupled to plasticity. A simple, and classical, technique is to assume that the plastic part of the model is written in term of the effective stress. The same applies to the modelling of rock masses.

Strain-softening, which is typical of concrete and rocks, induces spurious strain localisation. Integral or gradient damage model are solutions to circumvent this problem. A central issue is the calibration of the model parameters, and in particular the experimental determination of the internal length. This can be achieved from size effect test data with the help of inverse analysis of structural responses. It is important, however, to underline that a proper calibration *cannot* be achieved from the fit of a single experiment. Size effect experiments on geometrically similar specimens, or bending and pure tension tests are required.

Acknowledgements

Financial support from the European Commission through the MAECENAS project (FIS5-2001-00100), through the DIGA RTN network (HPRN-CT-2002-00220), and from the partnership between Electricité de France and the R&DO group through the MECEN project are gratefully acknowledged.

3.8 References

Askes H., 2000, Advanced spatial discretisation strategies for localised failure. mesh adaptivity and meshless methods, PhD Thesis, Delft University of Technology, Faculty of Civil Engineering and Geosciences.

Bazant, Z.P., and Pijaudier-Cabot, G., 1989, Measurement of the characteristic length of non local continuum, J. Engrg. Mech. ASCE, 115, 755-767.

Bazant, Z.P. and Planas, J., 1998, Fracture and size effect in concrete and other quasibrittle Materials, CRC Press.

Benallal A., Billardon R., Geymonat G., 1993, Bifurcation and rate-independent materials, Bifurcation and stability of dissipative systems, CISM Lecture Notes 327, Springer, 1-44.

de Borst R., Sluys L.J., Muhlhaus H.B., and Pamin J., 1993, Fundamental issues in finite element analyses of strain localisation, Engrg. Comput., 10, 99-121.

Carmeliet J., 1999, Optimal estimation of gradient damage parameters from localisation phenomena in quasi-brittle materials, Int.J. Mech. Cohesive Frict. Mats., 4, 1-16.

Etse, G. Willam, K., 1994, Fracture energy formulation for inelastic behavior of plain concrete. J. Engrg. Mech. ASCE, 120, 1983-2011.

Fichant S., La Borderie C., and Pijaudier-Cabot G., 1999, Isotropic and anisotropic descriptions of damage in concrete structures, Int.J. Mech. Cohesive Frict. Mats., 4, pp. 339-359.

Grassl, P. Lundgren, K. Gylltoft, K., 2002, Concrete in compression: a plasticity theory with a novel hardening law, International Journal of Solids and Structures 39:5205-5223.

Godard V., 2001, Modélisation mécanique non locale des matériaux fragiles: délocalisation de la déformation, Stage de DEA de Mécanique, Université Pierre et Marie Curie.

Godard V., 2003, Modélisation non locale à gradients de déformation, Documentation de référence Code_Aster R5.04.02, www.code-aster.org.

La Borderie C., 1991, Phénomènes unilatéraux dans un matériau endommageable, Thèse de Doctorat de l'Université Paris 6.

Le Bellego C., Dubé J-F., Pijaudier-Cabot G., B. Gérard, 2003, Calibration of nonlocal damage model from size effect tests, European Journal of Mechanics A/Solids, 22, 33 –46.

Le Bellego, C., Gérard B., and Pijaudier-Cabot G., 2000, Chemomechanical effects in mortar beams subjected to water hydrolysis, J. Engrg. Mech. ASCE, 126, 266-272.

Lemaitre J., 1992, A course on damage mechanics, Springer-Verlag.

Mazars J., 1984, Application de la mécanique de l'endommagement au comportement non linéaire et à la rupture du béton de structure, Thèse de Doctorat ès Sciences, Université Paris 6.

Nadai A., 1950, Theory of flow and fracture of solids, Vol. 1, second edition, Mc Graw Hill, New York, p. 572.

Peerlings R.H., de Borst R., Brekelmans W.A.M., de Vree J.H.P., 1996, Gradient enhanced damage for quasi-brittle materials, Int. J. Num. Meth. Engrg., 39, 3391-3403.

Peerlings R.H., de Borst R., Brekelmans W.A.M, Geers M.G.D., 1998, Gradient enhanced modelling of concrete fracture, Int.J. Mech. Cohesive Frict. Mats., 3, 323-343.

Picandet, V. Khelidj, A. Bastian, G., 2001, Effect of axial compressive damage on gas permeability of ordinary and high-performance concrete, Cement and Concrete Research, 31, 1525-1532.

Pijaudier-Cabot G. and Bazant Z.P., 1987, Nonlocal damage theory, J. of Engrg. Mech., ASCE, 113, 1512-1533.

Pijaudier-Cabot G., Mazars J., and Pulikowski J., 1991, Steel-concrete bond analysis with nonlocal continuous damage, J. of Structural Engineering, ASCE, 117, 862-882.

Sfer, D., Carol I. Gettu R., Etse, G., 2002, Study of the behavior of concrete under triaxial compression, J. Engrg. Mech. ASCE, 128, 156-163.

Sinha, B.P. Gerstle, K.H. Tulin, L.G., 1964, Stress strain relations for concrete under cyclic loading. Journal of the American concrete institute, 195-211.

Linear Micro-elasticity

Ioannis Vardoulakis[1]

[1] Faculty of Applied Mathematics and Physics, Department of Mechanics
National Technical University of Athens, Greece

Abstract. This chapter provides a brief introduction to the theory of linear Elasticity with micro-structure. The relations between constitutive and balance stresses are derived and the issues of uniqueness and boundary conditions are addressed.

4.1 Introduction

Fifty years after the first publication of the original work of the brothers Cosserat, Eugéne and François in 1909, the basic kinematic and static concepts of the "Cosserat" continuum were reworked in a milestone paper by late Professor Günther (1958). Günther's paper marks the rebirth of continuum micro-mechanics in the late 50's and early 60's. Following this publication, several hundred papers were published all over the world on that subject. A variety of names have been invented and given to theories of various degrees of rigor and complexity: They were called "Cosserat continua" or "micro-polar media", "oriented media", "continuum theories with directors", "multi-polar continua", "micro-structured" or "micro-morphic continua", etc (cf. Herrmann, 1972). The state-of-the-art at this time was reflected in the collection of papers presented at the historical IUTAM Symposium on the *"Mechanics of Generalized Continua"*, in Freudenstadt and Stuttgart in 1967.

On the subject of Cosserat Elasticity recommendable for their clarity and didactical value are the papers by H. Schaeffer (1962, 1967) and by Kessel (1964) in German and the paper of Koiter (1964) in English. Notable and of equal importance in relation to Gradient Elasticity is the milestone paper by Mindlin (1964) and two papers by Germain (1973a &b), the latter written partially in French and partially in English.

In this chapter we present a simple version of Mindlin's linear Elasticity theory with microstructure. Frantziskonis & Vardoulakis (1991), Exadaktylos (1998), Exadaktylos et al. (1996, 1998, 2001 a & b), Georgiadis (2003), Geogiadis et al. (1998, 2000, 2001, 2002, 2003), and Vardoulakis & Sulem (1995) have applied this theory in a number papers, where static and dynamic boundary-value problems have been addressed and solved analytically[1] and numerically by Amanatidou & Aravas (2002). Here the corresponding static, linear micro-elasticity theory is outlined. The basic reference for the micro-elasticity theory given herein, is the paper by Mindlin & Eshel (M&E, 1968). Central to our discussion is the distinction between the 'balance' stresses, which appear in the formulation of force- and moment-equilibrium equations, and the 'constitutive' stresses, which are introduced through the variation of the elastic strain energy density function. The existence of higher order stresses such as couple- and double-stresses is demonstrated and the relations between equilibrium and

---

[1] see also additional references to the current research literature at the end of this chapter.

constitutive stresses are derived. The uniqueness theorem proof presented here follows the line of the proof given by M&E and is a variant of the Kirchhoff uniqueness argument. The uniqueness theorem proof is based on the hypothesis for positive elastic strain energy density function, which in our case is guarantied as soon as some constitutive inequalities are satisfied; cf. Georgiadis et al. (2002). We close our discussion here by considering the issue of boundary conditions, which is central to any higher order continuum theory and we try to clarify the issue by discussing some simple boundary-value problems.

4.2 The Constitutive Equations of the "Casal-Mindlin" Micro-elasticity: The strain energy density function

As a starting point for our presentation we select here a rather unnoticed early publication by Casal (1961), which is referenced in the aforementioned papers by Germain. Casal's model is one-dimensional and starts from postulating the total elastic strain-energy in a tension bar with stiffness (EA) and length L

$$W = \frac{1}{2}\int_0^L (EA)\left(\varepsilon^2 + \ell(\nabla\varepsilon)^2\right)dx + \frac{1}{2}\left((EA)\ell'\varepsilon^2\right)\Big|_0^L \qquad (1)$$

In Eq. (1) one can easily recognize the classical term, pertaining to the contribution of strain, $\varepsilon = \nabla u$. On top of that, Eq. (1) introduces a strain-gradient effect in the volume-part as well as on the boundaries of the bar. Casal's tension bar is characterized by the introduction of two additional material constants on top of the Young's modulus E. These additional constants, ℓ and ℓ', have the dimension of length, and are responsible for the corresponding volume and surface strain-gradient dependent elastic energy contributions.

Motivated by Casal's original idea, we consider the following special form for the elastic strain-energy density function (Vardoulakis & Sulem 1995)

$$\hat{w} = \frac{1}{2}\lambda\varepsilon_{mm}\varepsilon_{nn} + G\varepsilon_{mn}\varepsilon_{nm} + \ell^2\left(\frac{1}{2}\lambda\hat{\kappa}_{kmm}\hat{\kappa}_{knn} + G\hat{\kappa}_{kmn}\hat{\kappa}_{knm}\right) +$$

$$\ell'_k\left(\frac{1}{2}\lambda\left(\hat{\kappa}_{kmm}\varepsilon_{nn} + \varepsilon_{mm}\hat{\kappa}_{knn}\right) + G\left(\hat{\kappa}_{kmn}\varepsilon_{nm} + \varepsilon_{mn}\hat{\kappa}_{knm}\right)\right) \qquad (2)$$

In Eq. (2), ε_{ij} and $\hat{\kappa}_{ijk}$ denote the infinitesimal strain and strain gradient, respectively

$$\varepsilon_{ij} = \varepsilon_{ji} = \frac{1}{2}(\partial_i u_j + \partial_j u_i) \ , \quad \hat{\kappa}_{ijk} = \hat{\kappa}_{ikj} = \partial_i\varepsilon_{jk} \qquad (3)$$

where $u_i = u_i(x_k,t)$ is the infinitesimal displacement field.

As compared to linear isotropic elasticity the present micro-elasticity theory is equipped by two additional constitutive parameters: the internal length ℓ and the director length ℓ'_k. If we set these to length parameters equal to zero we regain Hooke's law. Thus in Eq. (2) λ and G are identified as the Lamé constants of the material. This theory could be seen is a special case of a Mindlin-type 2nd gradient linear Elasticity theory with micro-structure[2], however as we will discuss in section 4.7.2, this is actually not the case, since the director ℓ'_k, will be specified in such a way, that the constitutive model equations remain isotropic.

According to the classification given by M&E the considered form of the elastic energy density function, Eq. (2), is falling into the category of energy functions of the 2nd form, which depend only on the strain and its gradient.

$$\hat{w} = \hat{w}(\varepsilon_{ij}, \hat{\kappa}_{ijk}) \qquad (4)$$

We recall the 1st form of elastic strain energy density function after M&E

$$\tilde{w} = \tilde{w}(\varepsilon_{ij}, \tilde{\kappa}_{ijk}) \quad , \quad \tilde{\kappa}_{ijk} = \partial_i \partial_j u_k \qquad (5)$$

and the 3rd form

$$\overline{w} = \overline{w}(\varepsilon_{ij}, \overline{\kappa}_{ij}, \overline{\overline{\kappa}}_{ijk}) \quad , \quad \overline{\kappa}_{ij} = \frac{1}{2}\varepsilon_{jlk}\partial_l \partial_i u_k \quad , \quad \overline{\overline{\kappa}}_{ijk} = \frac{1}{3}\left(\partial_i \partial_j u_k + \partial_j \partial_k u_i + \partial_k \partial_i u_j\right) \qquad (6)$$

As is shown in M&E all three forms for the strain energy density function are equivalent.

Using the first variation of the elastic energy density function we can introduce stress-like quantities, which are dual in energy to the strain and to the strain-gradient respectively,

$$\hat{w} = \hat{w}(\varepsilon_{ij}, \hat{\kappa}_{ijk}) \quad \Rightarrow \quad \delta\hat{w} = \hat{\tau}_{ij}\delta\varepsilon_{ij} + \hat{\mu}_{ijk}\delta\hat{\kappa}_{ijk} \qquad (7)$$

The stresses $\hat{\tau}_{ij}$ and $\hat{\mu}_{ijk}$ are called here **constitutive stresses** and are not to be confused with the balance stresses, which are introduced through the momentum and angular momentum balance laws. As we will see in this section, unlike in classical Elasticity Theory the above-defined constitutive stresses, Eqs. (7), do not coincide with the balance stresses. From Eqs. (2) and (7) we get the following constitutive equations for these

$$\hat{\tau}_{ij} = \frac{\partial \hat{w}}{\partial \varepsilon_{ij}} = \lambda\varepsilon_{kk}\delta_{ij} + 2G\varepsilon_{ij} + \ell'_k\left(\lambda\hat{\kappa}_{knn}\delta_{ij} + 2G\hat{\kappa}_{kij}\right) \qquad (8)$$

$$\hat{\mu}_{ijk} = \frac{\partial \hat{w}}{\partial \hat{\kappa}_{ijk}} = \ell^2\left(\lambda\hat{\kappa}_{inn}\delta_{jk} + 2G\hat{\kappa}_{ijk}\right) + \ell'_i\left(\lambda\varepsilon_{nn}\delta_{jk} + 2G\varepsilon_{jk}\right) \qquad (9)$$

---

[2] Notice that Mindlin's general elastic strain energy function has 903 independent coefficients!

4.3 Balance Stresses

As already mentioned balance stresses are defined independently of any constitutive equations, through the formulation of the balance laws of linear and angular momentum, which in the static limit are reduced to force and moment equilibrium conditions. In order to formulate these balance laws we must postulate the external forces, which act on the considered continuum. Accordingly for the considered micro-elastic continuum we assume the existence of the following set of external forces:

- body forces, $F_k dV$
- surface tractions, $t_k dS$
- surface couples with moment $m_k dS$
- self-equilibrating surface double tractions, $R_{ik} dS$

The set forces $\{F_k, t_k\}$ appear in the balance of linear and angular momentum, whereas the couples m_k appear only in the balance of angular momentum. The self-equilibrating double tractions R_{ik} do not appear in the balance laws.

The body forces $F_i(x_k)$ could be for example gravity-field forces, acting per unit volume of the material; e.g. $F_i = \rho g_i$, where $\rho(x_k, t)$ is the mass density field and g_i the acceleration of gravity.

Surface tractions are related to a non-symmetric **balance stress tensor** field $\tau_{ij}(x_k, t)$, and to the unit outward normal vector n_i on the considered surface element on which they act

$$t_i = n_k \tau_{ki} \tag{10}$$

In the static limit, balance of linear momentum is expressed by the following global force equilibrium condition,

$$\int_V F_i dV + \int_{\partial V} t_i dS = 0 \tag{11}$$

This condition together with Eq. (10) and Gauss' theorem yields the corresponding local equilibrium conditions for the balance stress tensor

$$\int_V (\partial_j \tau_{jk} + F_k) dV = 0 \quad \Rightarrow \quad \underline{\partial_j \tau_{jk} + F_k = 0} \tag{12}$$

Thus in the static limit, the principle of conservation of linear momentum reduces to the equilibrium equations for the non-symmetric balance stress tensor (Fig. 1).

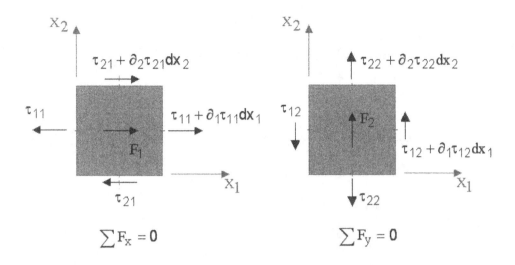

Figure 1. Force equilibrium on an infinitesimal material element

Similarly we define a couple-stress tensor field $\mu_{ij}(x_k,t)$, which is relates the surface couples m_i to the corresponding outward unit vector on the surface element on which they act

$$m_i = n_k \mu_{ki} \tag{13}$$

In the static limit balance of angular momentum is expressed by the following global moment equilibrium condition,

$$\int_V \varepsilon_{ijk} x_j F_k dV + \int_{\partial V} (\varepsilon_{ijk} x_j t_k + m_i) dS = 0 \tag{14}$$

where ε_{ijk} is the Levi-Civita permutation tensor. Eq. (14), together with Eq. (13) and Gauss' theorem yield the local moment equilibrium conditions for the balance couple-stress tensor (Fig. 2):

$$\int_V \varepsilon_{ijk} x_j \partial_l \tau_{lk} dV + \int_{\partial V} (\varepsilon_{ijk} x_j n_l \tau_{lk} + n_k \mu_{ki}) dS = 0$$

or

$$\int_V (\varepsilon_{ijk} x_j \partial_l \tau_{lk} - \partial_l(\varepsilon_{ijk} x_j \tau_{lk}) - \partial_k \mu_{ki}) dV = 0 \quad \Rightarrow$$

$$\partial_j \mu_{jk} + \varepsilon_{ijk} \tau_{ij} = 0 \tag{15}$$

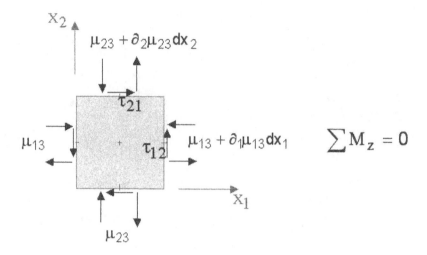

Figure 2. Moment equilibrium on an infinitesimal material element

In order to formulate the energy balance equation, we assume that the body forces and the surface tractions are working on the velocity field $v_i \approx \partial_t u_i$ and that the surface couples are working rate of the rotation vector,

$$\dot{w}_i = \frac{1}{2}\varepsilon_{ijk}\partial_j v_k \tag{16}$$

Finally we assume that the self-equilibrating surface double-tractions R_{ij} are defined by the 3rd order tensor field of the "balanced" double-stresses $\mu_{ijk}(x_1, t)$,

$$R_{ij} = n_k\mu_{kij} \tag{17}$$

These double tractions are assumed to be working on the corresponding (symmetric) strain-rate tensor

$$\dot{\varepsilon}_{ij} = \frac{1}{2}(\partial_i v_j + \partial_j v_i) \tag{18}$$

According to the above definitions the power of the external forces is defined by the following expression

$$W^{(ext)} = \int_V F_k v_k \, dV + \int_{\partial V} \left(t_k v_k + m_k \dot{w}_k + R_{ij} \dot{\varepsilon}_{ij} \right) dS \qquad (19)$$

We decompose additively the true stress in a symmetric part $\tau_{(ij)}$ and in an anti-symmetric part $\tau_{[ij]}$

$$\tau_{ij} = \tau_{(ij)} + \tau_{[ij]}$$

$$\tau_{(ij)} = \frac{1}{2}(\tau_{ij} + \tau_{ji}) \quad , \quad \tau_{(ij)} = \tau_{(ji)} \qquad (20)$$

$$\tau_{[ij]} = \frac{1}{2}(\tau_{ij} - \tau_{ji}) \quad , \quad \tau_{[ij]} = -\tau_{[ji]}$$

The moment equilibrium Eq. (15) becomes (Fig. 3),

$$\tau_{[ik]} = -\frac{1}{2}\varepsilon_{ikm}\partial_l \mu_{lm} \qquad (21)$$

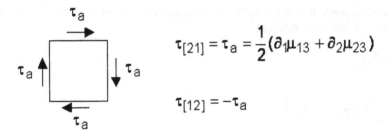

$$\tau_{[21]} = \tau_a = \frac{1}{2}(\partial_1 \mu_{13} + \partial_2 \mu_{23})$$

$$\tau_{[12]} = -\tau_a$$

Figure 3. The antisymmetric part of the balance stress tensor

Further we split the couple stress tensor into its spherical and deviatoric part

$$\mu_{ij} = \mu_{ij}^D + \frac{1}{3}\mu_{kk}\delta_{ij} \qquad (22)$$

and we introduce this in Eq. (21)

$$\tau_{[ik]} = -\frac{1}{2}\varepsilon_{ikm}\partial_l\mu_{lm}^D \tag{23}$$

Then by using the stress decomposition Eq. (19.1) we get

$$\tau_{ij} = \tau_{(ij)} - \frac{1}{2}\varepsilon_{ikm}\partial_l\mu_{lm}^D \tag{24}$$

and with that the force equilibrium Eq. (12) becomes

$$\partial_i\tau_{(ik)} + F_k = \frac{1}{2}\varepsilon_{ikm}\partial_i\partial_l\mu_{lm}^D \tag{25}$$

For elastic materials and for isothermal processes, the principle of energy conservation postulates that the power of external forces must be in equilibrium with the rate of the total elastic-strain energy

$$\dot{W}^{(el)} = W^{(ext)} \quad \Rightarrow \quad \int_V \dot{w}dV = \int_V F_k v_k dV + \int_{\partial V}\left(t_k v_k + m_k \dot{w}_k + R_{ij}\dot{\varepsilon}_{ij}\right)dS \tag{26}$$

with (cf. Eq. 7)

$$\dot{w} = \hat{\tau}_{ij}\dot{\varepsilon}_{ij} + \hat{\mu}_{ijk}\delta\dot{\kappa}_{ijk} \tag{27}$$

By using the divergence theorem and the definitions of kinematic variables from Eqs. (26) and (27) we obtain the following relations[3]:

$$\hat{\tau}_{jk} = \tau_{(jk)} + \partial_i\mu_{ijk} \tag{28}$$

$$\hat{\mu}_{ijk} = \mu_{ijk} + \frac{1}{2}\left(\varepsilon_{lik}\mu_{jl} + \varepsilon_{lij}\mu_{kl}\right) \tag{29}$$

These relations are connecting the constitutive stresses $\hat{\tau}_{jk}$, $\hat{\mu}_{ijk}$, which are dual in energy to the strain and its gradient, to the balance stresses τ_{jk}, μ_{ijk}, which are defined by the equations of force and moment equilibrium. We eliminate μ_{ijk} from Eqs. (28) and (29) and we get

$$\tau_{(jk)} = \hat{\tau}_{jk} - \partial_i\hat{\mu}_{ijk} + \frac{1}{2}\partial_i\mu_{jl}^D\varepsilon_{lik} + \frac{1}{2}\partial_i\mu_{kl}^D\varepsilon_{lij} \tag{30}$$

---

[3] The proof of Eq. (31) can be found in the paper by M&E. For the proof of Eqs. (32) to (34) see also in Appendix I. Appendices I, II and III of this section are based on the paper by M&E and have been worked out by Mrs Lilian Vougiouka, as a part of her Master of Civil Engineering Thesis at NTU Athens in November 2001.

We can prove the following relationships[4]

$$\mu_{ij}^D = \frac{2}{3}(\hat{\mu}_{ipq} + \hat{\mu}_{piq})\varepsilon_{jpq} \tag{31}$$

$$\overline{\mu}_{ij} = \mu_{ij}^D \tag{32}$$

$$\mu_{ijk} = \frac{1}{3}(\hat{\mu}_{ijk} + \hat{\mu}_{jki} + \hat{\mu}_{kij}) \tag{33}$$

$$\overline{\mu}_{il}\varepsilon_{ljk} + \overline{\mu}_{jl}\varepsilon_{lki} + \overline{\mu}_{kl}\varepsilon_{lij} = 0 \tag{34}$$

We define now the following symmetric tensor

$$\sigma_{jk} = \hat{\tau}_{jk} - \partial_i\hat{\mu}_{ijk} \tag{35}$$

Substituting Eqs. (30) to (35) into Eq. (25) we get that the equilibrium equations can be written in terms of the above introduced tensor σ_{ij},

$$\partial_j\sigma_{jk} + F_k = 0 \tag{36}$$

For this reason the stress-tensor σ_{ij} will be called here the **equilibrium stress tensor.**
    By using the constitutive Eqs. (8) and (9) for the constitutive stresses $\hat{\tau}_{ij}$ and $\hat{\mu}_{ijk}$ and the definition of the balance stress σ_{ij}, Eq. (34), we get that

$$\sigma_{jk} = (1 - \ell^2\nabla^2)(\lambda\varepsilon_{ii}\delta_{jk} + 2G\varepsilon_{jk}) \tag{37}$$

With this expression for the balance stress the equilibrium condition (36) yields,

$$\partial_j(1 - \ell^2\nabla^2)(\lambda\varepsilon_{ii}\delta_{jk} + 2G\varepsilon_{jk}) + F_k = 0 \tag{38}$$

or

$$\lambda\delta_{jk}\partial_j\partial_i(u_i - \ell^2\partial_l\partial_l u_i) + G(\partial_j\partial_j(u_k - \ell^2\partial_l\partial_l u_k) + \partial_j\partial_k(u_j - \ell^2\partial_l\partial_l u_j)) + F_k = 0 \tag{39}$$

---

[4] Recall that $\overline{\mu}_{ij} = \dfrac{\partial w}{\partial\overline{\kappa}_{ij}}$, where $\overline{w} = \overline{w}(\varepsilon_{ij}, \overline{\kappa}_{ij}, \overline{\overline{\kappa}}_{ijk})$ is M&E's 3rd form of the strain energy density function (Appendix II).

Exadaktylos & Vardoulakis (2001) showed that the general solution of Eq. (39) may be expressed in terms of generalized Neuber-Papkovich displacement functions.

Remarks

First we emphasize here that the symmetric equilibrium stress tensor σ_{ij} does not coincide with the non-symmetric balance stress tensor τ_{ij}. Both stress tensors fulfil the force equilibrium condition. Thus their difference is divergence free,

$$\partial_j(\tau_{jk} - \sigma_{jk}) = 0$$

Secondly we may re-interpret Eq. (37) as follows: We assume the existence of a local equilibrium
stress σ_{ij}

$$\partial_j\sigma_{jk} + F_k = 0$$

We assume that Hooke's law does not hold for the local equilibrium stress but an average stress, defined over a representative elementary volume:

$$<\sigma_{ij}> = \int_{(REV)} \sigma_{ij}dV = \lambda\varepsilon_{kk}\delta_{ij} + 2G\varepsilon_{ij}$$

We can easily show that for cubic (REV) with dimension a the following approximation formula holds:

$$<\sigma_{ij}>_{(REV)} = \left(1 + \frac{a^2}{24}\nabla^2\right)\sigma_{ij}(x_k)$$

By inverting the differential operator[5] we get Eq. (37):

$$\sigma_{ij} = \frac{1}{1 + \frac{a^2}{24}\nabla^2} <\sigma_{ij}>_{(REV)} \approx \left(1 - \frac{a^2}{24}\nabla^2\right)\left(\lambda\varepsilon_{kk}\delta_{ij} + 2G\varepsilon_{ij}\right)$$

We remark that the governing differential Eq. (38), that guaranties equilibrium, is written in terms of the strain tensor ε_{ij} and involves only the material length ℓ. Thus for the determination of the displacement field $u_k(x_i)$ we do not need to specify the spatial

---

[5] cf. Chapter 5, Appendix I.

distribution of the director field $\ell'_i(x_k)$, which is appearing in the strain-energy density function, Eq. (2). As we will see in section 4.7.2, the director field ℓ'_k needs only to be specified on the boundary of the considered micro-elastic body. This property of the considered micro-elasticity model is indeed reminiscent of the problem of surface tension T in fluids. For example in the study of capillary surface waves in fluids the information concerning the characteristic length of the problem, $\ell' = 2\pi\sqrt{T/(\rho g)}$, enters only through the dynamic boundary condition, that assigns on the surface of the fluid a fictitious pressure that is proportional to the mean curvature of that surface[6], $\Delta p = -T \nabla^2 \zeta(x,y)$.

4.4 Relations Between Balance- and Constitutive Stresses

We start from the decomposition of the true tensor in symmetric and antisymmetric part, Eq. (20). From the derivations presented in the previous section we get that,

$$\tau_{(jk)} = \sigma_{jk} + \frac{1}{2}\partial_i\mu^D_{jl}\varepsilon_{lik} + \frac{1}{2}\partial_i\mu^D_{kl}\varepsilon_{lij} \tag{40}$$

$$\tau_{[ik]} = -\frac{1}{2}\varepsilon_{ikm}\partial_l\mu_{lm} \tag{41}$$

$$\mu_{ik} = \mu^D_{ik}, \quad \underline{\mu_{nn} = 0} \tag{42}$$

The assumption that the spherical part of the couple-stress tensor (i.e. the mean torsion) is vanishing ($\mu_{nn} = 0$) is justified by the fact that it can be proven[7] that the solution of the generic boundary-value problem is independent of the value of μ_{nn}. Moreover it holds,

$$\mu^D_{ij} = \frac{2}{3}(\hat{\mu}_{ipq} + \hat{\mu}_{piq})\varepsilon_{jpq} \tag{43}$$

with $\hat{\mu}_{ijk}$ given by Eq. (9) in terms of the strain and its gradient. Thus we get the following relationship between the balance stress and the constitutive stresses,

$$\tau_{jk} = \hat{\tau}_{jk} - \frac{1}{3}\partial_i(\hat{\mu}_{ijk} + 2\hat{\mu}_{jki}) \tag{44}$$

or explicitly in terms of the strain and the strain gradient,

---

[6] cf. L. A. Segel, *Mathematics Applied to Continuum Mechanics*, Dover, 1977.
[7] see section 4.5

$$\tau_{jk} = (1 - \frac{1}{3}\dot{\partial}_i\ell'_i)(\lambda\varepsilon_{nn}\delta_{jk} + 2G\varepsilon_{jk}) - \frac{2}{3}\partial_i\ell'_j(\lambda\varepsilon_{nn}\delta_{ki} + 2G\varepsilon_{ki}) +$$

$$+ \frac{1}{3}\left\{\lambda\left[(2\ell'_i - \ell^2\partial_i)\delta_{jk} - 2(\ell'_j + \ell^2\partial_j)\delta_{ki}\right]\hat{\kappa}_{inn} + \right.$$

$$\left. + 2G\left[(2\ell'_i - \ell^2\partial_i)\hat{\kappa}_{ijk} - 2(\ell'_j + \ell^2\partial_j)\hat{\kappa}_{iki}\right]\right\}$$

(45)

Similarly for the balance couple stress tensor we get the following expressions: a) in terms of constitutive stresses:

$$\mu_{ik} = \frac{2}{3}(\hat{\mu}_{ipq} + \hat{\mu}_{piq})\varepsilon_{kpq}$$

(43)

or b) in terms of strains and strain gradients

$$\mu_{ik} = \frac{2}{3}\left\{\lambda(\ell^2\hat{\kappa}_{pnn} + \ell'_p\varepsilon_{nn})\delta_{iq} + 2G(\hat{\kappa}_{ipq} + \ell'_i\varepsilon_{pq} + \hat{\kappa}_{piq} + \ell'_p\varepsilon_{iq})\right\}\varepsilon_{kpq}$$

(44)

$$= \frac{2}{3}\left\{\lambda(\ell^2\partial_p + \ell'_p)\varepsilon_{nn}\delta_{iq} + 2G\left[(\partial_i + \ell'_i)\varepsilon_{pq} + (\partial_p + \ell'_p)\varepsilon_{iq}\right]\right\}\varepsilon_{kpq}$$

The balanced double stresses may be expressed again in terms of constitutive stresses and strains, as follows[8]:

$$\mu_{ijk} = \frac{1}{3}(\hat{\mu}_{ijk} + \hat{\mu}_{jki} + \hat{\mu}_{kij})$$

(45)

or

$$\mu_{ijk} = \frac{1}{3}\left\{\lambda\left[(\ell^2\partial_i + \ell'_i)\delta_{jk} + (\ell^2\partial_j + \ell'_j)\delta_{ki} + (\ell^2\partial_k + \ell'_k)\delta_{ij}\right]\varepsilon_{nn} + \right.$$

$$\left. 2G\left[(\partial_i + \ell'_i)\varepsilon_{jk} + (\partial_j + \ell'_j)\varepsilon_{ki} + (\partial_k + \ell'_k)\varepsilon_{ij}\right]\right\}$$

(46)

4.5 A Uniqueness Theorem

In this section the uniqueness of the solution of the general boundary value problem as stated in the previous sections is discussed. The equations for the static boundary value problem of the 2nd gradient elasticity, based on M&E's 2nd form of the elastic strain energy-density function $(\overline{w} = \hat{w} = \overline{\overline{w}})$ are:

---

[8] Appendix II

- *Equilibrium equations (balance of linear and angular momentum)*

$\{3\}$: $\partial_j \tau_{jk} + F_k = 0$ *(true stress equilibrium)* $\hspace{4cm}$ (47.1)

$\{3\}$: $\partial_j \mu_{jk} + \varepsilon_{ijk}\tau_{ij} = 0$ or $\tau_{[ij]} = -\frac{1}{2}\varepsilon_{ijk}\partial_l\mu_{lk}$ *(true couple stress equilibrium)* (47.2)

- *Compatibility equations*

$\{6\}$: $\varepsilon_{ij} = \varepsilon_{ji} = \frac{1}{2}\left(\partial_i u_j + \partial_j u_i\right)$ *(strain)* $\hspace{4cm}$ (47.3)

$\{3\}$: $w_i = \frac{1}{2}\varepsilon_{ijk}\partial_j u_k$ *(rotation)* $\hspace{4cm}$ (47.4)

$\{18\}$: $\hat{\kappa}_{ijk} = \hat{\kappa}_{ikj} = \partial_i\varepsilon_{jk} = \frac{1}{2}\left(\partial_i\partial_j u_k + \partial_i\partial_k u_j\right)$ *(strain gradient)* (47.5)

- *Constitutive equations*

$\{6\}$: $\hat{\tau}_{ij} = \dfrac{\partial \hat{w}}{\partial \varepsilon_{ij}} = \lambda\varepsilon_{kk}\delta_{ij} + 2G\varepsilon_{ij} + \ell'_k\left(\lambda\kappa_{knn}\delta_{ij} + 2G\kappa_{kij}\right)$ $\hspace{1.5cm}$ (47.6)

$\{18\}$: $\hat{\mu}_{ijk} = \dfrac{\partial \hat{w}}{\partial \hat{\kappa}_{ijk}} = \ell^2\left(\lambda\hat{\kappa}_{inn}\delta_{jk} + 2G\kappa_{ijk}\right) + \ell'_i\left(\lambda\varepsilon_{nn}\delta_{jk} + 2G\varepsilon_{jk}\right)$ (47.7)

- *Restrictions due to energy balance*

$\{6\}$: $\tau_{(jk)} = \hat{\tau}_{jk} - \partial_i\mu_{ijk} = \hat{\tau}_{jk} - \partial_i\hat{\mu}_{ijk} + \frac{1}{2}\varepsilon_{lik}\partial_i\mu^D_{jl} + \frac{1}{2}\varepsilon_{lij}\partial_i\mu^D_{kl}$ (47.8)

$\{8\}$: $\mu^D_{ij} = \frac{2}{3}(\hat{\mu}_{ipq} + \hat{\mu}_{piq})\varepsilon_{jpq}$ or $\mu^D_{ij} = \frac{2}{3}\hat{\mu}_{kpi}\varepsilon_{jkp}$ (47.9)

These are 71 equations for 72 dependent field variables, namely:

- (3) u_i, (3) w_i, (6) ε_{ij}, (18) $\hat{\kappa}_{ijk}$, (9) τ_{ij}, (9) μ_{ij}, (6) $\hat{\tau}_{ij}$, (18) $\hat{\mu}_{ijk}$.

The vector of unknown dependent variables is denoted by,

$$X = \{u_i, w_i, \hat{\kappa}_{ijk}, \tau_{ij}, \mu_{ij}, \hat{\tau}_{ij}, \hat{\mu}_{ijk}\}^T \hspace{4cm} (48)$$

The additional variable is the spherical part of the couple stress, which we call the mean torsion,

$$\mu = \frac{1}{3}\mu_{nn} \hspace{6cm} (49)$$

The mean torsion does not contribute to the elastic stain energy-density and is indeterminate within the framework of the considered micro-elasticity theories, represented by the above 71 equations (47.1-9).

For the proof of uniqueness of the solution of the general boundary-value problem, we will consider that there are two different solution vectors X' and X'' with μ' and μ'' arbitrary. Let their difference be $X = X' - X''$. If each set is a solution so is their difference. For this set X from Eq. (47.1) follows

$$\int_{t_0}^{t} dt \int_{V} \Pi \, dV = 0 \quad , \quad \Pi = \left(\partial_i \tau_{ij} + F_j\right) \dot{u}_j \tag{50}$$

where at time t_0 the body is unstrained and at time t the total loading has been accomplished. In Eq. (50) the triple integral is over the volume V with boundary surface ∂V of the considered body $\mathbf{B}$. With $\dot{u}_i$ we denote the velocity,

$$\dot{u}_i \approx \partial_t u_i \approx v_i \tag{51}$$

We observe that

$$\partial_i \tau_{ij} v_j = \partial_i (\tau_{ij} v_j) - \tau_{(ij)} v_{(j,i)} - \tau_{[ij]} v_{[j,i]} \tag{52}$$

where

$$v_{(j,i)} = \dot{\varepsilon}_{ij} = \frac{1}{2}\left(\partial_i v_j + \partial_j v_i\right) \quad , \quad v_{[j,i]} = \dot{\omega}_{ij} = \frac{1}{2}\left(\partial_i \dot{u}_j - \partial_j \dot{u}_i\right) \tag{53}$$

where, using the Eq. (47.2-4) and (47.8), we get:

$$\partial_i \tau_{ij} v_j = \partial_i (\tau_{ij} v_j) - (\hat{\tau}_{ij} - \partial_k \mu_{kij}) v_{(j,i)} + \frac{1}{2} e_{ijk} \partial_l \mu_{lk} v_{[j,i]} \tag{54}$$

$$= \partial_i (\tau_{ij} v_j) - (\hat{\tau}_{ij} - \partial_k \mu_{kij}) \dot{\varepsilon}_{ij} + \underline{\partial_l \mu_{lk} \dot{w}_k}$$

Also we have

$$A = \partial_l \mu_{lk} \dot{w}_k + \partial_k \mu_{kij} \dot{\varepsilon}_{ij} = \partial_i \left(\mu_{ij} \dot{w}_j + \mu_{ijk} \dot{\varepsilon}_{jk}\right) - \underline{\mu_{ij} \partial_i \dot{w}_j} - \mu_{ijk} \partial_i \dot{\varepsilon}_{jk} \tag{55}$$

and

$$B = \mu_{ij} \partial_i \dot{w}_j = \mu_{ij} \varepsilon_{jlk} \dot{\hat{\kappa}}_{lik} = \frac{1}{2} \mu_{ij} \varepsilon_{jlk} \dot{\hat{\kappa}}_{lik} + \frac{1}{2} \mu_{kj} \varepsilon_{jli} \dot{\hat{\kappa}}_{lki} \tag{56}$$

Eq. (55) becomes

$$A = \partial_i \left(\mu_{ij} \dot{w}_j + \mu_{ijk} \dot{\varepsilon}_{jk} \right) - B - \mu_{ijk} \dot{\hat{\kappa}}_{ijk}$$

$$= \partial_i \left(\mu_{ij} \dot{w}_j + \mu_{ijk} \dot{\varepsilon}_{jk} \right) - \left(\mu_{ijk} + \frac{1}{2} \varepsilon_{lij} \mu_{kl} + \frac{1}{2} \varepsilon_{lik} \mu_{jl} \right) \dot{\hat{\kappa}}_{ijk} \qquad (57)$$

$$= \partial_i \left(\mu_{ij} \dot{w}_j + \mu_{ijk} \dot{\varepsilon}_{jk} \right) - \hat{\mu}_{ijk} \dot{\hat{\kappa}}_{ijk}$$

where Eq. (29)[9] was used,

$$\hat{\mu}_{ijk} = \mu_{ijk} + \frac{1}{2} \varepsilon_{lik} \mu_{jl} + \frac{1}{2} \varepsilon_{lij} \mu_{kl} \qquad (29)$$

Substituting Eq. (54) with Eq. (57) into the integrand of Eq. (50.2) and by using Eq. (7.2) for the rate of elastic energy density

$$\dot{\hat{w}} = \hat{\tau}_{ij} \dot{\varepsilon}_{ij} + \hat{\mu}_{ijk} \dot{\hat{\kappa}}_{ijk} \qquad (7.2)$$

we get

$$\Pi = F_j v_j + \partial_i (\tau_{ij} v_j) - \left(\hat{\tau}_{jk} \dot{\varepsilon}_{jk} + \hat{\mu}_{ijk} \dot{\hat{\kappa}}_{ijk} \right) + \partial_i \left(\mu_{ij} \dot{w}_j + \mu_{ijk} \dot{\varepsilon}_{jk} \right)$$

$$= F_j v_j + \partial_i (\tau_{ij} v_j + \mu_{ij} \dot{w}_j + \mu_{ijk} \dot{\varepsilon}_{jk}) - \dot{\hat{w}} \qquad (58)$$

Using now the divergence theorem we get

$$\int_V \Pi \, dV = 0 \quad \Rightarrow \quad \int_V \dot{\hat{w}} dV = \int_V (F_k v_k) dV + \int_{\partial V} \left(\tau_{ij} v_j + \mu_{ij} \dot{w}_j + \mu_{ijk} \dot{\varepsilon}_{jk} \right) n_i \, dS \qquad (59)$$

Thus starting with the system of 71 equations we recovered the energy balance equation for the rates. By integrating over time we get the energy balance equation for the solution vector X,

$$\int_0^t dt \int_V \dot{\hat{w}} dV = \int_0^t dt \int_V (F_k v_k) dV + \int_0^t dt \int_{\partial V} \left(\tau_{ij} v_j + \mu_{ij} \dot{w}_j + \mu_{ijk} \dot{\varepsilon}_{jk} \right) n_i \, dS \qquad (60)$$

Since we have assumed that there the two different solutions hold for the same system of external forces, then

$$F_k = F_k' - F_k'' = 0 \qquad (61)$$

and the volume integral in Eq.(60) vanishes.

---

[9] Appendix I

We remark that in the energy balance Eq. (60) appear the velocity v_i and its derivatives $\partial_i v_j$, through the definitions of the spin vector and the strain-rate tensor, respectively. It can be shown that not all partial derivatives of a field prescribed along a boundary can be chosen independently[10]. Actually, if along a smooth boundary a component of the velocity field v_i is given, then at the same boundary only its normal derivative Dv_i, with $D = n_k \partial_k$ may be independently prescribed. Thus starting with the system of 71 equations we get the equation of virtual work for the difference solution vector X. We have to reduce the 12 dependent variables of the surface integral in Eq. (60) to 6, because only 6 are independent. So, if it is assumed that the velocity v_k and its normal derivative Dv_k are known on the boundary ∂V we write

$$\int_{\partial V} \left(\tau_{ij} v_j + \mu_{ij} \dot{w}_j + \mu_{ijk} \dot{\varepsilon}_{jk} \right) n_i \, dS = \int_{\partial V} \left(P_k v_k + Q_k Dv_k \right) dS \tag{62}$$

In Appendix III we show that indeed for the problem, as is formulated above by Eqs. (47), the forcing vectors P_k and Q_k may be expressed in terms of the constitutive stresses and the mean torsion as follows:

$$P_k = \left(\left(\hat{\tau}_{jk} - \partial_i \left(\hat{\mu}^t{}_{ijk} - \hat{\mu}^t{}_{jik} \right) - n_l n_m \partial_m \hat{\mu}^t{}_{jlk} \right) - \left(\hat{\mu}^t{}_{jik} - n_m n_n \delta_{ij} \hat{\mu}^t{}_{mnk} + \frac{1}{2} \varepsilon_{jik} \mu \right) D_i \right) n_j \tag{63}$$

$$Q_k = \hat{\mu}^t{}_{ijk} n_i n_j \tag{64}$$

where

$$D = n_k \partial_k \quad , \quad D_k = \partial_k - n_k D \tag{65}$$

and

$$\hat{\mu}^t{}_{ijk} = \frac{1}{3} \left(\hat{\mu}_{ijk} + 2 \hat{\mu}_{jki} \right) \tag{66}$$

Finally from Eqs. (60) and (62) we get

$$\int_0^t dt \int_V \dot{w} dV = \int_0^t dt \int_{\partial V_u} \left(P_k v_k + Q_k Dv_k \right) dS + \int_{\partial V_\sigma} \left(P_k v_k + Q_k Dv_k \right) dS \tag{67}$$

or

---

[10] Appendix IV

$$\int_V \hat{w}dV = \frac{1}{2}\left\{ \int_{\partial V_u}(P_k u_k + Q_k Du_k)dS + \int_{\partial V_\sigma}(P_k u_k + Q_k Du_k)dS \right\} \tag{68}$$

where ∂V_u is that subset of the boundary, where kinematic constraints are given, and ∂V_σ is the subset, where static constraints are given. However we have that,

- on $\partial V_u : u_k = u'_k - u''_k = 0$, $Du_k = Du'_k - Du''_k = 0$ $\tag{69a}$

- on $\partial V_\sigma : P_k = P'_k - P''_k = 0$, $Q_k = Q'_k - Q''_k = 0$ $\tag{69b}$

and we get finally that

$$\int_V \hat{w}dV = 0 \tag{70}$$

The elastic strain energy-density function has been proven to be positive as soon as the following constitutive inequalities hold[11] (Georgiadis et al. 2002),

$$(3\lambda + 2G) > 0, \ G > 0, \ \ell^2 > 0, \ -1 < \frac{\ell'}{\ell} < 1 \tag{71}$$

Thus, considering Eq. (2), the only solution that Eq. (60) admits is the trivial one,

$$\varepsilon_{ij} = 0 \text{ and } \hat{\kappa}_{ijk} = \partial_i \varepsilon_{jk} = 0 \tag{72}$$

If rigid body motions are excluded from the boundary constraints, then this results to

$$u_i = u'_i - u''_i = 0 \tag{73}$$

which proves that the two solutions are identical.

We remark that the constitutive stresses $\hat{\tau}_{jk}$, $\hat{\mu}_{ijk}$ can be computed exactly, whereas the true stresses $\tau_{jk}, \mu_{ij}, \mu_{ijk}$ can be computed only within an approximation of the mean torsion. We notice however that the value of the mean torsion does not influence the uniqueness of the solution of the considered boundary-value problem. Thus, as already mentioned above, we will assume that the spherical part of the couple-stress tensor is vanishing ($\mu = 0$); cf. Eq. (42).

---

[11] cf. definition of ℓ' in section 4.7.2.

4.6 Boundary Conditions and the Virtual Work Equation

Let us consider a micro-elastic body **B**, with volume V and boundary ∂V. In a classical continuum, the boundary is divided in two complementary parts s.t.

$$\partial V = \partial V_u \cup \partial V_\sigma \quad \text{and} \quad \partial V_u \cap \partial V_\sigma = \emptyset \tag{74}$$

The set ∂V_u is that set, where displacements u_i are prescribed; ∂V_σ is the complementary set, where tractions t_i are prescribed, such that at any point $x_k \in \partial V$ $u_i t_i = 0$. In a higher grade continuum, like the one we are studying here, in addition to stresses and displacements, higher order stresses and displacement derivatives are needed to be known on the boundary. In particular from Eq. (62) we get the virtual work of the external forces on the boundary ∂V_σ, where external forces are prescribed. By forcing the virtual displacement and its normal derivative to vanish on $\partial V_u = \partial V - \partial V_\sigma$,

$$\text{on } \partial V_u : \quad \delta u_i = 0 \quad \wedge \quad D\delta u_i = 0 \tag{75}$$

we get

$$\delta W_S^{(ext)} = \int_{\partial V} (P_k \delta u_k + Q_k D\delta u_k) n_i \, dS \tag{76}$$

In this expression P_k and Q_k are vectors, which work on the displacement and on its normal derivative. As shown in section 4.4 the forcing vectors depend on the constitutive stresses according to Eqs. (63) and (64). For easy reference we summarize here these conditions, considering the assumption of vanishing mean torsion:

$$P_k = \left(\left(\hat{\tau}_{jk} - \partial_i \left(\hat{\mu}^t_{ijk} - \hat{\mu}^t_{jik} \right) - n_l n_m \partial_m \hat{\mu}^t_{jlk} \right) - \left(\hat{\mu}^t_{jik} - n_m n_n \delta_{ij} \hat{\mu}^t_{mnk} \right) D_i \right) n_j$$

$$Q_k = \hat{\mu}^t_{ijk} n_i n_j$$

$$D = n_k \partial_k \quad , \quad D_k = \partial_k - n_k D$$

$$\hat{\mu}^t_{ijk} = \frac{1}{3}(\hat{\mu}_{ijk} + 2\hat{\mu}_{jki}) \tag{77}$$

$$\hat{\tau}_{ij} = \frac{\partial \hat{w}}{\partial \varepsilon_{ij}} = \lambda \varepsilon_{kk} \delta_{ij} + 2G\varepsilon_{ij} + \ell'_k \left(\lambda \hat{\kappa}_{knn} \delta_{ij} + 2G\hat{\kappa}_{kij} \right)$$

$$\hat{\mu}_{ijk} = \frac{\partial \hat{w}}{\partial \hat{\kappa}_{ijk}} = \ell^2 \left(\lambda \hat{\kappa}_{inn} \delta_{jk} + 2G\hat{\kappa}_{ijk} \right) + \ell'_i \left(\lambda \varepsilon_{nn} \delta_{jk} + 2G\varepsilon_{jk} \right)$$

From the virtual work of the external forces, Eq. (74), it follows that the vector P_k has clear physical meaning: P_k is the <u>total</u>, work-producing, boundary-traction; i.e. of the total force that does work on the boundary displacement. We will call P_k the *working traction or the potency*. We remark that the potency P_k differs from both traction vectors that can be build from the constitutive or the true stress tensors

$$P_k \neq \hat{\tau}_{ik} n_i \quad , \quad P_k \neq \tau_{ik} n_i \tag{78}$$

In that sense we use the potency P_k for defining boundary conditions for tractions, by identifying the notion of "force" with the notion of "work producing force". The other work producing quantity is the vector Q_k. Its physical meaning is less clear, but we will try to shed some light to it by working out a simple example.

Thus we conclude that at any boundary point of the considered micro-elastic body **B** and for the directions normal to- and in the tangential plane we can prescribe either static quantities or their energy conjugate kinematic quantities, by selecting from the set[12] $\{P_k, Q_k \mid u_k, Du_k\}$ such that at any point $x_k \in \partial V$ $u_i P_i = 0 \wedge Du_i Q_i = 0$.

On the basis of the above derivations we write the virtual work equation for the considered medium as follows (cf. Vardoulakis & Sulem, 1995)

$$\delta W^{(int)} = \delta W^{(ext)} \tag{79}$$

where the virtual work of internal forces is defined (cf. Eqs. (7) and (60)),

$$\delta W^{(int)} = \int_V (\hat{\tau}_{ij} \delta \varepsilon_{ij} + \hat{\mu}_{ijk} \delta \hat{\kappa}_{ijk}) dV \tag{80}$$

and the virtual work of external forces is (cf. Eqs. (62) and (75)),

$$\delta W^{(ext)} = \int_V F_i \delta u_i dV + \int_{\partial V} (P_k \delta u_k + Q_k D \delta u_k) dS \tag{81}$$

The virtual work equation (79) thus becomes

$$\int_V (\hat{\tau}_{ij} \delta \varepsilon_{ij} + \hat{\mu}_{ijk} \delta \hat{\kappa}_{ijk}) dV - \int_V F_i \delta u_i dV = \int_{\partial V} (P_k \delta u_k + Q_k D \delta u_k) dS \tag{82}$$

We remark that utilizing the virtual work Eq. (82)[13]: a) From the volume integrals we retrieve the equilibrium condition (36) for the equilibrium stress tensor , Eq. (35). b) From the remaining surfaces integrals we retrieve exactly the boundary conditions (77.1) and (77.2).

---

[12] As is known from Continuum Mechanics, other boundary conditions may be also conceived, like for example the "elastic support", where the traction at a point is a linear function of the displacement of that point, $P_k = ku_k$.

[13] cf. Vardoulakis & Sulem (1995), sect. 10.2.4.

4.7 Examples

The uniqueness theorem, presented in section 4.5, justifies among others the use of the so-called *"semi-inverse method"* of solving b.v. problems. According to this method a strain field with some general properties is selected before hand and then a solution is constructed, which fulfills all field and constitutive equations as well as sufficient boundary conditions. Then according to the above theorem such a solution must be unique.

4.7.1   One-dimensional Compression

As a first application of the present theory we will consider here the problem of one-dimensional compression[14] (Fig. 4) of a material specimen of initial height H.

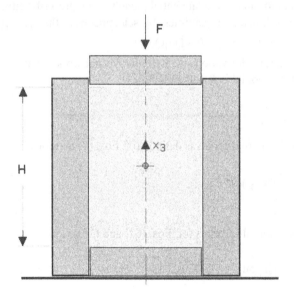

Figure 4. One-dimensional compression

The assumed strain field is,

$$[\varepsilon] = \begin{bmatrix} 0 & 0 & 0 \\ 0 & 0 & 0 \\ 0 & 0 & \varepsilon_{33} \end{bmatrix} , \quad \varepsilon_{33} = \hat{\varepsilon}(x_3) = \frac{du_3}{dx_3} \tag{83}$$

---

[14] In Soil Mechanics this deformation is called an oedometric compression.

Thus the only non-vanishing component of the strain gradient is,

$$\hat{\kappa}_{333} = \frac{d\hat{\varepsilon}}{dx_3} \qquad (84)$$

By neglecting the influence of the self weight, the equilibrium Eq. (37) reads

$$\frac{d\sigma_{33}}{dx_3} = 0 \quad \Rightarrow \quad \sigma_{33} = \text{const.} \qquad (85)$$

From Eq. (37) we get

$$\sigma_{33} = (\lambda + 2G)\left(1 - \ell^2 \frac{d^2}{dx_3^2}\right)\varepsilon_{33} \qquad (86)$$

Thus the equilibrium Eq. (85) yields that

$$\frac{d}{dx_3}\left(\varepsilon_{33} - \ell^2 \frac{d^2\varepsilon_{33}}{dx_3^2}\right) = 0 \qquad (87)$$

From that we get the following general solution for the strain-field:

$$\varepsilon_{33} = \varepsilon_0 + c_1 \sinh\left(\frac{x_3}{\ell}\right) + c_2 \cosh\left(\frac{x_3}{\ell}\right) \qquad (88)$$

We assume that the boundary potency is known; i.e.

$$\text{at} \quad x_3 = \pm H/2 \ (n_3 = \pm 1): \quad P_1 = P_2 = 0 \quad , \quad P_3 = -\bar{\sigma}n_3 \qquad (89)$$

where $\bar{\sigma}$ is computed from the total applied load F and from the cross-sectional area of a horizontal section of the specimen (Fig. 4),

$$\bar{\sigma} = \frac{F}{A} \qquad (90)$$

The boundary condition for the potency P_3, Eq. (77.1), yields:

$$P_3 = \hat{\tau}_{33} - \frac{d\hat{\mu}}{dx_3} = \sigma_{33} \quad \Rightarrow \quad \varepsilon_0 = -\frac{\bar{\sigma}}{\lambda + 2G} < 0 \quad (\textit{compressive primary strain}) \qquad (91)$$

For solving the problem at hand we must also invoke an additional boundary condition, which will allow for the determination of the remaining integration constants c_1 and c_2 of the strain field, Eq. (88). As explained in the previous section this additional boundary condition could be a statement either for the value of the normal derivative of the vertical displacement (i.e. for the strain ε_{33}) or for the value of the vertical component Q_3 of the corresponding forcing vector. Here we will select the former. This selection is justified as follows: We assume that the specimen is compressed by two "rigid" pistons. Within the classical continuum theory we would admit that at the interface, between specimen and rigid pistons, a weak displacement discontinuity develops, that will allow for the normal to the interface strain, inside the deformable body, to take immediately the value ε_0. In the contrary, within the enhanced continuum theory such an interface does not need to be a strain discontinuity line for the normal strain. Continuity of strain may be achieved by imposing the normal strain to be zero at that interface; i.e.

at $x_3 = \pm H/2$: $Du_3 = \varepsilon_{33} = 0 \implies$ (92)

$$c_1 \sinh\left(\frac{H}{2\ell}\right) + c_2 \cosh\left(\frac{H}{2\ell}\right) = -\varepsilon_0 \qquad\qquad c_1 = 0$$

$$\implies \qquad\qquad (93)$$

$$-c_1 \sinh\left(\frac{H}{2\ell}\right) + c_2 \cosh\left(\frac{H}{2\ell}\right) = -\varepsilon_0 \qquad\qquad c_2 = -\frac{\varepsilon_0}{\cosh\left(\dfrac{H}{2\ell}\right)}$$

With these values for the integration constants we get from Eq. (78) the following solution (Fig. 5)

$$\varepsilon = \varepsilon_0 \left(1 - \frac{\cosh\left(z^* \dfrac{H}{\ell}\right)}{\cosh\left(\dfrac{1}{2}\dfrac{H}{\ell}\right)} \right) \quad, \quad z^* = \frac{x_3}{H} \in [-\frac{1}{2}, \frac{1}{2}] \qquad\qquad (94)$$

We remark that the strain-field, Eq. (94), can be decomposed into two terms as

$$\varepsilon = \varepsilon^{(0)} + \varepsilon^{(2)} \qquad\qquad (95)$$

where $\varepsilon^{(0)}$ and $\varepsilon^{(2)}$ denote the part of the solution that is due o the 0th- (classical-continuum) and the 2nd-grade (enhanced continuum) terms of the underlying constitutive theory,

$$\varepsilon^{(0)} = \varepsilon_0 = \text{const} \quad , \quad \varepsilon^{(2)} = -\frac{\cosh\left(\dfrac{H}{\ell}z^*\right)}{\cosh\left(\dfrac{1}{2}\dfrac{H}{\ell}\right)}\varepsilon_0 \tag{96}$$

The 2nd grade term $\varepsilon^{(2)}$ perturbs significantly the classical constant-strain solution near to the boundaries. This can be seen easily, if we introduce the transformation,

$$\bar{x}_3 = x_3 - \frac{H}{2} \quad \Rightarrow \quad \bar{z} = \frac{\bar{x}_3}{H} = z^* - \frac{1}{2} \tag{97}$$

Then, for example close to the upper interface boundary, we get that the perturbed strain-field is varying quadratically

$$\varepsilon \approx 2\varepsilon_0\, e^{-\frac{1}{2}\frac{H}{\ell}}\left(\frac{\bar{x}_3}{\ell}\right)^2 \tag{98}$$

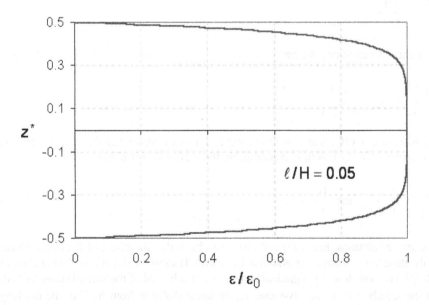

Figure 5. Strain distribution for $\ell/H = 0.05$

The corresponding displacement field is also a linear combination of a 0th and a 2nd grade term (Fig. 6),

$$u_3 = u_3^{(0)} + u_3^{(2)} \tag{99}$$

where

$$u_3^{(0)} = \varepsilon_0 \left(\frac{1}{2} + z^* \right) H \tag{100}$$

$$u_3^{(2)} = -\varepsilon_0 \left(\tanh\left(\frac{1}{2} \frac{H}{\ell} \right) + \frac{\sinh\left(\frac{H}{\ell} z^* \right)}{\cosh\left(\frac{1}{2} \frac{H}{\ell} \right)} \right) \ell \tag{101}$$

For a given load the compliance of the specimen can be expressed in terms of the engineering strain,

$$\bar{\varepsilon} = \frac{\Delta H}{H} = \frac{u_3(H)}{H} \tag{102}$$

In case of classical elasticity we get,

$$\bar{\varepsilon}^{(0)} = \frac{u_3^{(0)}}{H} = -\varepsilon_0 = \frac{F/A}{\lambda + 2G} \tag{103}$$

Thus micro-elasticity effects increase the compliance of the specimen, due to the extra terms in the strain energy-density function. Accordingly the scale effect is linear,

$$\bar{\varepsilon} \approx \bar{\varepsilon}^{(0)} \left(1 + 2\left(\frac{\ell}{H} \right) \right) \quad \text{for} : \frac{H}{\ell} \gg 1 \tag{104}$$

From the above solutions, Eqs. (95) and (99), we obtain that the effect of the micro-structure is felled only close to the boundaries and that the micro-elasticity model allows for the formation of a boundary layer, within which a gradual adjustment is achieved of the normal strain from its zero value at the interface to its steady-value ε_0 in some distance from it. The internal length ℓ determines the "thickness" of the boundary layer; i.e. the material constant ℓ determines how close to the boundary the perturbation of the far-field strain is still significant. In the present example the director field ℓ'_k does not affect the solution.

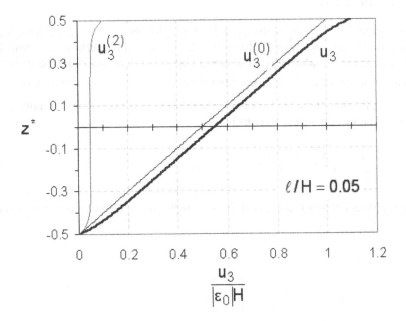

Figure 6. Vertical displacement distribution for, $\ell / L = 0.05$. The linear term, the non-linear (boundary-layer) solution and their normalized superposition, indicating the increased compliance of the structure due to the extra strain-gradient energy stored.

Remark

From the above derivation we can see clearly that the present micro-elasticity theory may be used to regularize the discontinuous solutions arising for example from the classical continuum formulation of problems in layered media. In these problems, classical continuum approach can only assure continuity of displacements and their derivatives parallel to the interface, a condition known as Maxwell's compatibility condition (see Vardoulakis & Sulem, 1995). In the case of micro-elastic media, however, we can always assure a smoother interface condition by requiring the continuity of the normal strain as well,

$$[u_i] = u_i^+ - u_i^- = 0 \ \text{(continuity of displacement)}$$

$$[\partial_i u_j] = 0 \ \text{(continuity of strain)}$$

4.7.2   The "Bolted" Layer

We consider an infinitely long layer of thickness H. We neglect body forces and assume that the surfaces of the layer are traction-free. However, we assume that the surfaces are "bolted" by a series of dense double-headed nails, as shown in Fig. 7.   The action of the pre-stressed bolts consists in the application of a co-linear force doublet $\{\vec{F}, -\vec{F}\}$, acting over a short distance $\tilde{\ell}$. During the pre-stressing of the bolts, the material in the vicinity of the bolted layers is strained. The work done by the bolts is,

$$W^{ext} = -p\frac{\Delta\tilde{\ell}}{2} - p\frac{\Delta\tilde{\ell}}{2} = -p\frac{1}{2}\varepsilon_{33}\tilde{\ell} - p\frac{1}{2}\varepsilon_{33}\tilde{\ell} = -p\tilde{\ell}\varepsilon_{33} \tag{105}$$

where p is the equivalent pressure, computed from the bolt forces distributed over the surface of the layer.

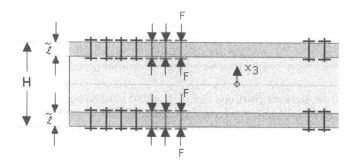

Figure 7. The bolted layer

Thus we may interpret the boundary condition for the forcing vector Q_k, Eq. (77.2) as follows:

at $x_3 = H/2$ $(n_3 = +1)$: $Q_3 = \hat{\mu}_{333} = (\lambda + 2G)(\ell^2\hat{\kappa}_{333} + \ell'_3(H/2)\varepsilon_{33})$

$$\tag{106}$$

$$W_{ext} = Q_3 Du_3 = Q_3 n_3 \partial_3 u_3 = Q_3 \varepsilon_{33}$$

at $x_3 = -H/2$ $(n_3 = -1)$: $Q_3 = \hat{\mu}_{333} = (\lambda + 2G)(\ell^2\hat{\kappa}_{333} + \ell'_3(-H/2)\varepsilon_{33})$

$$\tag{107}$$

$$W_{ext} = Q_3 Du_3 = Q_3 n_3 \partial_3 u_3 = -Q_3 \varepsilon_{33}$$

resulting to

$$x_3 = H/2: \quad (\lambda + 2G)(\ell^2 \hat{\kappa}_{333} + \ell'_3(H/2)\varepsilon_{33}) = -p\tilde{\ell} \qquad (108)$$

$$x_3 = -H/2: \quad (\lambda + 2G)(\ell^2 \hat{\kappa}_{333} + \ell'_3(-H/2)\varepsilon_{33}) = p\tilde{\ell} \qquad (109)$$

The considered example is typical of a boundary-value problem with two significant, but otherwise fully identical boundaries. This fact (i.e. the strip-like geometry) excludes the possibility of assuming a constant director field inside the volume of the body (Fig. 8a), because this model would distinguish among the upper and lower surface of the layer. Moreover, since for an observer sitting on one of these boundaries there is no reason to distinguish between right and left, we must assume that at boundary points the director field is always directed normal to it (Vardoulakis & Sulem, 1995);

$$\ell'_k = \ell' n_k \quad \forall x_i \in \partial V \qquad (110)$$

where n_k is the unit outward normal vector at the boundary[15]. Thus we are forced to assume that the director field is confined into two boundary layers, and that it is having the direction of the outward normal to the corresponding boundary (Fig. 8b).

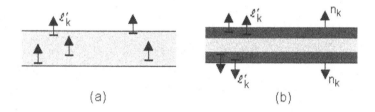

(a)   (b)

Figure 8. (a) Constant director field (b) Piece-wise constant director field

With these remarks from Eqs (108) to (110) we get,

$$x_3 = H/2: \quad \ell\frac{d\varepsilon_{33}}{dx_3} + \frac{\ell'}{\ell}\varepsilon_{33} = -r \qquad (111)$$

$$x_3 = -H/2: \quad \ell\frac{d\varepsilon_{33}}{dx_3} - \frac{\ell'}{\ell}\varepsilon_{33} = r \qquad (112)$$

---

[15] Of course, we could equally assume that the director is pointing inwards; this is taken care, however, by the sign of ℓ'.

where r is the dimensionless double force, applied at the boundary, scaled by the elastic oedometric modulus and the material length ℓ,

$$r = \frac{p}{\lambda + 2G} \frac{\tilde{\ell}}{\ell} \tag{113}$$

With the general solution, Eq. (82), and the observation that in this case $P_3 = 0$ ($\varepsilon_0 = 0$), from Eqs. (105) and (106) we get the following solution for the strain field,

$$\varepsilon_{33} \approx -\frac{r}{1 + \left(\dfrac{\ell'}{\ell}\right)} \frac{\cosh\left(\dfrac{x_3}{H}\dfrac{H}{\ell}\right)}{\cosh\left(\dfrac{1}{2}\dfrac{H}{\ell}\right)} \qquad \text{for}: \frac{H}{\ell} \gg 1 \tag{114}$$

As can be seen from Fig. 9, the strain is compressive and clearly localized close to the boundaries.

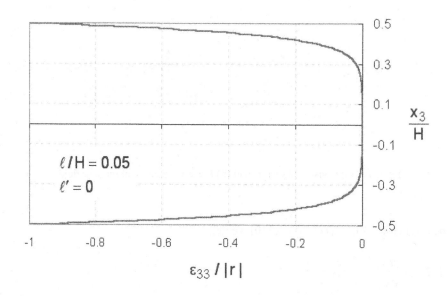

Figure 9. Strain distribution in a bolted layer

We remark that the net relative compression of the layer due to bolting is,

$$\bar{\varepsilon} = -\frac{1}{H}\int_{-H/2}^{H/2}\varepsilon_{33}dx_3 \approx 2\frac{p}{\lambda + 2G}\frac{\left(\dfrac{\tilde{\ell}}{H}\right)}{1+\left(\dfrac{\ell'}{\ell}\right)} \quad \text{for}: \frac{H}{\ell} \gg 1 \tag{115}$$

It is remarkable that the scale effect is expressed directly by the bold length, which enters into the problem through the boundary condition for double forces. The material lengths appear only through their ratio. According to Eqs. (114) and (115) the effect of the bolts is reduced or amplified depending on the magnitude and sign of the director length ℓ'. As already mentioned, meaningful, within the frame of the present micro-elasticity theory, are values: $-1 < \dfrac{\ell'}{\ell} < 1$; cf. Ineq. (71). Above example, indicates that most probably negative values of ℓ' should rather be excluded, since they lead to infinite compliance for limit $\ell'/\ell \to -1$.

References

1. Amanatidou, E. and Aravas, N. (2002). Mixed finite element formulations of strain-gradient elasticity. *Compt. Methods Appl. Mech. & Engrg*, 191:1723-1751.
2. Casal, P. (1961). La capilaritè, interne. *Cahier du Groupe Français d'Etudes de Rhèologie C.N.R.SVI* no. 3, 31-37.
3. Cosserat, E. and F. *Théorie des Corps Déformables*. A. Hermann et Fils, Paris (1909).
4. Exadaktylos, G. (1998). Gradient elasticity with surface energy: Mode-I crack problem. *Int. J. Solids Structures*, 35: 421-456.
5. Exadaktylos, G. and Vardoulakis, I. (1998). Surface instability in gradient elasticity. *Int. J. Solids Structures.*, 35: 2251-2281.
6. Exadaktylos, G. and Vardoulakis, I. (2001 a). Microstructure in linear elasticity and scale effects: a reconsideration of basic rock mechanics and rock fracture mechanics. *Tectonophysics*, 335: 81-109.
7. Exadactylos, G., Vardoulakis, I., and Aifantis, E. (1996). Cracks in gradient elastic bodies with surface energy. *Int. J.Fract.*, 79: 107-119.
8. Frantziskonis G. and Vardoulakis I. (1991). On the microstructure of surface instabilities. *Eur. J. Mech. A/Solids*, 11: 21-34 .
9. Georgiadis, H.G. (2003). The mode III crack problem in microstructured solids governed by dipolar gradient elasticity: Static and dynamic analysis. *J Appl. Mech.-T. ASME* 70: 517-530.
10. Georgiadis, H.G. and Vardoulakis, I. (1998). Anti-plane shear Lamb's problem treated by gradient elasticity with surface energy, *Wave Motion*, 28: 353-366.
11. Georgiadis , H.G. and Vardoulakis, I (2003). Dispersive generalized and high-frequency Rayleigh waves in a gradient-elastic half-space, in print.
12. Georgiadis , H.G., Vardoulakis, I. and Lykotrfitis G. (2000) Torsional surface waves in a gradient-elastic half-space, *Wave Motion*, 31: 333-348.

13. Georgiadis, H.G. and Velgaki, E.G. (2003). High-frequency Rayleigh waves in materials with micro-structure and couple-stress effects. *Int. J. Solids Struct.* 40: 2501-2520.
14. Germain, P. (1973a). La mèthode des puissances virtuelles en mècanique des milieux continus. Part I. *Journal de Mècanique*, 12: 235-274.
15. Germain, P. (1973b). The Method of virtual power in continuum mechanics. Part 2: Microstructure. *SIAM, J. Appl. Math*, 25: 556-575.
16. Günther, W. (1958). Zur Statik und Kinematik des Cosseratschen Kontinuums. *Abhandlungen der Braunscheigschen Wissenschaftlichen Gesellschaft*, Göttingen, Verlag E. Goltze, 101: 95-213.
17. Hermann, G. (1972). Some applications of micromechanics. *Experimental Mechanics*, 12: 235-238.
18. Kessel, S. (1964). Lineare Elastizitätstheorie des anisotropen Cosserat-Kontinuums. *Abhandlungen der Braunscheigschen Wissenschaftlichen Gesellschaft*, Göttingen, Verlag E. Goltze, **16**, 1-22.
19. Koiter, W. T. (1964). Couple-stresses in the Theory of Elasticity. I & II. *Proc. Ned. Akad. Uet.*, 67: 17-29 & 30-44.
20. Mindlin, R.D., (1964). Microstructure in linear elasticity. *Arch. Rat. Mech. Anal.*, 10: 51-77.
21. Mindlin, R.D. and Eshel, N.N. (1968). On first strain-gradient theories in linear Elasticity. *Int. J. Solids Structures*, 4:109-124.
22. Schaeffer, H. (1962). Versuch einer Elastizitätstheorie des zweidimensionalen ebenen COSSERAT-Kontinuums. *Miszellannenn der Angewandten Mechanik*. Akademie Verlag, Berlin.
23. Schaeffer, H. (1967). Das Cosserat-Kontinuum. *ZAMM*, 47/8: 485-498.
24. Vardoulakis, I., Exadactylos,G. and Aifantis, E. (1995). Gradient elasticity with surface energy.Mode III crack problem. *Int. J. Solids and Structures*, 33: 4531-4559.
25. Vardoulakis, I. and Georgiadis, H.G. (1997). SH surface waves in a homogeneous gradient-elastic half-space with surface energy, *Journal of Elasticity*, 47: 147-165.
26. Vardoulakis, I. and Sulem, J. *Bifurcation Analysis in Geomechanics*, Blackie Academic and Professional, 1995.

Recent Research Literature

1. Askes, H., Suiker, A.S.J. and Sluys, L.J. (2002). A classification of higher-order strain-gradient models - linear analysis. *Arch. Appl. Mech.*, 72: 171-188.
2. Exadaktylos, G. (1999). Some basic half-plane problems of the cohesive elasticity theory with surface energy. *Acta Mech.*, 133: 175-198.
3. Fannjiang, A.C., Chan Y.S. and Paulino, G.H. (2002). Strain gradient elasticity or antiplane shear cracks: A hypersingular integrodifferential equation approach. *SIAM J Apll. Math.* 62: 1066-1091.
4. Kalpakides, V.K. and Agiasofitou, A.K. (2002). On material equations in second gradient electroelasticity. *J Elasticity* 67: 205-227.
5. Luzon, F, Ramirez, L., Sanchez-Sesma, F.J., et al. (2003). Propagation of SH elastic waves in deep sedimentary basins with an oblique velocity gradient. *Wave Motion*, 38: 11-23.
6. Manolis, G.D. (2000). Some basic solutions for wave propagation in a rod exhibiting non-local elasticity. *Eng. Anal. Bound. Elem.*, 24: 503-508.

7.   Paulino, G.H., Fannjiang, A.C., and Chan. Y.S. (2003). Gradient elasticity theory for mode III fracture in functionally graded materials - Part I: Crack perpendicular to the material gradation. *J Appl. Mech. – T. ASME* 70: 531-542.
8.   Polizzotto C., (2001). Nonlocal elasticity and related variational principles . *Int. J. Solids Struct.*, 38: 7359-7380.
9.   Polizzotto, C. (2002). Thermodynamics and continuum fracture mechanics for nonlocal-elastic plastic materials. *Eur. J. Mech. A- Solids*, 21: 85-103.
10. Polizzotto,   C. (2003). Unified thermodynamic framework-for nonlocal/gradient continuum theories. *Eur. J. Mech. A-Solid,* 22: 651-668.
11. Polyzos, D., Tsepoura, K.G., Tsinopoulos, S., et al. (2003). A boundary element method for solving 2-D and 3-D static gradient elastic problems - Part I: Integral formulation. *Comput. Method. Appl. Mech.,* 192: 2845-2873 .
12. Tsepoura K.G., Papargyri-Beskou, S. and Polyzos, D. (2002). A boundary element method for solving 3D static gradient elastic problems with surface energy. *Comput. Mech.,* 29: 361-381.
13. Tsepoura, K.G., Papargyri-Beskou, S., Polyzos, D. et al. (2002). Static and dynamic analysis of a gradient-elastic bar in tension. *Arch. Appl. Mech.,* 72: 483-497.
14. Tsepoura, K.G. and Polyzos, D. (2003). Static and harmonic BEM solutions of gradient elasticity problems with axisymmetry. *Comput. Mech.,* 32: 89-103.
15. Zhou, D. and Jin, B. (2003). Boussinesq-Flamant problem in gradient elasticity with surface energy. *Mech. Res. Comun.* 30: 463-468.

Appendix I : *Relations between true and constitutive stresses*

Energy balance is expressed by the equation between the total elastic energy rate and the power of external forces,

$$\dot{W}^{(el)} = W^{(ext)} \tag{A1.1}$$

Using Eq. (7.2) for the elastic strain energy density function we get

$$\dot{W}^{(el)} = \int_V (\hat{\tau}_{jk}\dot{\varepsilon}_{jk} + \hat{\mu}_{lik}\dot{\hat{\kappa}}_{lik})dV \tag{A1.2}$$

Using the divergence theorem and the definitions of the kinematic variables, Eqs. (3), the power of external forces becomes:

$$W^{(ext)} = \int_V \left\{ (F_j + \partial_i \tau_{ij})v_j + \tau_{ij}\partial_i v_j + \partial_i \mu_{ij}\dot{w}_j + \mu_{ij}\partial_i \dot{w}_j + \partial_i \mu_{ijk}\dot{\varepsilon}_{jk} + \mu_{ijk}\partial_i \dot{\varepsilon}_{jk} \right\}dV \tag{A1.3}$$

We remark that:

1. $F_j + \partial_i \tau_{ij} = 0$

due to the equilibrium Eq. (13), and

2. $\tau_{ij} \partial_i v_j = \tau_{(ij)} \dot{\varepsilon}_{ij} + \tau_{[ij]} \dot{w}_{ij} = \tau_{(ij)} \dot{\varepsilon}_{ij} + \tau_{[ij]} e_{jki} \dot{w}_k$ (A1.4)

3. $\mu_{ij} \partial_i \dot{w}_j = \mu_{ij} \varepsilon_{jlk} \dot{\hat{\kappa}}_{lik} = \frac{1}{2} \mu_{ij} \varepsilon_{jlk} \dot{\hat{\kappa}}_{lik} + \frac{1}{2} \mu_{kj} \varepsilon_{jli} \dot{\hat{\kappa}}_{lki}$ (A1.5)

4. $\mu_{ijk} \partial_i \dot{\varepsilon}_{jk} = \mu_{ijk} \dot{\hat{\kappa}}_{ijk} = \mu_{lik} \dot{\hat{\kappa}}_{lik}$ (A1.6)

Introducing the above expressions into Eq. (A1.3) and using Eqs. (A1.1) and (A1.2) we get the following equation:

$$\int_V (\hat{\tau}_{jk} \dot{\varepsilon}_{jk} + \hat{\mu}_{lik} \dot{\hat{\kappa}}_{lik}) dV =$$
$$\int_V \left\{ (\tau_{[ki]} \varepsilon_{ijk} + \partial_i \mu_{ij}) \dot{w}_j + (\tau_{(jk)} + \partial_i \mu_{ijk}) \dot{\varepsilon}_{jk} + (\frac{1}{2} \mu_{ij} \varepsilon_{jlk} + \frac{1}{2} \mu_{kj} \varepsilon_{jli} + \mu_{lik}) \dot{\hat{\kappa}}_{lik} \right\} dV$$
(A1.7)

We remark that:

5. $\partial_j \mu_{jk} + \varepsilon_{ijk} \tau_{ij} = 0, \quad \Rightarrow \quad \partial_i \mu_{ij} + \varepsilon_{ijk} \tau_{[ki]} = 0 \quad (\varepsilon_{ijk} \tau_{(ki)} = 0)$

due to the moment equilibrium Eq. (16), and therefore Eq. (A1.7) becomes

$$\int_V (\hat{\tau}_{jk} \dot{\varepsilon}_{jk} + \hat{\mu}_{lik} \dot{\hat{\kappa}}_{lik}) dV =$$
$$\int_V \left\{ (\tau_{(jk)} + \partial_i \mu_{ijk}) \dot{\varepsilon}_{jk} + (\frac{1}{2} \mu_{ij} \varepsilon_{jlk} + \frac{1}{2} \mu_{kj} \varepsilon_{jli} + \mu_{lik}) \dot{\hat{\kappa}}_{lik} \right\} dV$$
(A1.8)

We restrict first the kinematics to the particular class of strain-rate fields which are homogeneous inside a volume $V' \subseteq V$ (i.e. strain fields with $\dot{\hat{\kappa}}_{lik} = \partial_l \dot{\varepsilon}_{ik} = 0$) and which vanish outside V'. If we apply Eq. (A1.8) for these fields then,

$$\int_{V'} \{\hat{\tau}_{jk} - (\tau_{(jk)} + \partial_i \mu_{ijk})\} \dot{\varepsilon}_{jk} dV = 0 \quad \Rightarrow \quad \hat{\tau}_{jk} = \tau_{(jk)} + \partial_i \mu_{ijk}$$
(28)

With this result we get then from Eq. (A1.8) also that

$$\hat{\mu}_{ijk} = \mu_{ijk} + \frac{1}{2}\mu_{jl}\epsilon_{lik} + \frac{1}{2}\mu_{kl}\epsilon_{lij}$$

(29)

Appendix II: *Evaluation of the deviatoric part of the true couple stress and of the expression of the true double stress*

We start with

$$\mu_{ij}^{D} = \frac{2}{3}(\hat{\mu}_{ipq} + \hat{\mu}_{piq})\epsilon_{jpq}$$

(31)

The proof of Eq. (31) can be found in the paper by M&E.

We consider the 1st form of elastic strain energy density function after M&E

$$w = \tilde{w}(\epsilon_{ij}, \tilde{\kappa}_{ijk}) \quad , \quad \tilde{\kappa}_{ijk} = \partial_i\partial_j u_k$$

(A2.1)

and we define the corresponding constitutive stresses

$$\tilde{\mu}_{ijk} = \frac{\partial\tilde{w}}{\partial\tilde{\kappa}_{ijk}} = \frac{\partial\hat{w}}{\partial\hat{\kappa}_{pqr}}\frac{\partial\hat{\kappa}_{pqr}}{\partial\tilde{\kappa}_{ijk}} = \hat{\mu}_{pqr}\frac{\partial\hat{\kappa}_{pqr}}{\partial\tilde{\kappa}_{ijk}}$$

(A2.2)

By using the relation of the kinematic variables between the 1st and 2nd form we get

$$\hat{\kappa}_{pqr} = \frac{1}{2}(\tilde{\kappa}_{pqr} + \tilde{\kappa}_{prq}) = \frac{1}{2}\tilde{\kappa}_{ijk}(\delta_{ip}\delta_{jq}\delta_{kr} + \delta_{ip}\delta_{jr}\delta_{kq})$$

(A2.3)

Then Eq. (A2.2) becomes

$$\tilde{\mu}_{ijk} = \hat{\mu}_{pqr}\frac{1}{2}(\delta_{ip}\delta_{jq}\delta_{kr} + \delta_{ip}\delta_{jr}\delta_{kq}) = \frac{1}{2}(\hat{\mu}_{ijk} + \hat{\mu}_{ikj})$$

(A2.4)

We consider also the **3rd form** of the elastic energy density function

$$w = \overline{w}(\epsilon_{ij}, \overline{\kappa}_{ij}, \overline{\overline{\kappa}}_{ijk})$$

(A2.5)

with,

$$\overline{\kappa}_{ij} = \frac{1}{2}\epsilon_{jlk}\partial_l\partial_i u_k$$

(A2.6)

$$\overline{\overline{\kappa}}_{ijk} = \frac{1}{3}\left(\partial_i\partial_j u_k + \partial_j\partial_k u_i + \partial_k\partial_i u_j\right) \tag{A2.7}$$

and we define the corresponding constitutive stresses

$$\overline{\overline{\mu}}_{ijk} = \frac{\partial \overline{w}}{\partial \overline{\overline{\kappa}}_{ijk}} = \frac{\partial \hat{w}}{\partial \hat{\kappa}_{pqr}}\frac{\partial \hat{\kappa}_{pqr}}{\partial \overline{\overline{\kappa}}_{ijk}} \tag{A2.8}$$

$$\overline{\mu}_{ij} = \frac{\partial \overline{w}}{\partial \overline{\kappa}_{ij}} = \frac{\partial \tilde{w}}{\partial \tilde{\kappa}_{pqr}}\frac{\partial \tilde{\kappa}_{pqr}}{\partial \overline{\kappa}_{ij}} = \tilde{\mu}_{pqr}\frac{\partial \tilde{\kappa}_{pqr}}{\partial \overline{\kappa}_{ij}} \tag{A2.9}$$

We are using further the relation between the kinematic variables of the 1st and 3rd energy form

$$\tilde{\kappa}_{pqr} = \overline{\overline{\kappa}}_{pqr} + \frac{2}{3}\overline{\kappa}_{pl}\varepsilon_{lqr} + \frac{2}{3}\overline{\kappa}_{ql}\varepsilon_{lpr} = \overline{\overline{\kappa}}_{pqr} + \frac{2}{3}\overline{\kappa}_{ij}(\delta_{ip}\delta_{lj}\varepsilon_{lqr} + \delta_{iq}\delta_{lj}\varepsilon_{lpr}) =$$
$$= \overline{\overline{\kappa}}_{pqr} + \frac{2}{3}\overline{\kappa}_{ij}(\delta_{ip}\varepsilon_{jqr} + \delta_{iq}\varepsilon_{jpr}) \tag{A2.10}$$

Then Eq. (A2.9) gives

$$\overline{\mu}_{ij} = \frac{2}{3}(\tilde{\mu}_{iqr}\varepsilon_{jqr} + \tilde{\mu}_{pir}\varepsilon_{jpr}) = \frac{2}{3}(\tilde{\mu}_{iqr}\varepsilon_{jqr} + \tilde{\mu}_{ipr}\varepsilon_{jpr})$$
$$= \frac{4}{3}\tilde{\mu}_{ipr}\varepsilon_{jpr} = \frac{4}{3}\tilde{\mu}_{ipq}\varepsilon_{jpq} \tag{A2.11}$$

From Eqs. (A2.4) and (A2.11) we obtain

$$\overline{\mu}_{ij} = \frac{2}{3}(\hat{\mu}_{ipq} + \hat{\mu}_{piq})\varepsilon_{jpq} \tag{A2.12}$$

and with that and Eq.(31) we have proven Eq. (32)

$$\overline{\mu}_{ij} = \mu_{ij}^{D} \tag{32}$$

The energy balance equation for the 3rd energy form reads,

$$W^{(el)} = W^{(ext)} \quad \Rightarrow \quad \int_V \dot{\overline{w}} dV = \int_V F_k v_k dV + \int_{\partial V}\left(t_k v_k + m_k \dot{w}_k + R_{ij}\dot{\varepsilon}_{ij}\right) dS \tag{A2.13}$$

We remark also the following equations

$$\mu_{ij}\partial_i\dot{w}_j = (\mu_{ij}^{D} + \frac{1}{3}\delta_{ij}\mu_{kk})\overline{\kappa}_{ij} = \mu_{ij}^{D}\dot{\overline{\kappa}}_{ij} \tag{A2.14}$$

$$\mu_{ijk}\partial_i\dot{\varepsilon}_{jk} = \mu_{ijk}\dot{\hat{\kappa}}_{ijk} = \mu_{ijk}(\dot{\overline{\kappa}}_{ijk} - \tfrac{1}{3}\dot{\kappa}_{jl}\varepsilon_{kil} - \tfrac{1}{3}\dot{\kappa}_{kl}\varepsilon_{jil})$$

$$= \mu_{ijk}\dot{\overline{\kappa}}_{ijk} - \tfrac{1}{3}\mu_{ijk}(\delta_{pj}\delta_{ql}\varepsilon_{kil} + \delta_{pk}\delta_{ql}\varepsilon_{jil})\dot{\overline{\kappa}}_{pq} \qquad (A2.15)$$

$$= \mu_{ijk}\dot{\overline{\kappa}}_{ijk} - \tfrac{2}{3}\mu_{ipk}\varepsilon_{kiq}\dot{\overline{\kappa}}_{pq}$$

Using the expression of the rate of elastic energy density

$$\dot{w} = \tau_{ij}\dot{\varepsilon}_{ij} + \overline{\mu}_{ij}\dot{\overline{\kappa}}_{ij} + \overline{\overline{\mu}}_{ijk}\dot{\overline{\kappa}}_{ijk} \qquad (A2.16)$$

and replacing all the above Eqs. (A2.14-16) into Eq. (A2.13) and using the divergence theorem we get:

$$\int_V (\overline{\tau}_{jk}\dot{\varepsilon}_{jk} + \overline{\mu}_{pq}\dot{\overline{\kappa}}_{pq} + \overline{\overline{\mu}}_{ijk}\dot{\overline{\kappa}}_{ijk})dV =$$

$$= \int_V \left\{ (F_j + \partial_i\tau_{ij})v_j + \tau_{ij}\partial_i v_j + \partial_i\mu_{ij}\dot{w}_j + \mu_{ij}\partial_i\dot{w}_j + \partial_i\mu_{ijk}\dot{\varepsilon}_{jk} + \mu_{ijk}\partial_i\dot{\varepsilon}_{jk} \right\}dV \qquad (A2.17)$$

$$= \int_V \left\{ (\tau_{[ik]}\varepsilon_{kji} + \partial_i\mu_{ij})\dot{w}_j + (\tau_{(jk)} + \partial_i\mu_{ijk})\dot{\varepsilon}_{jk} + (\mu^D_{pq} - \tfrac{2}{3}\mu_{ipk}\varepsilon_{kiq})\dot{\overline{\kappa}}_{pq} + \mu_{ijk}\dot{\overline{\kappa}}_{ijk} \right\}dV$$

From this expression and

$$\tau_{[ik]} = -\tfrac{1}{2}\varepsilon_{ikm}\partial_l\mu^D_{lm}$$

Eq. (A2.17) becomes

$$\int_V (\overline{\tau}_{jk}\dot{\varepsilon}_{jk} + \overline{\mu}_{pq}\dot{\overline{\kappa}}_{pq} + \overline{\overline{\mu}}_{ijk}\dot{\overline{\kappa}}_{ijk})dV =$$

$$= \int_V \left\{ (\tau_{(jk)} + \partial_i\mu_{ijk})\dot{\varepsilon}_{jk} + (\mu^D_{pq} - \tfrac{2}{3}\mu_{ipk}\varepsilon_{kiq})\dot{\overline{\kappa}}_{pq} + \mu_{ijk}\dot{\overline{\kappa}}_{ijk} \right\}dV \qquad (A2.18)$$

Using a similar argument as in Appendix I we obtain finally from (A2.18) the following local equations

$$\overline{\tau}_{jk} = \tau_{(jk)} + \partial_i\mu_{ijk} \qquad (A2.19)$$

$$\overline{\mu}_{pq} = \mu^D_{pq} - \tfrac{2}{3}\mu_{ipk}e_{kiq} \qquad (A2.20)$$

$$\overline{\overline{\mu}}_{ijk} = \mu_{ijk}$$

$$(A2.21)$$

Finally by using its definition (cf. M&E) we get

$$\overline{\overline{\mu}}_{ijk} = \frac{1}{3}(\hat{\mu}_{ijk} + \hat{\mu}_{jki} + \hat{\mu}_{kij}) \tag{A2.22}$$

Thus from Eqs. (A2.21) and (A2.22) we get

$$\mu_{ijk} = \frac{1}{3}(\hat{\mu}_{ijk} + \hat{\mu}_{jki} + \hat{\mu}_{kij}) \tag{33}$$

For the proof of Eq. (34) we remark first that[16]

$$\varepsilon_{mnp}\varepsilon^{rst} = \delta^{rst}_{mnp} = \begin{vmatrix} \delta^r_m & \delta^r_n & \delta^r_p \\ \delta^s_m & \delta^s_n & \delta^s_p \\ \delta^t_m & \delta^t_n & \delta^t_p \end{vmatrix} \tag{A2.23}$$

$$\delta_{ij} = \delta^{ij} = \delta^j_i = \begin{cases} 1 & \text{if}: i = j \\ \\ 0 & \text{if}: i \neq j \end{cases}, \quad \varepsilon_{ijk} = \varepsilon^{ijk} \quad \text{etc.} \tag{A2.24}$$

In particular we get

$$\delta^{rsp}_{mnp} = \delta^{rs}_{mn} = \begin{vmatrix} \delta^r_m & \delta^r_n \\ \delta^s_m & \delta^s_n \end{vmatrix} \tag{A2.25}$$

With Eq. (27), Eq. (29) becomes

$$\mu^D_{il}\varepsilon_{ljk} + \mu^D_{jl}\varepsilon_{lki} + \mu^D_{kl}\varepsilon_{lij} = 0 \tag{A2.26}$$

where μ^D_{ij} is given by Eq. (31). Thus by using Eq. (26), (A2.23-5) we conclude by inspection that Eq. (A2.26) is correct.

---

[16] McConnell A.J. *Applications of Tensor Analysis*, Dover, 1957.

Appendix III: *On the uniqueness theorem*

We consider the virtual work equation

$$\int_{\partial V}\left(\tau_{ij}\delta u_j + \mu_{ij}\delta w_j + \mu_{ijk}\delta\varepsilon_{jk}\right)n_i\,dS = \int_{\partial V}\left(P_k\delta u_k + Q_k D\delta u_k\right)dS \tag{A3.1}$$

- The first term of the left hand side of Eq. (A3.1) becomes:

$$\tau_{jk} = \tau_{(jk)} + \tau_{[jk]} = \hat{\tau}_{jk} - \partial_i\mu_{ijk} - \frac{1}{2}\varepsilon_{ijk}\partial_l\mu_{li} =$$

$$= \hat{\tau}_{jk} - \frac{1}{3}\partial_i\left(\hat{\mu}_{ijk} + \hat{\mu}_{jki} + \hat{\mu}_{kij}\right) - \frac{1}{2}\varepsilon_{ijk}\partial_l\mu_{li}^D - \frac{1}{2}\varepsilon_{ijk}\delta_{li}\partial_l\mu \tag{A3.2}$$

where μ is the mean torsion

$$\mu = \frac{1}{3}\mu_{nn} \tag{A3.3}$$

Using Eq. (47.9)

$$\tau_{jk} = \hat{\tau}_{jk} - \frac{1}{3}\partial_i\left(\hat{\mu}_{ijk} + \hat{\mu}_{jki} + \hat{\mu}_{kij}\right) - \frac{1}{3}\partial_l\left(\hat{\mu}_{jkl} - \hat{\mu}_{kjl}\right) - \frac{1}{2}\varepsilon_{ijk}\delta_{li}\partial_l\mu$$

$$= \hat{\tau}_{jk} - \frac{1}{3}\partial_i\left(\hat{\mu}_{ijk} + 2\hat{\mu}_{jki}\right) - \frac{1}{2}\varepsilon_{ijk}\delta_{li}\partial_l\mu \tag{A3.4}$$

and therefore

$$\int_{\partial V}n_j\tau_{jk}\delta u_k\,dS = \int_{\partial V}n_j\left\{\hat{\tau}_{jk} - \frac{1}{3}\partial_i\left(\hat{\mu}_{ijk} + 2\hat{\mu}_{jki}\right) - \frac{1}{2}\varepsilon_{ijk}\delta_{li}\partial_l\mu\right\}\delta u_k\,dS \tag{A3.5}$$

- The second term of the left hand side of Eq. (A3.1) becomes:

$$n_i\mu_{ij}\delta w_j = n_i\mu_{ij}^D\delta w_j + n_i\mu\delta_{ij}\delta w_j \tag{A3.6}$$

Analyzing the displacement gradient into normal and tangential derivative

$$w_j = \frac{1}{2}\varepsilon_{jkl}\partial_k\delta u_l = \frac{1}{2}\varepsilon_{jkl}\left(n_k D\delta u_l + D_k\delta u_l\right) \tag{A3.7}$$

we get:

$$n_i \mu_{ij}^D \delta w_j = \frac{1}{2} n_i \mu_{ij}^D \varepsilon_{jkl} \left(n_k D \delta u_l + D_k \delta u_l \right)$$ (A3.8)

$$n_i \mu \delta_{ij} \delta w_j = \frac{1}{2} n_i (\mu \delta_{ij}) \varepsilon_{jkl} \left(n_k D \delta u_l + D_k \delta u_l \right)$$ (A3.9)

We remark that

$$\frac{1}{2} n_i \mu_{ij}^D \varepsilon_{jkl} D_k \delta u_l = \frac{1}{2} D_k \left(n_i \mu_{ij}^D \varepsilon_{jkl} \delta u_l \right) - \frac{1}{2} D_k (n_i) \mu_{ij}^D \varepsilon_{jkl} \delta u_l - \frac{1}{2} n_i D_k (\mu_{ij}^D) \varepsilon_{jkl} \delta u_l =$$
$$= \frac{1}{2} D_k \Phi_k - \frac{1}{2} D_k (n_i) \left(\frac{2}{3} (\hat{\mu}_{kli} - \hat{\mu}_{lki}) \right) \delta u_l - \frac{1}{2} n_i D_k \left(\frac{2}{3} (\hat{\mu}_{kli} - \hat{\mu}_{lki}) \right) \delta u_l$$ (A3.10)

with

$$\Phi_k = n_i \mu_{ij}^D \varepsilon_{jkl} \delta u_l$$ (A3.11)

for which the following identity holds:

$$D_k \Phi_k = D_p n_p n_k \Phi_k - n_q \varepsilon_{qpm} \partial_p \left(\varepsilon_{mlk} n_l \Phi_k \right)$$ (A3.12)

With

$$A_m = \varepsilon_{mlk} n_l \Phi_k$$ (A3.13)

we get that over a closed surface

$$\oint_S n_q \varepsilon_{qpm} \partial_p A_m dS = 0$$ (A3.14)

Using the above identity and Eq. (A3.10), the integral of Eq. (A3.8) results to the following expression:

$$W_\mu = \int_{\partial V} n_i \mu_{ij}^D \delta w_j = \int_{\partial V} \left\{ \frac{1}{2} n_i n_k \left(\frac{2}{3} (\hat{\mu}_{kli} - \hat{\mu}_{lki}) \right) \right\} D \delta u_l dS +$$
$$+ \int_{\partial V} \left(D_p n_p n_k \Phi_k - \frac{1}{2} D_k (n_i) \left(\frac{2}{3} (\hat{\mu}_{kli} - \hat{\mu}_{lki}) \right) - \frac{1}{2} n_i D_k \left(\frac{2}{3} (\hat{\mu}_{kli} - \hat{\mu}_{lki}) \right) \right) \delta u_l dS$$

or

$$W_\mu = \int_{\partial V} \left\{ \frac{1}{3} n_i n_k \left(\hat{\mu}_{kli} - \hat{\mu}_{lki} \right) \right\} D\delta u_l dS +$$

$$+ \int_{\partial V} \left\{ \frac{1}{3} D_p n_p n_k n_i \left(\hat{\mu}_{kli} - \hat{\mu}_{lki} \right) - \frac{1}{3} D_k (n_i)\left(\hat{\mu}_{kli} - \hat{\mu}_{lki} \right) - \frac{1}{3} n_i D_k \left(\hat{\mu}_{kli} - \hat{\mu}_{lki} \right) \right\} \delta u_l dS \tag{A3.15}$$

The integral of Eq. (A3.9) becomes:

$$\int_{\partial V} n_i \delta_{ij} \mu \delta w_j = \int_{\partial V} \left\{ \frac{1}{2} n_i (\mu \delta_{ij}) \varepsilon_{jkl} \left(n_k D\delta u_l + D_k \delta u_l \right) \right\} dS =$$

$$= \int_{\partial V} \left(\frac{1}{2} n_i (\mu \delta_{ij}) \varepsilon_{jkl} n_k \right) D\delta u_l dS +$$

$$+ \int_{\partial V} \left\{ \frac{1}{2} D_k \left(n_i (\mu \delta_{ij}) \varepsilon_{jkl} \delta u_l \right) - \frac{1}{2} D_k (n_i)(\mu \delta_{ij}) \varepsilon_{jkl} \delta u_l - \frac{1}{2} n_i D_k \left((\mu \delta_{ij}) \varepsilon_{jkl} \right) \delta u_l \right\} dS \tag{A3.16}$$

$$= \int_{\partial V} \left\{ \frac{1}{2} D_p n_p n_k n_i (\mu \delta_{ij}) \varepsilon_{jkl} \delta u_l - \frac{1}{2} D_k (n_i)(\mu \delta_{ij}) \varepsilon_{jkl} \delta u_l - \frac{1}{2} n_i D_k \left((\mu \delta_{ij}) \varepsilon_{jkl} \right) \delta u_l \right\} dS$$

$$= \int_{\partial V} \left\{ -\frac{1}{2} D_k (n_i)(\mu \delta_{ij}) \varepsilon_{jkl} - \frac{1}{2} n_i D_k \left((\mu \delta_{ij}) \varepsilon_{jkl} \right) \right\} \delta u_l dS$$

And therefore the integral of the second term of Eq.(A3.1) yields

$$\int_{\partial V} n_i \mu_{ij} \delta w_j dS = \int_{\partial V} \left\{ \frac{1}{3} n_i n_j \left(\hat{\mu}_{jki} - \hat{\mu}_{kji} \right) \right\} D\delta u_k dS +$$

$$+ \int_{\partial V} \left\{ \frac{1}{3} D_p n_p n_i n_j \left(\hat{\mu}_{jki} - \hat{\mu}_{kji} \right) - \frac{1}{3} D_j (n_i)\left(\hat{\mu}_{jki} - \hat{\mu}_{kji} \right) - \frac{1}{3} n_i D_j \left(\hat{\mu}_{jki} - \hat{\mu}_{kji} \right) \right\} \delta u_k dS - \tag{A3.17}$$

$$- \int_{\partial V} \left\{ \frac{1}{2} D_k (n_i)(\mu \delta_{ij}) \varepsilon_{jkl} + \frac{1}{2} n_i D_k \left((\mu \delta_{ij}) \varepsilon_{jkl} \right) \right\} \delta u_l dS$$

- The **third term** of the left hand side of Eq. (A3.1) is analyzed as follows:

$$n_i \mu_{ijk} \delta \varepsilon_{jk} = n_i \mu_{ijk} \partial_j \delta u_k = n_i \mu_{ijk} n_j D\delta u_k + n_i \mu_{ijk} D_j \delta u_k =$$

$$= n_i \mu_{ijk} n_j D\delta u_k + D_j (n_i \mu_{ijk} \delta u_k) - D_j n_i \mu_{ijk} \delta u_k - n_i D_j \mu_{ijk} \delta u_k \tag{A3.18}$$

and its integral becomes

$$\int_{\partial V} n_i \mu_{ijk} \delta\varepsilon_{jk} dS = \int_{\partial V} \frac{1}{3} n_i \left(\hat{\mu}_{ijk} + \hat{\mu}_{jki} + \hat{\mu}_{kij}\right) n_j D\delta u_k dS +$$

$$+ \int_{\partial V} \left\{ \frac{1}{3} D_p n_p n_i \left(\hat{\mu}_{ijk} + \hat{\mu}_{jki} + \hat{\mu}_{kij}\right) n_j \right\} \delta u_k dS - \qquad \text{(A3.19)}$$

$$- \int_{\partial V} \left\{ \frac{1}{3} D_j n_i \left(\hat{\mu}_{ijk} + \hat{\mu}_{jki} + \hat{\mu}_{kij}\right) + \frac{1}{3} n_i D_j \left(\hat{\mu}_{ijk} + \hat{\mu}_{jki} + \hat{\mu}_{kij}\right) \right\} \delta u_k dS$$

Finally by substituting Eqs. (A3.5), (A3.17) and (A3.19) into the left hand side of Eq. (A3.1) we get

$$\int_{\partial V} \left(\tau_{ij}\delta u_j + \mu_{ij}\delta w_j + \mu_{ijk}\delta\varepsilon_{jk}\right) n_i \, dS$$

$$= \int_{\partial V} \frac{1}{3} n_i \left(\hat{\mu}_{ijk} + 2\hat{\mu}_{jki}\right) n_j D\delta u_k dS + \int_{\partial V} \left\{ n_j \hat{\tau}_{ij} - \frac{1}{3} n_j \partial_i \left(\hat{\mu}_{ijk} + 2\hat{\mu}_{jki}\right) + \frac{1}{3} D_p n_p n_i \left(\hat{\mu}_{ijk} + 2\hat{\mu}_{jki}\right) n_j - \right.$$

$$\left. - \frac{1}{3} D_j n_i \left(\hat{\mu}_{ijk} + 2\hat{\mu}_{jki}\right) + \frac{1}{3} n_i D_j \left(\hat{\mu}_{ijk} + 2\hat{\mu}_{jki}\right) \right\} \delta u_k dS - \qquad \text{(A3.20)}$$

$$- \int_{\partial V} \left\{ \frac{1}{2}(\mu\delta_{ij})\varepsilon_{jlk} D_l n_i + \frac{1}{2} n_i D_l (\mu\delta_{ij})\varepsilon_{jlk} + \frac{1}{2} n_j \varepsilon_{ijk}\delta_{li}\partial_l\mu \right\} \delta u_k dS$$

and

$$\frac{1}{2} n_i D_l (\mu\delta_{ij})\varepsilon_{jlk} + \frac{1}{2} n_j \varepsilon_{ijk}\delta_{li}\partial_l\mu = \frac{1}{2} n_j (D_l - \partial_l\mu)\varepsilon_{jlk} = \frac{1}{2} n_j n_l \varepsilon_{jlk} D\mu = 0 \qquad \text{(A3.21)}$$

And therefore

$$\int_{\partial V} \left(\tau_{ij}\delta u_j + \mu_{ij}\delta w_j + \mu_{ijk}\delta\varepsilon_{jk}\right) n_i \, dS = \int_{\partial V} \frac{1}{3} n_i \left(\hat{\mu}_{ijk} + 2\hat{\mu}_{jki}\right) n_j D\delta u_k dS +$$

$$\int_{\partial V} \left\{ n_j \hat{\tau}_{ij} - \frac{1}{3} n_j \partial_i \left(\hat{\mu}_{ijk} + 2\hat{\mu}_{jki}\right) + \frac{1}{3} D_p n_p n_i \left(\hat{\mu}_{ijk} + 2\hat{\mu}_{jki}\right) n_j - \right. \qquad \text{(A3.22)}$$

$$\left. - \frac{1}{3} D_j n_i \left(\hat{\mu}_{ijk} + 2\hat{\mu}_{jki}\right) + \frac{1}{3} n_i D_j \left(\hat{\mu}_{ijk} + 2\hat{\mu}_{jki}\right) \right\} \delta u_k dS - \int_{\partial V} \left\{ \frac{1}{2}(\mu\delta_{ij}) e_{jlk} D_l n_i \right\} \delta u_k dS$$

And by substituting Eq. (A3.22) into Eq.(A3.1) we get Eq.(52).

We can also formulate the relations between P_k and Q_k in terms of the "true" forces:

$$P_k = t_k + \frac{1}{2} e_{jlk} \left(D_p n_p n_l - D_l\right) m_j + \left(D_p n_p n_j - D_j\right) R_{jk} \qquad \text{(A.3.23)}$$

$$Q_k = \frac{1}{2} e_{jlk} n_l m_j + n_j R_{jk} \qquad \text{(A3.24)}$$

Appendix IV: *On a field and its normal derivative along a curve*

Let us consider for simplicity a two-dimensional situation and let in the plane (x, y) a curve (Γ), which is described by its analytic function

$$(\Gamma): \quad y = f(x) \tag{A4.1}$$

This curve (Γ), is that part of the boundary of a body $(\mathbf{B})$ where some or all displacement components are given as boundary conditions, say

$$u_\alpha(x, f(x)) = \bar{u}_\alpha(x) \tag{A4.2}$$

Let us consider now two neighbouring points A and B on that curve (Γ) with the co-ordinates

$$A(x_A, y_A) \quad , \quad B(x_B = x_A + \Delta x, y_B = y_A + \Delta y) \tag{A4.3}$$

where

$$\frac{\Delta y}{\Delta x} \approx \left(\frac{df}{dx}\right)_A \tag{A4.4}$$

For the computation of the values of the function $u_\alpha(x, y)$ at the interior point $C(x_B, y_A) \in \mathbf{B}$, we expand this function in Taylor series around the points A and B and we keep only the two first terms (Fig. A4.1). We remark that we can compute the value at point C by either going from A directly to C on the horizontal line $y = y_A$ or indirectly by first going to point B along the curve (Γ) and then moving vertically along the line $x = x_B$ from B to C; i.e.:

$$u_\alpha(C) \approx u_\alpha(A) + \left(\frac{\partial u_\alpha}{\partial x}\right)_A \Delta x \tag{A4.5}$$

Since

$$\left(\frac{\partial u_\alpha}{\partial y}\right)_B = \left(\frac{\partial u_\alpha}{\partial y}\right)_A + \left(\frac{\partial}{\partial x}\left(\frac{\partial u_\alpha}{\partial x}\right)\right)_A + \frac{\partial}{\partial y}\left(\frac{\partial u_\alpha}{\partial y}\right)_A \frac{df}{dx} \Delta x \tag{A4.6}$$

we get from Eq. (A4.5) that

$$u_\alpha(C) \approx u_\alpha(B) - \left(\frac{\partial u_\alpha}{\partial y}\right)_B \Delta y \approx u_\alpha(A) + \left(\frac{\partial u_\alpha}{\partial x}\right)_A \Delta x + \left(\frac{\partial u_\alpha}{\partial y}\right)_A \Delta y - \left(\frac{\partial u_\alpha}{\partial y}\right)_B \Delta y$$

$$= u_\alpha(A) + \left(\frac{\partial u_\alpha}{\partial x}\right)_A \Delta x + O(\Delta x^2) \tag{A4.7}$$

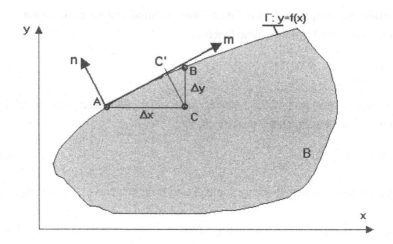

Thus the value of the function $u_\alpha(x, y)$ at point C does not depend on the path,

$$u_\alpha(C) \approx u_\alpha(A) + \left(\frac{\partial u_\alpha}{\partial x}\right)_A \Delta x \approx u_\alpha(B) - \left(\frac{\partial u_\alpha}{\partial y}\right)_B \Delta y \qquad (A4.8)$$

Moreover we observe that the value of $\left(\frac{\partial u_\alpha}{\partial y}\right)_A$ and of $\left(\frac{\partial u_\alpha}{\partial x}\right)_A$ are interrelated.

Indeed from

$$u_\alpha(B) \approx u_\alpha(A) + \left(\frac{\partial u_\alpha}{\partial x}\right)_A \Delta x + \left(\frac{\partial u_\alpha}{\partial y}\right)_A \Delta y \qquad (A4.9)$$

we get

$$\left(\frac{d\bar{u}_\alpha}{dx}\right)_A \approx \frac{u_\alpha(B) - u_\alpha(A)}{\Delta x} \approx \left(\frac{\partial u_\alpha}{\partial x}\right)_A + \left(\frac{\partial u_\alpha}{\partial y}\right)_A \left(\frac{df}{dx}\right)_A \qquad (A4.10)$$

Thus for the continuation of the considered function $u_\alpha(x, y)$ from a boundary point $A \in (\Gamma)$ to a point C, in the interior of body **B,** we need:

1.   The data for the function $u_\alpha(x,y)$ on the boundary, Eq. (A4.2).

2.   The information concerning its gradient in <u>only one</u> direction.

In particular we may chose point B to coincide with the normal projection of point C on the curve, $B \equiv C'$. From the figure we can see immediately that,

$$u_\alpha(C) \approx u_\alpha(A) + \left(\frac{d\bar{u}_\alpha}{dx}\right)_A \Delta m + \left(\frac{\partial u_\alpha}{\partial n}\right)_{C'} \Delta n \approx u_\alpha(C') + \left(\frac{\partial u_\alpha}{\partial n}\right)_{C'} \Delta n \qquad (A4.11)$$

where m and n are the directions tangential and normal to the boundary in the vicinity of the considered points. This means that for the determination of the function $u_\alpha(x,y)$ in the interior domain, starting from boundary data we must:

a) specify the function along the boundary curve $u_\alpha(x,f(x)) = \bar{u}_\alpha(x)$, which will allow us to compute

$$u_\alpha(C') = u_\alpha(A) + \left(\frac{d\bar{u}_\alpha}{dx}\right)_A \Delta m \qquad (A4.12)$$

b) we must know the value of the normal derivative of the considered function along the boundary

$$Du_\alpha(A) = \left(\frac{\partial u_\alpha}{\partial n}\right)_A \qquad (A4.13)$$

so that

$$u_\alpha(C) \approx u_\alpha(A) + \left(\frac{d\bar{u}_\alpha}{dx}\right)_A \Delta m + Du_\alpha \Delta n \qquad (A4.15)$$

2nd Gradient Flow Theory of Plasticity

Ioannis Vardoulakis[1]

[1] Faculty of Applied Mathematics and Physics, Department of Mechanics
National Technical University of Athens, Greece

Abstract. This chapter provides a brief introduction to the 2nd Gradient flow theories of plasticity with emphasis on strain softening and shear-banding.

5.1 Introduction

Localization of deformation leads to a change of scale of the problem, so that phenomena occurring at the scale of the grain cannot be ignored anymore in the mechanical modeling process of the macroscopic behavior of the material. Thus in order to describe correctly localization phenomena it appears necessary to resort to continuum models with micro-structure, such as Cosserat or Gradient models (cf. Vardoulakis & Sulem, 1995). The description of statics and kinematics of continuous media with microstructure has been studied systematically by many authors in the past (cf. Chapter 4). This work was revived in the last two decades of the 20th century, in order to address the problem of localization in Geomaterials.

Here Mindlin's continuum approach is used as the background for the presentation of the most basic results that pertain to more recent works on 2nd gradient elasto-plasticity theories, which can be used for addressing localization phenomena properly.

5.2 A Cohesion-Softening Model for an Ideal Quasi-Brittle Solid

5.2.1 Local Continuum Formulation

As a brief introduction to the problem at hand we describe here a simple mathematical model for quasi brittle solids, by sing a "kinematic hardening" approach (Mroz, 1973, Vardoulakis & Frantziskonis, 1992). Accordingly the Cauchy stress tensor σ_{ij} is split formally into a reduced stress τ_{ij} and into a back-stress α_{ij},

$$\sigma_{ij} = \tau_{ij} + \alpha_{ij} \tag{1}$$

These stress tensors are further decomposed into deviatoric and spherical parts

$$\sigma_{ij} = s_{ij} + p\delta_{ij} \quad , \quad \alpha_{ij} = a_{ij} + q\delta_{ij} \quad , \quad \tau_{ij} = t_{ij} + p_\tau \delta_{ij} \tag{2}$$

For simplicity we assume that plastic yielding is described by a Drucker-Prager yield surface, which is expressed in terms of the reduced stress as follows:

$$F = F(\tau_{ij}) = T_\tau + p_\tau f \tag{3}$$

where p_τ and T_τ are related to the 1st and the 2nd deviatoric invariants of the reduced stress tensor

$$p_\tau = \frac{1}{3}I_{1\tau} = \frac{1}{3}\tau_{kk} \quad , \quad T_\tau = \sqrt{J_{2t}} = \sqrt{\frac{1}{2}t_{ij}t_{ji}} \tag{4}$$

and the "friction" coefficient f is assumed to be constant,

$$f = const. > 0 \tag{5}$$

We assume also that the back-stress is isotropic,

$$a_{ij} = 0 \quad \Rightarrow \quad \alpha_{ij} = q\delta_{ij} \tag{6}$$

and we get

$$p_\tau = p - q \quad , \quad t_{ij} = s_{ij} \quad \Rightarrow \quad T_\tau \equiv T = \sqrt{J_{2s}} = \sqrt{\frac{1}{2}s_{ij}s_{ji}} \tag{7}$$

With these assumptions Eq. (3) for the yield surface becomes

$$F = T + f(p - q) \tag{8}$$

As will be explained below the strength parameter q is assumed to be a decreasing function of the plastic hardening parameter Ψ :

$$q = Q(\Psi) > 0 \downarrow \forall \Psi \geq 0 \Rightarrow q_0 = q(0) > 0 \tag{9}$$

From Fig. 1(a) we see clearly that the evolving yield surface $F(\sigma_{ij}, \Psi) = 0$, Eq. (8), describes the behavior of a constant-friction, cohesion-softening model.

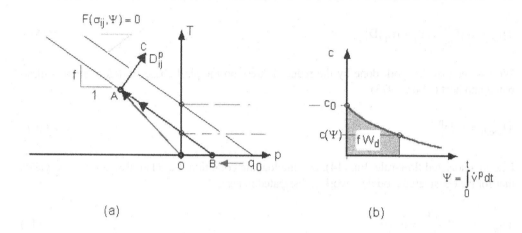

Figure 1. Cohesion softening model

Plastic loading is defined by the conditions

$$F = 0 \quad \Rightarrow \quad T + f(p - q) = 0 \tag{10}$$

and

$$\dot{F} = \frac{\partial F}{\partial \tau_{ij}} \dot{\tau}_{ij} = 0 \Rightarrow \quad F_{ij}(\dot{\sigma}_{ij} - \dot{\alpha}_{ij}) = 0 \tag{11}$$

where

$$F_{ij} = \frac{\partial F}{\partial \tau_{ij}} \tag{12}$$

In the considered case of a D.-P. model we have the following simple expressions for the gradient F_{ij}

$$F_{ij} = \frac{1}{\sqrt{2}} m_{ij} + \frac{1}{3} f \delta_{ij} \quad , \quad m_{ij} = \frac{s_{ij}}{\sqrt{2}\,T} \quad , \quad m_{ij} m_{ji} = 1 \tag{13}$$

We assume that the material is dilatant, and we adopt normality as a plastic flow-rule. This assumption may be also seen as a condition for maximum dilatancy,

$$D_{ij}^p = F_{ij} \dot{\Psi} \quad \text{with} : \dot{\Psi} \geq 0 \tag{14}$$

The total dissipation is computed as the work of the stress on the plastic rate of deformation

$$D_{loc} = \sigma_{ij} D_{ij}^p = (\tau_{ij} + \alpha_{ij}) D_{ij}^p \tag{15}$$

We assume that the work done by the reduced stress on the plastic rate of deformation is dissipated into heat (Mroz, 1973)

$$D_{loc,h} = \tau_{ij} D_{ij}^p \tag{16}$$

Due to associated flow-rule, Eq. (14), and the loading condition, Eq. (11), this assumption yields that for the considered model <u>no work</u> is dissipated in heat,

$$D_{loc,h} = \tau_{ij} \dot{\Psi} \frac{\partial F}{\partial \tau_{ij}} = \dot{F} = 0 \tag{17}$$

This can be seen clearly in Fig. 1 (a), where the reduced-stress vector, depicted there by the vector $\vec{\tau} = (\overrightarrow{BA})$, is normal to the plastic rate of deformation, depicted by the vector $\vec{D}^p = (\overrightarrow{AB})$. This means in turn that the considered elasto-plastic model describes an ideal quasi-brittle material, for which all dissipation is due to "damage" only

$$D_{loc} = D_{loc,d} = \alpha_{ij} D_{ij}^p \tag{18}$$

To this "cold"-type plasticity theory is given sometimes the name "material damage" theory, meaning that energy dissipated during loading is spend mainly for internally damaging the material by e.g. the generation of micro-cracks. Material damage is indirectly observable, since it is usually accompanied by a clear and localizable acoustic emission. Within the mathematical frame of the present constitutive model material damage is observed macroscopically as cohesion softening (Fig. 1 b).

In order to explore further mathematically these assumptions we remark first

$$F_{ij} = \frac{\partial F}{\partial \tau_{ij}} = \frac{\partial F}{\partial \sigma_{kl}} \frac{\partial \sigma_{kl}}{\partial \tau_{ij}} = \frac{\partial F}{\partial \sigma_{ij}} \tag{19}$$

With that from the flow rule, Eq. (14) it follows that the tensors σ_{ij} and D_{ij}^p are co-axial. Thus we may apply Maxwell's energy decomposition on the expression for the dissipation, by splitting into a volumetric and a deviatoric part

$$D_{loc} = \sigma_{ij} D_{ij}^p = p\dot{v}^p + T\dot{g}^p \tag{20}$$

In Eq. (20) $\dot{v}^P$ and $\dot{g}^P$ are the plastic volumetric- and shearing strain-rate, respectively,

$$\dot{v}^P = D^P_{kk}, \quad \dot{g}^P = \sqrt{2D^{p'}_{ij} D^{p'}_{ij}}, \quad D^p_{ij} = D^{p'}_{ij} + \frac{1}{3}D^p_{kk} \tag{21}$$

Due to the assumed normality flow-rule we have that the ratio of plastic volumetric- to plastic shear strain-rate is given by the friction coefficient

$$\frac{\dot{v}^P}{\dot{g}^P} = f = \text{const.} \tag{22}$$

Using the loading condition, Eq. (10) we get from Eq. (20)

$$D_{loc} = \dot{v}^P\left(T\frac{\dot{g}^P}{\dot{v}^P} + p\right) = \dot{v}^P\left(T\frac{1}{f} + p\right) = q\dot{v}^P \tag{23}$$

This result is consistent with the above Eqs. (6) and (18),

$$D_{loc} = q\delta_{ij}D^p_{ij} = \alpha_{ij}D^p_{ij} = D_{loc,d} \Rightarrow D_{loc,d} = q\dot{v}^P \tag{24}$$

Thus, if we select the cumulative plastic volumetric strain as "hardening" parameter,

$$\Psi = \int_0^t \dot{v}^P dt \tag{25}$$

then the strength parameter q can be computed from the cumulative damage energy, as indicated in Fig. 1(b),

$$W_d = \int_0^t D_{loc,d} dt = \int_0^{v^P} q(\Psi)d\Psi \Leftrightarrow q = \frac{dW_d}{dv^P} \tag{26}$$

The observed dilatancy is describing the porosity increase of the material due to micro-cracking. To illustrate this statement we recall the rate equation between the plastic volumetric strain and the porosity ϕ of the material (cf. Vardoulakis & Sulem, 1995)

$$\dot{v}^P \approx \frac{\dot{\phi}}{1-\phi} \tag{27}$$

In Eq. (27) elastic volumetric strains are neglected.

These observations allow us to assume that the strength parameter q should be seen in general as a decreasing function of the porosity of the material; i.e.

$$q = Q(\phi) \tag{28}$$

with

$$H_t = \frac{dQ}{d\phi} < 0 \tag{29}$$

Remark

The simplest model of strength reduction due to porosity increase a "damage" power law of the form,

$$q = q_0 (1 - \phi)^n$$

resulting to

$$H_t = -nq_0 (1 - \phi)^{n-1}$$

For $n = 1$ we get a Kachanov-type model with constant rate of softening,

$$q = q_0 (1 - \phi) \quad \Rightarrow \quad H_t = -q_0$$

The major defect of such constitutive models remains the fact that these are strain-softening models. The application of such a softening models leads to mathematically ill-posed boundary-value problem formulations. This mathematical ill-posedeness can be easily verified by means of a linear stability analysis. Such an analysis shows that in the limit a perturbation of vanishing wave-length grows with infinite pace (Schaeffer, 1990). This theoretical result indicates that damage in the softening regime has the tendency to localize in a narrow zone, which macroscopically is interpreted as a "fracture" surface. The classical continuum approach, pursued up to this point is not capable to simulate the localization of the deformation in a finite strip, leading thus to zero-fracture-zone thickness and to infinite Ljapunov exponents for the time evolution of the instability.

5.2.2   The Non-Local Assumption

A remedy for the defect of the classical continuum with strain softening is to modify the constitutive Eq. (28). Accordingly we replace this "local" continuum assumption by a non-local one (Vardoulakis & Aifantis, 1989): Through Eq. (28) we assumed in fact that the strength is a function of the local value of the porosity. Accordingly we reformulate the above statement by stating instead that the strength is a function of the mean value of the porosity over a small but finite characteristic volume, V_c, whose center of gravity coincides with the position of the considered material point,

$$q = Q(<\phi>) \tag{30}$$

$$<\phi> = \frac{1}{V_c} \int_{V_c} \phi \, dV \tag{31}$$

For the evaluation of the integral in Eq. (31) we put the origin of the co-ordinate system at the considered material point. Let $x_i = h_i$ be the coordinates of an arbitrary point inside the volume of integration V_c. By using a truncated Taylor series expansion we get,

$$\phi(x_i) = \phi(0) + \left.\frac{\partial \phi}{\partial x_i}\right|_0 h_i + \frac{1}{2}\left.\frac{\partial^2 \phi}{\partial x_i \partial x_j}\right|_0 h_i h_j + O\left(h_i^3\right) \tag{32}$$

We assume that V_c is a cube, whose sides have the length $\overline{\ell}_c$.

From Eqs. (31) and (32) we get the following well-known integration rule from Numerical Analysis[1],

$$<\phi> \approx \phi + \frac{\overline{\ell}_c^2}{24} \nabla^2 \phi \tag{33}$$

We observe that "local homogeneity" means that in Eq. (33) the term with the Lapalcian is negligible, thus yielding the classical result

$$<\phi> = \phi \tag{34}$$

Thus in a locally homogeneous setting the value of a field at point coincides with its mean value over an elementary volume that is centered at this point. Eq. (34) is equivalent with a linear-in-space variation of the field $\phi(x_i)$. In a locally non-homogeneous setting, Eq. (33) is

---

[1] cf. Henrici, P. *Elements of Numerical Analysis*, Wiley, 1964.

used instead, for the approximation of the mean value over the sampling volume. In this approximation we assumed that the characteristic length

$$\ell_c = \bar{\ell}_c / (2\sqrt{6})$$ (35)

is a small parameter, as compared to any other geometric length of the problem at hand.
    With these assumptions the non-local constitutive Eq. (30) yields

$$q \approx Q\left(\phi + \ell_c^2 \nabla^2 \phi\right) \approx Q(\phi) + \frac{dQ}{d\phi} \ell_c^2 \nabla^2 \phi$$ (36)

or with Eq. (29) to the following expression,

$$q \approx Q(\phi) - \ell_c^2 |H_t| \nabla^2 \phi \quad (H_t < 0)$$ (37)

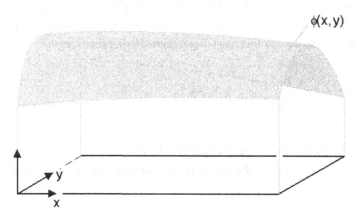

Figure 2. Negative mean curvature of the porosity distribution function in regions of damage localization

The physical meaning of this non-local modification of the constitutive law for the strength is the following (Fig.2): It is expected that within a zone of damaged material the porosity (due to micro-cracking) assumes large values. This is expressed by a negative "mean-curvature condition", (or "localization")

$$\nabla^2 \phi < 0$$ (38)

Thus the non-local term in Eq. (37) acts in a way to counterbalance effect of softening.
    From Eq. (37) by material time differentiation we get,

$$\dot{q} \approx \left(H_t + \frac{dH_t}{d\phi} \ell_c^2 \nabla^2 \phi \right) \dot{\phi} + H_t \ell_c^2 \nabla^2 \dot{\phi} \approx H_t \left(\dot{\phi} + \ell_c^2 \nabla^2 \dot{\phi} \right) \tag{39}$$

or due to Eqs. (27) and (25)

$$\overline{H}_t \dot{\Psi} < \dot{q} \approx \overline{H}_t \left(\dot{\Psi} + \ell_c^2 \nabla^2 \dot{\Psi} \right) < 0 \tag{40}$$

where

$$\overline{H}_t = (1 - \phi) \frac{dQ}{d\phi} < 0 \tag{41}$$

From the loading condition, Eq. (11), the constitutive equations for the Cauchy stress and the back stress

$$\dot{\sigma}_{ij} = C^e_{ijkl} D^e_{kl} = C^e_{ijkl} (D_{kl} - \dot{\Psi} F_{kl}) \tag{42}$$

$$\dot{\alpha}_{ij} = \dot{q} \delta_{ij} \tag{43}$$

and the evolution law for the strength parameter q, Eq. (40) we obtain consistency condition,

$$\dot{F} = F_{ij} \dot{\tau}_{ij} = 0 \Rightarrow \quad F_{ij} \left(\dot{\sigma}_{ij} - \dot{\alpha}_{ij} \right) = 0$$

$$F_{ij} C^e_{ijkl} D_{kl} - F_{ij} C^e_{ijkl} F_{kl} \dot{\Psi} - F_{ij} \overline{H}_t (\dot{\Psi} + \ell_c^2 \nabla^2 \dot{\Psi}) \delta_{ij} = 0$$

$$B_{kl} D_{kl} = H \dot{\Psi} + F_{kk} \overline{H}_t \ell_c^2 \nabla^2 \dot{\Psi}$$

or

$$\dot{\Psi} - \ell^2 \nabla^2 \dot{\Psi} = \frac{1}{H} B_{kl} D_{kl} \tag{44}$$

where

$$B_{kl} = F_{ij} C^e_{ijkl} \tag{45}$$

$$H_0 = B_{kl} F_{kl} \tag{46}$$

$$H = H_0 + F_{kk} \overline{H}_t \quad (\overline{H}_t = -|\overline{H}_t|) \tag{47}$$

and

$$\ell^2 = \frac{|H_t|}{H} F_{kk} \ell_c^2 \tag{48}$$

We require that the plastic modulus is positive, thus imposing a restriction to the softening modulus not to exceed the snap-back threshold value H_0 (Nguyen & Bui, 1974):

$$H > 0 \quad \Rightarrow \quad F_{kk}|\bar{H}_t| < H_0 \tag{49}$$

With

$$F_{kk} = f \quad , \quad 0 < f < 0.87 \tag{50}$$

and

$$H_0 = G\left(1 + \frac{2}{3}\frac{1+\nu}{1-2\nu}f^2\right) \tag{51}$$

Ineq. (49) is satisfied as soon as the softening rate is not exceeding the elastic shear modulus

$$|H_t| < G \tag{52}$$

Thus the evolving material length-parameter, defined above through Eq. (48), is always real,

$$\ell^2 > 0 \tag{53}$$

For homogeneous ground plastic-strain states, the term with the Laplacian on the l.h.s. of Eq. (44) vanishes identically and the consistency condition collapses to that of the classical flow theory of plasticity; i.e. to a monomial equation, which can be readily solved in terms of the plastic multiplier $\dot{\Psi}$. In general direct elimination of $\dot{\Psi}$ will not be possible and one has to carry the consistency condition as an additional field equation together with the balance equations, and to treat $\dot{\Psi}$ as an additional degree of freedom (Mühlhaus & Aifantis,1991)[7]. An approximate procedure however proposed by Vardoulakis et al. (1989 and 1992) allows for the elimination of $\dot{\Psi}$ from the set of governing equations and simplifies the problem drastically (see Appendix I),

$$\dot{\Psi} = \frac{1}{1-\ell^2\nabla^2}\left(\frac{1}{H}B_{kl}D_{kl}\right) \quad \Rightarrow \quad \dot{\Psi} \approx \frac{<1>}{H}B_{kl}\left(1 + \ell^2\nabla^2\right)D_{kl} \tag{54}$$

where $<\cdot>$ denote the Föppl-Macauley brackets, so that $\dot{\Psi} \geq 0$; i.e. with

$$H > 0 \tag{55}$$

we define the switch function:

$$<1> = \begin{cases} 1 & \text{if} : F = 0 \text{ and } B_{kl}\left(1 + \ell^2 \nabla^2\right) D_{kl} > 0 \\ 0 & \text{if} : F < 0 \text{ or } F = 0 \text{ and } B_{kl}\left(1 + \ell^2 \nabla^2\right) D_{kl} \leq 0 \end{cases} \tag{56}$$

Above approximation, Eq. (54), allows to derive explicitly the rate constitutive equations of gradient flow theory of plasticity. Indeed, from Eqs. (42) and (54) we get

$$\dot{\sigma}_{ij} = C^e_{ijkl} D^e_{kl} \quad \Rightarrow$$

$$\dot{\sigma}_{ij} = C^e_{ijkl} D_{kl} - C^e_{ijmn} F_{mn} \frac{<1>}{H} B_{kl}\left(1 + \ell^2 \nabla^2\right) D_{kl}$$

$$= C^e_{ijkl} D_{kl} - \frac{<1>}{H} C^e_{ijmn} F_{mn} B_{kl} D_{kl} - \frac{<1>}{H} C^e_{ijmn} F_{mn} B_{kl} B_{kl} \ell^2 \nabla^2 D_{kl}$$

or

$$\dot{\sigma}_{ij} = C^{ep}_{ijkl} D_{kl} - C^p_{ijkl} \ell^2 \nabla^2 D_{kl} \tag{57}$$

where

$$C^{ep}_{ijkl} = C^e_{ijkl} - C^p_{ijkl} \quad , \quad C^p_{ijkl} = \frac{<1>}{H} C^e_{ijmn} F_{mn} F_{pq} C^e_{pqkl} \tag{58}$$

Notice that the constitutive Eqs. (57) constitute a singular perturbation of the ones of classical flow theory.

5.3 Isotropic Softening Plasticity

The previous considerations allow us to produce the following approach to a 2nd gradient plasticity model for isotropically softening solid. We start with a generalization of the yield function, by assuming that it is of the form,

$$F = \hat{F}(\sigma_{ij}, <\Psi>) \tag{59}$$

where

$$< \Psi >= \frac{1}{V_c} \int_{V_c} \Psi dV \tag{60}$$

is the average of the hardening parameter over the characteristic volume V_c. For example in Critical State Soil Plasticity Theory we assume that the yield function depends on the current state of the voids ratio, $\Psi = e$. In that case Eq. (59) would reflect the fact that porosity is to be evaluated over a characteristic volume V_c, and that the assumption $< e >= e$ simply reflects a hypothesis about the local homogeneity of the porosity distribution. With

$$< \Psi >\approx \Psi + \ell_c^2 \nabla^2 \Psi \tag{61}$$

we get from Eq. (59),

$$F \approx \hat{F}(\sigma_{ij} \Psi) + \frac{\partial \hat{F}}{\partial \Psi} \ell_c^2 \nabla^2 \Psi \tag{62}$$

For a softening solid the hardening modulus is negative,

$$H_t = -\frac{\partial \hat{F}}{\partial \Psi} < 0 \quad \Rightarrow \quad |H_t| = -H_t = \frac{\partial \hat{F}}{\partial \Psi} > 0 \tag{63}$$

and Eq. (62) becomes,

$$F \approx \hat{F}(\sigma_{ij} \Psi) + |H_t| \ell_c^2 \nabla^2 \Psi \tag{64}$$

This yield function is now used together with constitutive equations of elasto-plasticity theory,

$$D_{ij} = D_{ij}^e + D_{ij}^p \tag{65.1}$$

$$D_{ij}^e = C_{ijkl}^{e\,-1} \dot{\sigma}_{kl} \tag{65.2}$$

$$D_{ij}^p = \dot{\Psi} \frac{\partial Q}{\partial \sigma_{ij}} \quad , \quad \dot{\Psi} \geq 0 \tag{65.3}$$

such that plastic strains are generated whenever the stress is and stays on the current yield surface.

With these remarks the consistency condition becomes,

$$\dot{F} = 0 \quad \Rightarrow \quad F \approx \hat{F}(\sigma_{ij} \Psi) + |H_t| \ell_c^2 \nabla^2 \dot{\Psi} + O(\nabla^2 \Psi) = 0$$

or

$$\frac{\partial \hat{F}}{\partial \sigma_{ij}} \dot{\sigma}_{ij} + \frac{\partial \hat{F}}{\partial \Psi} \dot{\Psi} + |H_t| \ell_c^2 \nabla^2 \dot{\Psi} + O(\nabla^2 \Psi) = 0$$

or

$$\frac{\partial F}{\partial \sigma_{ij}} C_{ijkl}^e \left(D_{kl} - \dot{\Psi} \frac{\partial Q}{\partial \sigma_{ij}} \right) - H_t \dot{\Psi} + |H_t| \ell_c^2 \nabla^2 \dot{\Psi} + O(\nabla^2 \Psi) = 0$$

or

$$(1 - \ell^2 \nabla^2) \dot{\Psi} \approx \frac{1}{H} B_{kl} D_{kl} \qquad (66)$$

where

$$F_{ij} = \frac{\partial \hat{F}}{\partial \sigma_{ij}} \quad , \quad Q_{ij} = \frac{\partial Q}{\partial \sigma_{ij}}$$

$$B_{kl} = F_{ij} C_{ijkl}^e \qquad (67)$$

$$H_0 = F_{ij} C_{ijkl}^e Q_{kl}$$

$$H = H_0 + H_t > 0$$

and

$$\ell^2 = \frac{|H_t|}{H} \ell_c^2 > 0 \qquad (68)$$

Within a good approximation, Eq. (66) can be inverted, yielding thus again the Eq. (54) and the loading condition, Eq. (56). The corresponding rate equations of 2nd gradient non-associated elasto-plasticity theory are:

$$\dot{\sigma}_{ij} = C_{ijkl}^{ep} D_{kl} - C_{ijkl}^p \ell^2 \nabla^2 D_{kl} \qquad (69)$$

$$C_{ijkl}^{ep} = C_{ijkl}^e - C_{ijkl}^p \quad , \quad C_{ijkl}^p = \frac{<1>}{H} C_{ijmn}^e F_{mn} Q_{pq} C_{pqkl}^e \qquad (70)$$

5.4 The Mindlin-Continuum Formulation

The constitutive Eqs. (57) or (69) are the rate-equations of a 2nd gradient elasto-plastic solid. In view of these constitutive equations we introduce decomposition of the Cauchy stress rate

$$\dot{\sigma}_{ij} = \dot{\sigma}_{ij}^{(0)} + \dot{\sigma}_{ij}^{(2)} \tag{71}$$

The first term on the r.h.s. of Eq. (71) is called the 0th grade constitutive stress-rate and set equal to the constitutive stress-rate of the classical continuum

$$\dot{\sigma}_{ij}^{(0)} = C_{ijkl}^{ep} D_{kl} \tag{72}$$

The second term is called the 2nd grade constitutive stress-rate and it is assumed to satisfy an internal balance equation,

$$\dot{\sigma}_{ij}^{(2)} + \partial_k \dot{m}_{kij} = 0 \tag{73}$$

where $\dot{m}_{ijk}$ has the dimensions of a double-stress-rate and is defined through the following constitutive equation of gradient type,

$$\dot{m}_{ijk} = C_{jkmn}^{p} \ell^2 \partial_i D_{mn} \tag{74}$$

We see that, within a reasonable approximation, from Eqs. (71) to (74) we recover, the original constitutive Eqs. (57) or (69) of 2nd Gradient elasto-plasticity.

We consider a virtual kinematical set that is derived from a virtual velocity field δv_i,

$$\delta D_{ij} = \frac{1}{2}\left(\partial_i \delta v_j + \partial_j \delta v_i\right) \quad , \quad \delta c_{ijk} = \partial_i \delta D_{jk} \tag{75}$$

We assume that the stress-rate tensor $\dot{\sigma}_{ij}^{(0)}$ is working on the rate of deformation tensor and the double stress-rate tensor $\dot{m}_{ijk}$ is working on the gradient of the rate of deformation tensor. Accordingly we define the total second-order virtual power of internal forces as follows (cf. Chapter 4, Eqs. (80) to (82)):

$$\delta_2 P^{(int)} = \int_V \left(\dot{\sigma}_{ij}^{(0)} \delta D_{ij} + \dot{m}_{ijk} \delta c_{ijk}\right) dV \tag{76}$$

In case of locally homogeneous deformations, the second term inside the integral of Eq. (76) is vanishing, and the stress tensor $\dot{\sigma}_{ij}^{(0)}$ collapses to the Cauchy stress-rate tensor ($\dot{\sigma}_{ij} = \dot{\sigma}_{ij}^{(0)}$). In this case the expression for the 2nd order stress power collapses also to the one holding for the classical continuum.

In a similar way we define the virtual power of the rates of the external forces (cf. Chapter 3, Eq. (62))

$$\delta_2 P^{(ext)} = \int_V \dot{F}_k \delta v_k \, dV + \int_{\partial V} \left(\dot{P}_k \delta v_k + \dot{Q}_k D \delta v_k \right) dS \tag{77}$$

where

$$D = n_k \partial_k \quad , \quad D_k = \partial_k - n_k D \tag{78}$$

denotes the normal-to-the-boundary derivative. As explained in Chapter 3 (Appendix IV), the surface integral in Eq. (77) has the following meaning: If we prescribe at boundary point the velocity component $v_{(i)}$ we have also the freedom to specify independently is normal derivative $Dv_{(i)}$.

We require, that the virtual work equation holds as an expression of equilibrium

$$\delta_2 P^{(int)} = \delta_2 P^{(ext)}$$

or explicitly,

$$\int_V \left(\dot{\sigma}_{ij}^{(0)} \delta D_{ij} + \dot{m}_{ijk} \partial_i \delta D_{jk} \right) dV = \int_V \dot{F}_k \delta v_k \, dV + \int_{\partial V} \left(\dot{P}_k \delta v_k + \dot{Q}_k D \delta v_k \right) dS \tag{79}$$

Using Gauss' theorem this integral equation becomes,

$$\int_V \left(\dot{\sigma}_{ij}^{(0)} - \partial_k \dot{m}_{kij} \right) \delta D_{ij} dV + \int_{\partial V} n_i \dot{m}_{ijk} \delta D_{jk} \, dS = \int_V \dot{F}_k \delta v_k \, dV + \int_{\partial V} \left(\dot{P}_k \delta v_k + \dot{Q}_k D \delta v_k \right) dS$$

or

$$\int_V \left(\partial_i \left(\dot{\sigma}_{ij} \delta v_j \right) - \partial_i \dot{\sigma}_{ij} \delta v_j \right) dV + \int_{\partial V} n_i \dot{m}_{ijk} \delta D_{jk} \, dS = \int_V \dot{F}_k \delta v_k \, dV + \int_{\partial V} \left(\dot{P}_k \delta v_k + \dot{Q}_k D \delta v_k \right) dS$$

or

$$\int_{\partial V} n_i \dot{\sigma}_{ij} \delta v_j \, dS - \int_V \partial_i \dot{\sigma}_{ij} \delta v_j \, dV + \int_{\partial V} n_i \dot{m}_{ijk} \delta D_{jk} \, dS$$

$$= \int_V \dot{F}_k \delta v_k \, dV + \int_{\partial V} \left(\dot{P}_k \delta v_k + \dot{Q}_k D \delta v_k \right) dS \tag{80}$$

where

$$\dot{\sigma}_{ij} = \dot{\sigma}_{ij}^{(0)} - \partial_k m_{kij} \tag{81}$$

In the virtual work Eq. (80) the volume integrals yield that indeed the stress-rate $\dot{\sigma}_{ij}$ is a balance stress-rate tensor,

$$\partial_i \dot{\sigma}_{ij} + \dot{F}_j = 0 \tag{82}$$

This result means that the virtual work Eq. (80), is in fact a weak form of equilibrium.

As shown in detail in Mindlin & Eshel (1968) and in Vardoulakis & Sulem (1995) the surface integrals in Eq. (80) yield the boundary conditions for the surface potencies, $\dot{P}_k$ and $\dot{Q}_k$

$$\dot{\sigma}_{ki} n_{ki} - D_1(\dot{m}_{kli} n_k) + (D_p n_p) \dot{m}_{kli} n_k n_l = \dot{P}_i \tag{83.1}$$

$$\dot{m}_{kli} n_k n_l = \dot{Q}_i \tag{83.2}$$

where the stress rates are given by the constitutive equations (71), (72) and (74).

We notice finally that the virtual Eq. (79) together with the constitutive Eqs. (74) and (75), can be used as the basis of Finite Element formulations for the numerical analysis of localization problems (Zervos et al., 2001 a &b)

$$\int_V \left(C_{ijkl}^{ep} D_{kl} \delta D_{ij} + C_{jkmn}^{p} \ell^2 \partial_i D_{mn} \partial_i \delta D_{jk} \right) dV = \int_V \dot{F}_k \delta v_k \, dV + \int_{\partial V} \left(\dot{P}_k \delta v_k + \dot{Q}_k D \delta v_k \right) dS \tag{84}$$

Remarks

1. As we saw in Chapter 43, the introduction of higher gradients into the constitutive relation necessarily leads to the definition of higher-order boundary conditions. A consistent mathematical model for strain-softening elasto-plastic material should be equipped with an elasticity that possesses the same order in higher gradients as the plasticity (Zervos et al. 2001 a & b). This is because one does not know before hand where unloading will take place, and boundary conditions should be defined everywhere to be of the same order. This remark results in a modification of previous 2nd gradient elasto-plastic models, since

we recognize that consistent to gradient plasticity is a gradient elasticity. Accordingly we should set that the background elasticity is non-local as well

$$\dot{\sigma}_{ij} = C^e_{ijkl}(1 - \ell^2_e \nabla^2) D^e_{ij}$$

where $\ell_e \ll \ell$ is a suitable elastic material length.

2. As already outlined above, the appearance of strain-rate gradients into the constitutive equations raises naturally the problem of higher order boundary conditions. Since strain-gradients become important in phenomena of strain localization, one realizes that the simplest class of boundary value problems, which involve strain gradients and the effect of higher-order boundary conditions is that of interface localizations. A discussion of such a model problem can be found in Vardoulakis et al. (1992).

5.5 Shear- Banding

We consider a plane-strain problem and for a small strain formulation. In that case the differential equations that govern continued equilibrium from a given configuration $C^{(t)}$ are

$$\partial_1 \dot{\sigma}_{11} + \partial_2 \dot{\sigma}_{21} = 0$$

$$\partial_1 \dot{\sigma}_{12} + \partial_2 \dot{\sigma}_{22} = 0 \tag{86}$$

We assume that in the configuration $C^{(t)}$ the fields of initial stress and hardening parameter vary slowly in space. Accordingly, the components of the plastic stiffness tensor C^p_{ijkl} may be treated as constants. For convenience, the governing equations are written in the coordinate system of the principal axes of the stress tensor in $C^{(t)}$. Under fully loading conditions we obtain the following set of rate-constitutive equations

$$\dot{\sigma}_{11} = C^u_{1111}\dot{\varepsilon}_{11} + C^u_{1122}\dot{\varepsilon}_{22} - \ell^2 C^p_{1111}\nabla^2\dot{\varepsilon}_{11} - \ell^2 C^p_{1122}\nabla^2\dot{\varepsilon}_{22}$$

$$\dot{\sigma}_{22} = C^u_{2211}\dot{\varepsilon}_{11} + C^u_{2222}\dot{\varepsilon}_{22} - \ell^2 C^p_{2211}\nabla^2\dot{\varepsilon}_{11} - \ell^2 C^p_{2222}\nabla^2\dot{\varepsilon}_{22} \tag{87}$$

$$\dot{\sigma}_{12} = 2G\dot{\varepsilon}_{12}$$

where the components of the stiffness tensors are defined according to Eq. (58) or (70) for $<1> = 1$. Thus with, $C_{ijkl}^u = C_{ijkl}^e + C_{ijkl}^p$ we denote the components of the corresponding upper-bound linear comparison solid[2] (cf. Vardoulakis, 1994).

Introducing these expressions into the equilibrium equations results finally into the following system of partial differential equations for the components of the velocity vector

$$\ell^2 C_{1111}^p \nabla^2 \left(\partial_{11}^2 v_1 \right) + \ell^2 C_{1122}^p \nabla^2 \left(\partial_{21}^2 v_2 \right) - C_{1111}^u \partial_{11}^2 v_1 - G \partial_{22}^2 v_1 - \left(C_{1122}^u + G \right) \partial_{12}^2 v_2 = 0 \qquad (88.1)$$

$$\ell^2 C_{2211}^p \nabla^2 \left(\partial_{12}^2 v_1 \right) + \ell^2 C_{2222}^p \nabla^2 \left(\partial_{22}^2 v_2 \right) - \left(C_{2211}^u + G \right) \partial_{12}^2 v_1 - G \partial_{11}^2 v_2 - C_{2222}^u \partial_{22}^2 v_2 = 0 \qquad (88.2)$$

These are 4th order partial differential equations for the velocity field and constitute a singular perturbation of the classical ones, which are of 2nd order.

Above p.d.e.'s. will be investigated here for the special case where solutions are sought that correspond to the localization of deformation into narrow zones of intense shear, the so-called shear-bands. According to Fig. 3, the (x_1, x_2)-coordinate system is chosen in such a fashion that the x_1-axis coincides with the minor minimum principal stress σ_1 in C.

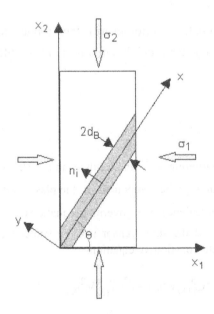

Figure 3. Shear band in an element test

[2] Shanley's or "upper-bound", linear comparison-solid is the solid, which corresponds to continuous loading. We notice that Raniecki & Bruhns (1981) defined a "lower bound", linear comparison-solid as well. These linear comparison solids provide upper and lower bounds for the load (strain) at which shear-band bifurcation is possible.

Let us assume that a shear-band is forming that is inclined with respect to the x_1-axis at an angle θ. A new coordinate system is introduced with its axes parallel and normal to the shear band

$$x = x_1 n_2 - x_2 n_1 \quad ; \quad y = x_1 n_1 + x_2 n_2 \tag{89}$$

where

$$n_1 = -\sin\theta \quad ; \quad n_2 = \cos\theta \tag{90}$$

is the unit vector that is normal to the shear band axis.

By assuming that all field properties related to the forming shear band do not depend on the longitudinal x-coordinate and by setting $(\bullet)' \equiv d/dy$, above Eqs. (88) reduce then to the following system of ordinary differential equations:

$$\ell^2 C^p_{1111} n_1^2 v_1^{(4)} + \ell^2 C^p_{1122} n_1 n_2 v_2^{(4)} - \Gamma_{11} v_1'' - \Gamma_{12} v_2'' = 0$$

$$\ell^2 C^p_{2211} n_1 n_1 v_1^{(4)} + \ell^2 C^p_{2222} n_2^2 v_2^{(4)} - \Gamma_{21} v_1'' - \Gamma_{22} v_2'' = 0 \tag{91}$$

In Eqs. (91) the tensor Γ_{ik} coincides with the acoustic tensor of the so-called upper-bound linear comparison solid for the direction n_i (cf. Vardoulakis & Sulem, 1995),

$$\Gamma_{ik} = C^u_{ijkl} n_j n_l \tag{92}$$

We search for periodic solutions of the system of Eqs. (79), which have the form (Fig. 4)

$$v_i = -\zeta_i \sin(Qy) \quad (i = 1,2) \tag{93}$$

and fulfill the following boundary conditions,

$$y = \pm d_B : \quad v_i = \mp \zeta_i \tag{94}$$

where $2d_B$ is the shear-band thickness. Thus the "wave number" Q in Eq. (93) is inversely proportional to the shear-band thickness,

$$Q = \frac{\pi}{2d_B} \tag{95}$$

For this velocity field the system of ordinary differential Eqs. (91) reduces to the following algebraic system of equations,

$$\begin{bmatrix} b_{11} & b_{12} \\ b_{21} & b_{22} \end{bmatrix} \begin{bmatrix} \zeta_1 \\ \zeta_2 \end{bmatrix} = \begin{bmatrix} 0 \\ 0 \end{bmatrix} \tag{96}$$

where

$$[b_{ij}] = \begin{bmatrix} \ell^2 C_{1111}^p n_1^2 Q^4 + \Gamma_{11} Q^2 & \ell^2 C_{1122}^p n_1 n_2 Q^4 + \Gamma_{12} Q^2 \\ \ell^2 C_{2211}^p n_1 n_2 Q^4 + \Gamma_{21} Q^2 & \ell^2 C_{2222}^p n_2^2 Q^4 + \Gamma_{22} Q^2 \end{bmatrix} \tag{97}$$

For non-trivial solutions in terms of the amplitude vector ζ_i the above system of Eqs. (96) results to the following condition for the shear-band thickness,

$$a_0(n_1, n_2)(Q\ell)^2 + a_1(n_1, n_2) = 0 \tag{98}$$

The coefficients of the above "dispersion equation" are quadratic forms of the components of the unit normal vector

$$a_0 = C_{1111}^p \Gamma_{22} n_1^2 - \left(C_{2211}^p \Gamma_{12} + C_{1122}^p \Gamma_{21} \right) n_1 n_2 + C_{2222}^p \Gamma_{11} n_2^2 = 0$$

$$\tag{99}$$

$$a_1 = \det(\Gamma_{ik})$$

In view of Eqs. (98) and (99) we make the following remarks:

1. The condition

    $$a_1 = \det(\Gamma_{ik}) = 0 \tag{100}$$

    coincides with the classical bifurcation condition (Vardoulakis & Sulem, 1995) and, if evaluated at the first occurring bifurcation state C_B, it yields the shear-band orientation.

1) The quadratic form $a_1(n_1, n_2)$ changes sign at the bifurcation state of the classical continuum,

    $$a_1(n_1, n_2) = \begin{cases} \geq 0 \text{ for} : \Psi \leq \Psi_B \\ < 0 \text{ for} : \Psi > \Psi_B \end{cases} \tag{101}$$

2) The quadratic form $a_0(n_1, n_2)$ determines the type of the governing partial differential equations. In order to regularize the original problem we require that

$$a_0(n_1, n_2) > 0 \tag{102}$$

With this condition the system of partial differential equations (76) is always elliptic, as opposed to the original system of governing equations ($\ell = 0$), which is of changing type,
namely turning from elliptic to hyperbolic at the point of shear-band bifurcation. Following
these observations we conclude that:

1. Prior to shear-band bifurcation there is no real solution for the shear-band thickness; i.e.
    there exist no localization solution.
1) At the bifurcation state the shear-band thickness is infinite as compared to the material
    length $\ell > 0$, and is rapidly decreasing in the post-bifurcation regime (Fig. 4),

$$(Q\ell) = \sqrt{-\frac{a_1(n_1, n_2)}{a_0(n_1, n_2)}} \tag{103}$$

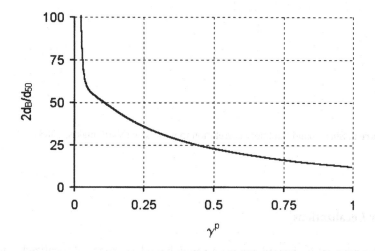

Figure 4. Example of estimated shear-band thickness as function of softening rate in a sand sp
ecimen tested in plane-strain compression ($\ell_c = d_{50} = 0.33$ mm) (Vardoulakis & Aifantis, 1989)

The above illustrated shear-band analysis is used for estimating the internal length parameter
ℓ_c. Empirical evidence is suggesting that shear-band thickness correlates to the mean grain
size, as shown below in Fig. 5.

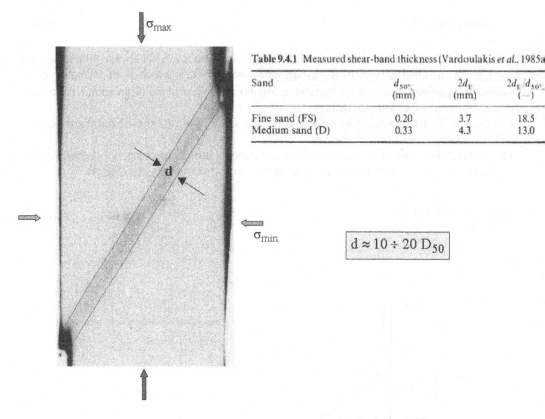

Figure 5. Shear-band thickness correlation to grain size (Vardoulakis, 1988).

5.6 Boundary Localizations[3]

When a granular material is sheared against a *rough* boundary, zones of localized deformation are observed at the interface. This happens for example along the walls of silos and at the interface between soil and pile driven into it. Fig. 6 below shows an x-ray plate of a shear-interface test performed in the biaxial apparatus (Tejhman & Wei Wu, 1995).

Following the paper by Bogdanova-Bontcheva & Lippmann (1975) and Unterreiner & Vardoulakis (1994), controlled interface shear tests on granular materials were also performed recently by Lerat et al. (1997), in a newly developed Ring-Shear Apparatus, Figs. 7 and 8. The boundary localization phenomenon in the Ring-Shear Apparatus was simulated by Zervos et al. (2000) by using a contact dynamics program, Fig. 9. These numerical tests have shown that,

---

[3] see also Vardoulakis et al. (1992)

for a 2-D Schneebeli material, at the interface localization the rolling contacts are at least twice as frequent as sliding contacts (Fig. 10).

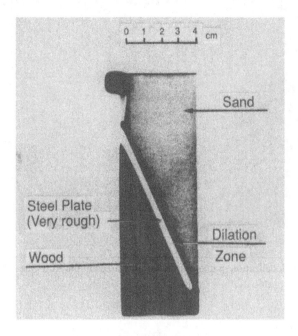

Figure.6. Interface softening dilatancy localization for a rough steel plate interfacing with sand in the biaxial apparatus (Tejhman & Wei Wu,1995).

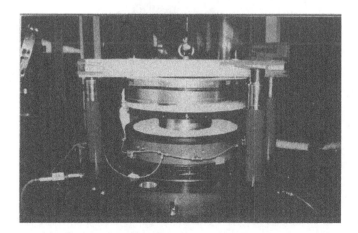

Figure 7. Plane-strain Couette apparatus for sand (Corfdir, A., Lerat, P. and Vardoulakis, I. (2003). A cylinder shear apparatus. Geotechnical Testing Journal, submitted.)

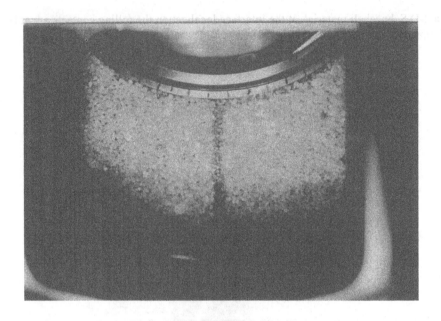

Figure 8. Interface localization in granular material realized in the Ring Shear Apparatus [2]

Figure 9. Contact dynamics simulation of interfacial localization (Zervos et al. 2000)

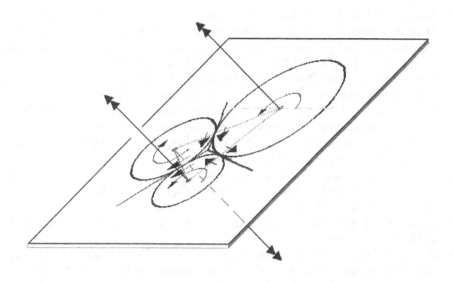

Figure 10. In this configuration we see that for each sliding contact two rolling contacts are necessary, an effect usually called coarse graining. This figure illustrates the necessity of extending the classical Cosserat granular model of Mühlhaus & Vardoulakis (1987) so as to include bipolar effects as well.

References

1. Bogdanova-Boncheva, N. and Lipmann, H. (1975). Rotationsymmetrisches ebenes Fliessen eines granularen Modellmaterials. *Acta Mechanica*, 21: 93-113.
2. Lerat, P., Schlosser, F., and Vardoulakis, I. (1997). Nouvel appareil de ciseillement pour l'étude des interfaces matériau granulaire-structure. *14th ICSMFE*.
3. Mindlin, R.D. and Eshel, N.N. (1968). On first strain-gradient theories in linear Elasticity. *Int. J. Solids Structures*, 4:109-124.
4. Mroz, Z. *Mathematical Models of Inelastic Behavior*. University of Waterloo Press, 1973 .
5. Mühlhaus, H.-B. and Aifantis, E.C. (1991). A variational principle for gradient plasticity. *Int. J. Solids Structures*, 28: 845-857.
6. Mühlhaus H.-B. and Vardoulakis, I. (1987). The thickness of shear bands in granular materials. *Géotechnique*, 37: 271-283.
7. Nguyen, Q.S. and Bui H.D. (1974). Sur les matériaux élastoplastiques à écrouissage positif ou négatif. *Journal de Méchanique*, 3: 322-432.
8. Raniecki, B. and Bruhns, O.T. (1981).Bounds to bifurcation stress in solids with non-associated plastic flow law at finite strain. *J. Mech. Phys. Solids*, 29: 153-172.
9. Schaeffer, D. (1990). Instability and ill-posedness in the deformation of granular materials. Int. *J. Num. Anal. Meth. In Geomechanics*, 14: 253-278.
10. Tejhman , J. and Wei Wu. (1995). Experimental and numerical study of sand-stell interfaces. *Int. J. Num. Anal. Meth. Geomech.*, 19: 513-536.

11. Unterreiner, F. and Vardoulakis, I. *Interfacial localisation in granular media. Computer Methods in Geomechanics* (Ed. Siriwardane & Zaamn), Balkema, 1711-1715, 1994.
12. Vardoulakis, I. (1988). Stability and bifurcation in geomechanics. *Numerical Methods in Geomechanics* (Innsbruck 1988), Swoboda (ed.), Balkema, 155-167.
13. Vardoulakis, I. (1994). Potentials and limitations of softening models in geomechanics (The role of second order work). *Europ. J. of Mechanics, A/Solids*, 13: 195-226.
14. Vardoulakis I. and Aifantis E. (1989). Gradient dependent dilatancy rule and its implications on shear banding and liquefaction. *Ingenieur Archiv*, 59: 197-208.
15. Vardoulakis, I. and Frantziskonis, G. (1992). Micro-structure in kinematic-hardening plasticity. *Eur. J. Mech. A Solids*, 11: 467-486.
16. Vardoulakis, I., Shah, K.R. and Papanastasiou P. (1992). Modelling of tool-rock interfaces using gradient-dependent flow theory of plasticity. *Int. J. Rock Mech. Min. Sci. & Geomech. Abstr.*, 29: 573-582.
17. Vardoulakis I. and Sulem J.. *Bifurcation Analysis in Geomechanics*, Blackie Academic and Professional,1995.
18. Zervos, A., Vardoulakis, I., Jean, M. and Lerat, P. (2000). Numerical investigation of granular kinematics. *Mechanics of Cohesive-Frictional Materials*, 5: 305-324.
19. Zervos, A., Papanastasiou P. and Vardoulakis, I. (2001a). A finite Element displacement formulation for gradient elastoplasticity. *Int. J. Num. Meth. Engrng.*, 50: 1369-1388.
20. Zervos, A., Papanastasiou P. and Vardoulakis, I. (2001b) Modeling of localization and scale effect in thick-walled cylinders with gradient elastoplasticity. *Int. J. Solids Strucures*, 38: 5081-5095.

Appendix I: *The Schrödinger operator*

We consider Eq. (54)

$$(1 - \ell^2 \nabla^2)\dot{\Psi} = \Lambda \quad , \quad \Lambda = \frac{1}{H} B_{kl} D_{kl} \tag{A.1}$$

Notice that in the literature the operator

$$S \approx 1 - \ell^2 \nabla^2 \tag{A.2}$$

is known in the literature as the Schrödinger operator.

We assume for simplicity the 1D-case

$$\nabla^2 \approx \frac{\partial^2}{\partial y^2} \tag{A.3}$$

where y is for example the direction along which the fields vary rapidly; i.e. the direction normal to the localization direction (Fig. 2).

We consider the Fourier transforms of the functions $\dot{\Psi}(y)$ and $\Lambda(y)$ as well as of their derivatives up to order two exist. In that case the functions $\dot{\Psi}$, $\dot{\Psi}'$ and $\dot{\Psi}''$ are continuous and $\dot{\Psi} \to 0$ as $|y| \to \infty$; the same is assumed to hold also for the function Λ. Thus we set,

$$Y(\alpha) = E\{\dot{\Psi}(y)\} = \int_{-\infty}^{+\infty} \dot{\Psi}(y) e^{-i\alpha y} dy \quad , \quad L(\alpha) = E\{\Lambda\} \int_{-\infty}^{+\infty} \Lambda(x) e^{-i\alpha y} dy \tag{A.4.1}$$

$$\dot{\Psi}(y) = E^{-1}\{Y(\alpha)\} = \frac{1}{2\pi} \int_{-\infty}^{+\infty} Y(\alpha) e^{i\alpha y} d\alpha \quad , \quad \Lambda(y) = E^{-1}\{L(\alpha)\} \int_{-\infty}^{+\infty} L(\alpha) e^{i\alpha y} d\alpha \tag{A.4.2}$$

If we apply the Fourier Transformation on both sides of Eq. (A.1) we get,

$$Y - \ell^2 \alpha^2 Y = L \quad \Rightarrow \quad Y = \frac{L}{1 - (\alpha \ell)^2} = \left(1 + (\alpha \ell)^2\right) L + O\left((\alpha \ell)^4\right) \tag{A.5}$$

By applying the inverse Fourier transform on the last equation we get,

$$E^{-1}\{Y\} \approx E^{-1}\{L\} + \ell^2 E^{-1}\{\alpha^2 L\}$$

or

$$\dot{\Psi} \approx \Lambda + \ell^2 \frac{d^2 \Lambda}{dy^2} \tag{A.6}$$

Thus for slow varying constitutive properties from Eq. (A.6) we get Eq. (54).

Slope Failure and the loss of controllability of the incremental response of a soil specimen subject to an arbitrary loading programme

Roberto Nova[*][1]

[*]Department of Structural Engineering, Milan University of Technology (Politecnico), Milan, Italy

Abstract An analysis of the instability conditions of slopes, based on the concept of latent instability, is presented first. It is shown that, for special conditions, multiple strain rate solutions are possible under the same loading increment. Drained and undrained cases for sand and clay slopes are considered. For rapid loading of sandy slopes, the latent instability analysis allows the reasons of catastrophic collapses of very flat slopes to be explained. Latent instability can occur even under more general loading programmes in the hardening regime, provided the flow rule is non-associate and the stiffness matrix is not positive-definite. Failures due to loss of load control in undrained tests or shear or compaction band occurrence in axisymmetric tests are examples.

1 Introduction

Slope stability is one of the main topics of Soil Mechanics. Actually, the term 'stability' is used improperly, from a mechanical stand-point. To assess whether a slope is stable or not, traditionally, only the equilibrium state at impending collapse is investigated, in fact, without any consideration of the slope response to perturbations.

From a conceptual view point, the 'stability' analysis of the slope of Figure 1a is analysed as that of the rigid block of Figure 1b. The soil wedge ABC of weight W is assumed to slide downwards along the shear band AC. This is a zone of very small width (few grain diameters), where strains are localised, delineating between the moving wedge ABC and the still soil below the band.

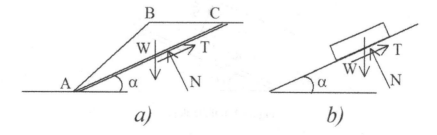

Figure 1. Simple stability analyses: a) of a slope; b) of a brick on an inclined plane.

[1] This paper is written within the framework of DIGA network (supported by the European Commission under contract n. HPRN-CT2002-00220) and ALERT Geomaterials.

Equilibrium conditions demand that:

$$T = N \tan \alpha \qquad (1.1)$$

For a granular soil, such as sand or gravel, the maximum shear force available is:

$$T = N \tan \phi \qquad (1.2)$$

where ϕ is the friction angle. By comparing (1.1) and (1.2), it is obvious that limit equilibrium conditions, and therefore the onset of 'instability' for the slope in the traditional sense, are reached when

$$\tan \alpha = \tan \phi \qquad (1.3)$$

Such an analysis can be easily extended to soils characterised by cohesion and friction and to subaqueous or saturated slopes in drained conditions, by using the effective stress concept. Curved or compound failure lines can be also considered. In any case, the limit slope angle is directly linked to the value of the soil friction angle. Even when water flows downwards along the slope seeping through a sand layer, impending collapse occurs when

$$\tan \alpha = \frac{\gamma'}{\gamma_{sat}} \tan \phi \qquad (1.4)$$

where γ' and γ_{sat} are the buoyant and saturated unit weight of the soil, respectively, the former being roughly equal to one half of the latter.

Since the friction angle for granular materials is larger than 30°, catastrophic failures of submarine slopes, normally inclined at 5° or less, cannot be explained by limit analysis. A different approach is necessary.

2 Conditions for drained shear band occurrence in plane strain

Consider, for the sake of simplicity, an infinite layer such as that shown in Figure 2.

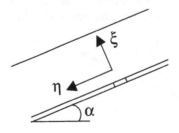

Figure 2. Infinite slope.

Let χ be the axis orthogonal to the figure and η and ξ two orthogonal axes in the plane of the figure, the former inclined as the slope. Take an element of the slope with faces orthogonal to each axis. In terms of effective stresses σ'_{ij}, its constitutive law can be written as

$$
\left\{
\begin{array}{c}
\dot{\sigma}'_{\xi} \\
\dot{\sigma}'_{\eta} \\
\dot{\sigma}'_{\chi} \\
\dot{\tau}_{\xi\eta} \\
\dot{\tau}_{\xi\chi} \\
\dot{\tau}_{\eta\chi}
\end{array}
\right\}
=
\begin{bmatrix}
D_{11} & D_{12} & D_{13} & D_{14} & D_{15} & D_{16} \\
D_{21} & D_{22} & D_{23} & D_{24} & D_{25} & D_{26} \\
D_{31} & D_{32} & D_{33} & D_{34} & D_{35} & D_{36} \\
D_{41} & D_{42} & D_{43} & D_{44} & D_{45} & D_{46} \\
D_{51} & D_{52} & D_{53} & D_{54} & D_{55} & D_{56} \\
D_{61} & D_{62} & D_{63} & D_{64} & D_{65} & D_{66}
\end{bmatrix}
\left\{
\begin{array}{c}
\dot{\varepsilon}_{\xi} \\
\dot{\varepsilon}_{\eta} \\
\dot{\varepsilon}_{\chi} \\
\dot{\gamma}_{\xi\eta} \\
\dot{\gamma}_{\xi\chi} \\
\dot{\gamma}_{\eta\chi}
\end{array}
\right\}
\tag{2.1}
$$

where D_{ij} are the elements of the stiffness matrix of the soil.

By imposing the plane strain condition and exploiting the symmetries, Equation (2.1) can be reduced to:

$$
\left\{
\begin{array}{c}
\dot{\sigma}'_{\xi} \\
\dot{\sigma}'_{\eta} \\
\dot{\tau}_{\xi\eta}
\end{array}
\right\}
=
\begin{bmatrix}
D_{11} & D_{12} & D_{14} \\
D_{21} & D_{22} & D_{24} \\
D_{41} & D_{42} & D_{44}
\end{bmatrix}
\left\{
\begin{array}{c}
\dot{\varepsilon}_{\xi} \\
\dot{\varepsilon}_{\eta} \\
\dot{\gamma}_{\xi\eta}
\end{array}
\right\}
\tag{2.2}
$$

The limit condition, in the traditional sense, can be seen as the state of stress for which unlimited strain rates can take place even without any change in external stresses. From Equation 2.2 this is equivalent to

$$
\det\left[D_{ij}\right]=0
\tag{2.3}
$$

Another possibility exists however. Assume that in a shear band directed as the slope, as in Figure 2, the state of stress and strain is different from the rest of the soil body. Of course, equilibrium and compatibility must be fulfilled anyway.
Therefore

$$
\Delta\dot{\sigma}'_{\xi}=\Delta\dot{\tau}_{\xi\eta}=\Delta\dot{\varepsilon}_{\eta}=0
\tag{2.4}
$$

where Δ indicates the difference between the value a quantity takes within the band and the value of the same quantity has outside it.

Assume further that the state of stress and strain before the load increment is the same inside and outside the band and that the increment direction is such that plastic strains occur in both zones. The stiffness matrix is the same, therefore, inside and outside the band. From Equations 2.2 and 2.4

$$
\left\{
\begin{array}{c}
0 \\
\Delta\dot{\sigma}'_{\eta} \\
0
\end{array}
\right\}
=
\begin{bmatrix}
D_{11} & D_{12} & D_{14} \\
D_{21} & D_{22} & D_{24} \\
D_{41} & D_{42} & D_{44}
\end{bmatrix}
\left\{
\begin{array}{c}
\Delta\dot{\varepsilon}_{\xi} \\
0 \\
\Delta\dot{\gamma}_{\xi\eta}
\end{array}
\right\}
\tag{2.5}
$$

The system is a homogeneous system which admits, in general, the trivial solution

$$
\Delta\dot{\sigma}'_{\eta}=\Delta\dot{\varepsilon}_{\xi}=\Delta\dot{\gamma}_{\eta}=0
\tag{2.6}
$$

If however

$$D_{11} D_{44} - D_{14} D_{41} = 0 \tag{2.7}$$

there exist an infinity of solutions of the type

$$
\begin{cases}
\Delta \dot{\varepsilon}_\xi = \psi \\[2mm]
\Delta \dot{\gamma}_{\xi\eta} = -\dfrac{D_{11}}{D_{14}} \psi \\[2mm]
\Delta \dot{\sigma}'_\eta = \left(D_{21} - \dfrac{D_{24} D_{11}}{D_{14}} \right) \psi
\end{cases}
\tag{2.8}
$$

where ψ is an arbitrary scalar. In particular, under zero external loading, the state of stress and strain in the layer is unchanged but for a shear band zone where the increments are given by Equation 2.8 and can be arbitrary large.

Note that $\Delta \dot{\sigma}'_\eta$ and $\Delta \dot{\sigma}'_\chi$, that we can obtain by imposing the nullity of $\Delta \dot{\varepsilon}_\chi$, are self-equilibrated stress increments, and neither equilibrium nor compatibility are violated.
Equation 2.7 is therefore the condition for the occurrence of a shear band of localised strains.

When either condition 2.3 or 2.7 are met, the second order work (density)

$$d^2 W = \dot{\sigma}'_{ij} \dot{\varepsilon}_{ij} \tag{2.9}$$

is zero. Both conditions are associated to the possibility of multiple solutions corresponding to zero load increment. They are both conditions for latent instability, therefore, analogous to that of an Euler beam.

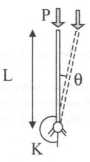

Figure 3. Latent instability of an Euler beam.

Consider in fact the beam of length L shown in Figure 3. For any value of the load, P, the vertical configuration is in equilibrium. However, the equilibrium may be stable or unstable depending on the value of the rotational spring stiffness, K. In particular, a dynamic analysis shows that the configuration is stable under any, small, perturbation if PL<K. It is unstable, when PL>K and while it is in a state of neutral equilibrium when PL=K.

This critical value of the load can be determined also by means of a static analysis. It can be shown in fact that for conservative systems the dynamic and the static analyses give the same critical load. Assume to give a small geometrical perturbation to the beam: a rotation around the hinge of a small angle ϑ. The moment equilibrium equation with respect to the hinge is:

$$(K - PL)\vartheta = 0 \tag{2.10}$$

Such an equation is fulfilled by $\vartheta = 0$ for any value of P. However, when

$$K - PL = 0 \tag{2.11}$$

any value of ϑ is compatible with the fulfilment of equilibrium. The critical load for which this occurs is known as the load of latent (Ziegler (1968)) or leaning instability for the beam. It is clear in fact that the beam can diverge from the horizontal configuration without changing the external load. Latent instability is therefore associated to the loss of uniqueness of the geometrical configuration that fulfils equilibrium. It is also associated to the nullity of the 'extended' stiffness, i.e. to the nullity of the coefficient of ϑ in Equation 2.10.

If a structure with more than one degree of freedom is considered, the condition for latent instability is given by the nullity of the determinant of an 'extended' stiffness matrix. An eigensolution in terms of displacements (or rotations) is therefore possible. The formal similarity between such a result and Equations 2.3 and 2.7 is evident. Therefore, we shall call henceforth latent instability condition the occurrence of the nullity of the determinant of the stiffness matrix or of one of its leading minors. In either case, in fact, an eigen-solution in terms of strain rates is possible.

If the stiffness matrix is symmetric, condition 2.3 coincides with the condition of loss of positive definiteness of the stiffness matrix. A leading minor of a symmetric positive definite matrix cannot be nil: as a consequence, Equation 2.7 cannot be fulfilled 'before' Equation 2.3, i.e. for a lower load level. A shear band cannot occur before the limit state is achieved.

Soils are characterised by non-symmetric stiffness matrices, however (Darve (1978), Nova and Wood (1979)) As a consequence, the stiffness matrix is not symmetric and the condition for the occurrence of a shear band in plane strain can be met before the condition for general collapse is achieved. The traditional concept of 'failure' has to be revised, therefore, since multiple solutions and unlimited strains can occur even though the determinant of the stiffness matrix is positive, i.e. before the traditional failure condition is met.

3   Conditions for undrained shear band occurrence in plane strain

The analysis of the previous section can be repeated in the case of very rapid loading. In this case, in the band at least, we may admit that there is no volume change, instantaneously. It follows, from Equation 2.4 and the plane strain conditions, that the no volume change condition is equivalent to a further constraint

$$\Delta \dot{\varepsilon}_\xi = 0 \tag{3.1}$$

and Equation 2.5 is modified accordingly. Together with the trivial solution (Equation 2.6), an infinity of eigensolutions of the type

$$\begin{cases} \Delta\dot{\varepsilon}_\xi = 0 \\ \Delta\dot{\gamma}_{\xi\eta} = \psi \\ \Delta\dot{\sigma}'_\eta = D_{24}\psi \end{cases} \qquad (3.2)$$

can be obtained when

$$D_{44} = 0 \qquad (3.3)$$

Equation 3.3 is therefore the condition for the occurrence of an undrained shear band in plane strain. Also in this case, the second order work is zero at this stage.

As in the drained case, the resulting $\Delta\dot{\sigma}'_\eta$ and $\Delta\dot{\sigma}'_\chi$ are self-equilibrated stresses. However from Equation 2.2 it comes out that

$$\Delta\dot{\sigma}'_\xi = D_{14}\psi \qquad (3.4)$$

Under zero external loading, the total stress $\Delta\sigma_\xi$ must be zero for equilibrium. This means that, within the band, a sudden increase of pore water pressure occurs

$$\Delta\dot{u} = -D_{14}\psi \qquad (3.5)$$

when condition 3.3 is fulfilled.

Di Prisco et al. (1995) have shown that spontaneous undrained collapse of a slope in sand with a plane strain friction angle of 37° occurs when the slope is inclined only of 24.5°. A consistent reduction of the critical angle for which latent instability occurs is obtained, therefore.

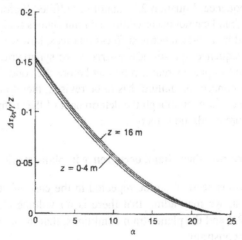

Figure 4. Normalised intensity of the shear perturbation becessary to cause the formation of an undrained shear band in an infinite subaqueous sandy slope as a function of its inclination – after di Prisco et al. (1995)

If a small horizontal perturbation is applied, due to seismic loading, for instance, the critical angle decreases with increasing the intensity of the perturbation. Figure 4 shows the results obtained by di Prisco et al. (1995). It is clear that a perturbation caused by a horizontal acceleration as small as 5% of the gravity acceleration is enough to cause the undrained collapse of a slope characterised by an inclination of 10° only.

Crosta et al. (1997) have further shown that, in case seepage conditions are present, (due for instance to the presence of water table and the non-miscibility of fresh and salt water) such a critical angle can decrease to very small values. These are of the same order of magnitude of those recorded on coastal sea shores.

4   A reappraisal of total stress analysis in clayey soils

When clayey soils are considered, the stability analysis is usually made in terms of total stresses. In that case, the material strength is expressed as

$$\tau = s_u \qquad\qquad\qquad\qquad (4.1)$$

where s_u is a parameter, known as undrained strength, usually evaluated by means of a vane test. In such a test, as in the shear band dividing the rigid wedge ABC of Figure 1b from the still soil, simple shear loading conditions apply, Figure 5a. Note that the shear band need not being a straight line, and the wedge motion can be rotatory around a centre at a finite distance, as in Figure 5b, instead of translatory. The shear band is in fact very thin with respect to the radius of the circular failure surface and the condition $\Delta\varepsilon_n = 0$ can be locally imposed anyway.

The undrained strength is determined as the peak shear value that can be applied to a soil specimen, Figure 5c. It is apparent from Equation 2.2 and the no volume change condition, that such a peak is reached when Equation 3.3 is fulfilled.

The latent instability analysis coincides, therefore, with the traditional limit equilibrium total stress analysis, which thus appears as a proper stability analysis.

Note however that in order to have correct predictions with the ordinary method s_u must be evaluated with a simple shear apparatus and not with a triaxial test, since the value of s_u would be different.

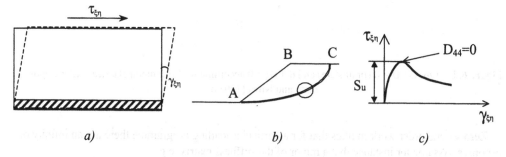

Figure 5. Undrained simple shear test: a) specimen deformation; b) collapse mechanism of a slope; c) stress strain relationship in simple shear and definition of undrained strength.

5   Determination of unstable states in general loading conditions

Shear bands such as those leading to slope instability occur also in laboratory specimens (Desrues 1984, Drescher et al., 1990). Other types of instability can take place, however, since homogeneous bifurcations can occur under special loading conditions. The second order work is in fact zero first when the stiffness matrix becomes positive semi-definite

$$\dot{\sigma}'_{ij}\dot{\varepsilon}_{ij} = D_{ijhk}\dot{\varepsilon}_{hk}\dot{\varepsilon}_{ij} = 0 \tag{5.1}$$

what happens when the determinant of the symmetric part of the stiffness matrix is zero. For the non-symmetric nature of the stiffness matrix, by virtue of the Ostrowksy and Taussky (1951) theorem, when this occurs, the determinant of the entire stiffness matrix is positive, which means that the soil is still in the hardening regime.

By taking a convenient constitutive model, e.g. Nova (1988), we can quantify when limit state and loss of positive semi-definiteness take place. Such loci are depicted in Figure 6 for loose Hostun sand. It is clear that the stress level for which the second order work can be zero for a particular stress path is very small, even lower that the stress level characteristic of the geostatic state at rest. For clay, the deviation from normality is less marked than for sand and the locus for loss of positive definiteness approaches, but it is still different, from the limit locus.

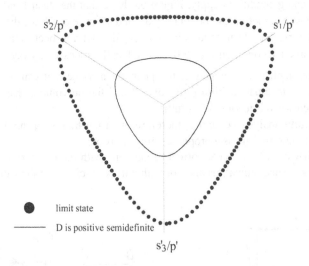

Figure 6. Loci for det **D** = 0 (limit state) and det **D**$_S$ = 0 according to 'Sinfonietta classica' (after Imposimato and Nova (1998)).

Zero second order work implies that for a particular loading programme there are an infinity of responses. Assume for instance that a minor of the stiffness matrix, e.g.

$$\det D_{\alpha\alpha} = 0 \tag{5.2}$$

where

$$\left\{ \begin{matrix} \dot{\sigma}'_\alpha \\ \dot{\sigma}'_\beta \end{matrix} \right\} = \begin{bmatrix} D_{\alpha\alpha} & D_{\alpha\beta} \\ D_{\beta\alpha} & D_{\beta\beta} \end{bmatrix} \left\{ \begin{matrix} \dot{\varepsilon}_\alpha \\ \dot{\varepsilon}_\beta \end{matrix} \right\} \qquad (5.3)$$

is the partitioned stiffness matrix.

To zero stress increment it is obviously associated a zero strain increment. However if we control partly stress ($\dot{\sigma}'_\alpha$) and partly strain ($\dot{\varepsilon}_\beta$), it can be derived (Imposimato and Nova, 1998) that a zero variation of the controlling quantities is associated to an infinity of responses of the type:

$$\dot{\sigma}'_\beta = \eta D_{\beta\alpha} \dot{\varepsilon}_\alpha * \qquad (5.4)$$

where $\dot{\varepsilon}_\alpha *$ is the normalised eigenvector of matrix $D_{\alpha\alpha}$ and η is an arbitrary scalar.

For instance assuming to control σ'_2, σ'_3 and $\dot{\varepsilon}_1$, Figure 7 shows the loci for which Equation 5.2 is fulfilled. In this figure it is shown that two particular stress paths (b and c) cannot be continued beyond the point for which such loci are reached.

Consider now the analysis in undrained conditions, for which we control the volume change and for instance two stress parameters, e.g. $\dot{\sigma}'_1 - \dot{\sigma}'_2$ and $\dot{\sigma}'_1 - \dot{\sigma}'_3$. It can be shown (Imposimato and Nova, 1997) that, when the volumetric compliance under isotropic loading [$C_{pp} = 0$] is zero, unlimited spontaneous increase in pore water pressure is possible under no external load increment ($\dot{\sigma}_1 = \dot{\sigma}_2 = \dot{\sigma}_3 = \dot{v} = 0$), with a consequent generation of unlimited strains. The locus for which this occurs is shown in Figure 8. In particular, in axisymmetric conditions, the locus collapses into a straight line passing through the origin, for which a peak occurs in undrained conditions. Such a line is also known as Lade's instability line (Lade (1992)).

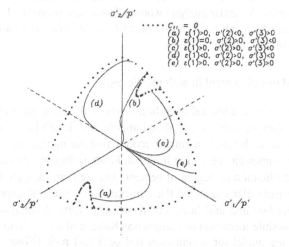

Figure 7. Stress paths in the deviatoric plane for different loading programs with one principal strain rate and two principal stress rates are controlled. Homogeneous bifurcation loci are starred. Note that loci for which $\det C_{\beta\beta} = 0$ always coincide with loci for which $\det D_{\alpha\alpha} = 0$ (after Imposimato and Nova, 1998).

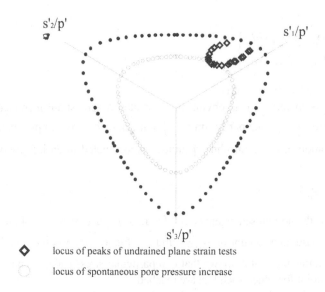

◇   locus of peaks of undrained plane strain tests

○   locus of spontaneous pore pressure increase

Figure 8. Locus of spontaneous pore pressure increase in undrained condition according to 'Sinfonietta classica' (deviatoric plane) and locus of stress points for which a locally undrained shear band and the peak in axial load in undrained plane strain tests occurs (Imposimato and Nova, 1997).

If we now consider a biaxial undrained stress path, we find that the locus of the peaks of the deviatoric stress, which coincides with the locus for which an undrained shear band can be formed, is given by a curve also shown in Figure 8, after Imposimato and Nova (1997). Depending on the initial state after consolidation, such a peak will be reached sooner or later and therefore far or close to the limit locus. A similar analysis would allow to determine the locus for occurrence of a drained shear band, which can be seen as a special case of loss of controllability of the loading programme (Imposimato and Nova, 1997).

6   Softening and loss of control in oedometric tests

Loss of control can occur, in special conditions, even in the most traditional of the geotechnical tests: the confined compression, so called oedometric, test. It will be shown in fact, by means of an elastoplastic strainhardening constitutive model, that during this test, an overconsolidated or a cemented soil specimen can reach the limit condition, soften and then harden again, provided the initial overconsolidation or degree of cementation are large enough. Furthermore, it will be shown that, for a particular choice of the constitutive parameters, appropriate for strong but brittle intergranular bonding and large initial porosity, vertical collapse of the specimen can take place, with possible formation of compaction bands and loss of load control.

Consider first the model for a cemented soil or a soft rock (Nova, 1992) where the bond strength is characterized by two fixed parameters p_t and p_m. The size of the elastic domain is controlled by p_s, that is a hardening variable, depending on the plastic strains experienced by the material. A picture of the initial elastic domain is shown in Figure 9. Assume, for the sake of simplicity, that the flow rule is associated and that hardening depends only on volumetric strains, as in Cam Clay (Schofield and Wroth, 1968). Compaction therefore corresponds to an increase in p_s and conversely dilation to its decrease.

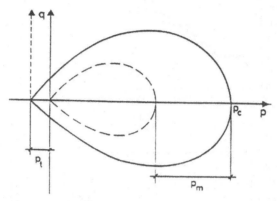

Figure 6. Initial elastic domain for a cemented material – after Nova (1992).

A typical stress path experimentally obtained in an oedometric test on cemented geomaterials, such as chalk (Leddra, 1988) or calcarenite (Coop and Atkinson, 1993), is shown in Figure 10a. The model predicts a similar trend. At the beginning the behavior is elastic, since the origin belongs to the initial elastic domain. The stress path is therefore given by a straight line, whose slope depends on the Poisson's ratio, only. The yield locus is reached at a point for which the horizontal component of the vector normal to it is directed in the negative sense of the hydrostatic axis, see Figure 10b. Plastic dilation is predicted, therefore, and consequently, for the assumed hardening law, softening of the material occurs.

From the standpoint of the theory of plasticity, the specimen 'fails' at this point. The stress path is directed within the initial elastic domain and the hardening variable decreases together with its size, Figure 10e. The deviator stress increases, however, as well as the external loading. Full controllability of the test is possible, therefore, even under load control.

It is quite interesting to note that, despite the deviator stress increases, the hardening modulus is negative, so that softening occurs, Figure 10d. On the other hand, softening does not imply loss of controllability of the test, for the special loading programme we are following, since vertical stress increases monotonically, Figure 10c. On the contrary, if the load programme would be stopped at any moment between points A and B of Figure 10 and a different load increment, say an increase in vertical load at constant horizontal stress, would be imposed, the specimen would instantaneously collapse.

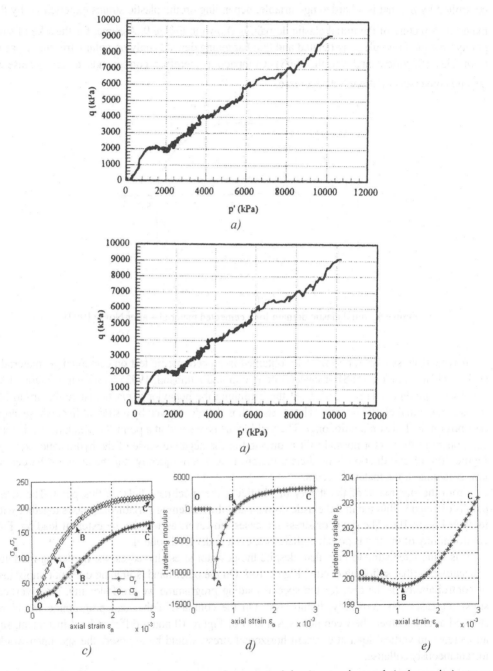

Figure 10. Oedometric test on a cemented geomaterial: a) experimental (calcarenite) stress path (after Coop and Atkinson); b) calculated stress path; c) stress-strain law; d) value of the hardening modulus e)value of the hardening variable (after Castellanza, 2002).

As point B is reached, the normal to the yield locus is directed towards the positive sign of the isotropic axis. Plastic compression takes place as well as hardening, for the assumed hardening and plastic flow rules. Not withstanding that, the deviator stress decreases. The stress path is obliged to follow more or less the shape of the yield locus, since the elastic strains are small and the zero lateral strain condition imposes that the plastic ones are of the same order of magnitude. Only when it is achieved a stress condition for which the normality rule determines plastic strain rates that are compatible with the oedometric condition, strains can occur freely, the stress path can leave the yield locus shape and the deviator stress can increase again.

Remarkably, similar results can be obtained by using a basic model as the original Cam Clay for a preconsolidated material with an overconsolidation ratio high enough to render the specimen 'dryer than critical', i.e. heavily overconsolidated.

Consider now a material whose bonds are brittle. Yielding is therefore associated to a decrease of the value of the variables p_t and p_m. In this case it is possible to have loss of load control even in the oedometric test. Figure 11 shows in fact the calculated results for a material characterized by convenient constitutive parameters. In this case the hardening parameter is given by the sum of three contributions (Gens and Nova, 1993):

$$H = H_s + H_t + H_m \tag{6.1}$$

each of which is related to the corresponding hardening variable p_s, p_t, p_m. While the first contribution can be either positive or negative, the other two are always negative. Softening can occur even in the region where the material is compacting, therefore, Figure 11c. The axial stress strain law shows a marked peak, implying that load control is impossible, beyond the peak. The calculated results were in fact obtained by imposing full kinematic control.

In such a test, in a specimen of finite size, but uniform state of stress and strain, at peak a bifurcation can take place without violating equilibrium nor compatibility. The specimen can be seen in fact as composed of layers, which are subject to the same vertical stress decrease but different horizontal stress variation, one layer undergoing elastic unloading while the adjacent one suffers elastoplastic loading, as shown in Figure 12. Such layers have been actually observed in soft rock specimens under compression (Olsson, 1999). Rudnicki (2002), following the same path of reasoning of Rudnicki and Rice (1975) for shear bands, has given the mathematical condition for the occurrence of compaction bands, i.e:

$$D_{ijhk} n_i n_j n_h n_k = 0 \tag{6.2}$$

where D_{ijhk} is the stiffness tensor and n_i are the director cosines with respect to the reference axes of the normal n to the plane of the band.

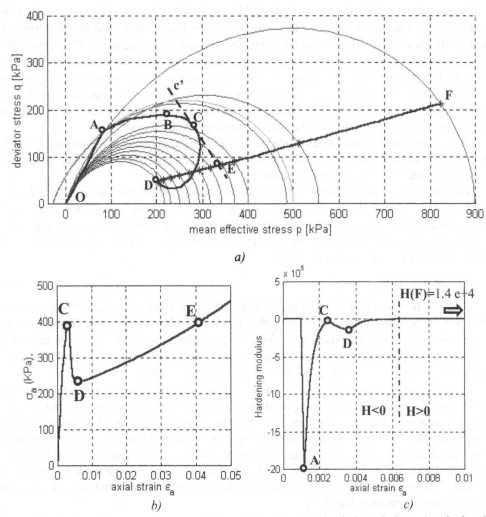

Figure 11. Oedometric test on a cemented geomaterial with bond degradation: a) calculated stress path; b) axial stress strain relationship; c) value of the hardening modulus (after Castellanza, 2002)

According to tensor transformation rules, the l.h.s. of Equation 6.2 gives the stiffness coefficient D_{nnnn}. Equation 6.2 becomes therefore:

$$D_{nnnn} = 0 \tag{6.3}$$

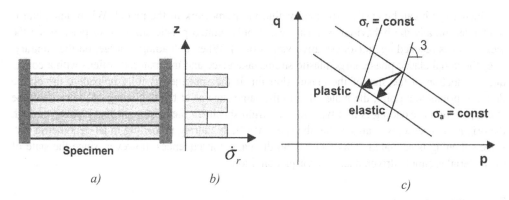

Figure 12. Bifurcation in an oedometric test on a cemented geomaterial with bond degradation a) compaction bands b) stress rates in the layers (after Nova et al., 2003).

The nullity of this stiffness coefficient, that coincides in this case with the occurrence of a peak in the axial stress-strain relationship (horizontal compaction band), is therefore the condition for bifurcation.

Softening can occur in unexpected conditions, therefore. As a final example, consider the results of an oedometric test on a specimen of loose sand, shown in Figure 13. During virgin loading, the ratio between horizontal and vertical stresses remains constant, so that the stress path in the plane p', q is a straight line passing through the origin. For moderate unloading, hysteretic unloading-reloading loops are apparent. If the material is fully unloaded, however, the reloading stress path is again a straight line, i.e. the material behaves as if it was virgin anew.

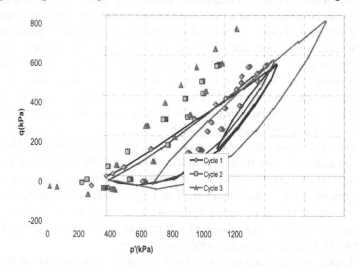

e 13. Experimental stress path in an oedometric test on a loose sand specimen, with cycles of unloading-reloading. After Castellanza (2002)

Such a result can be easily explained within the framework of the model. When unloading is moderate, initially the behaviour is elastic while only limited plastic strains take place when the yield locus is reached again (for negative values of q). When unloading is large, on the contrary, after the initial elastic phase, large plastic strains take place and the material softens with a consequent reduction of the size of the elastic domain. If the specimen is fully unloaded, the elastic domain shrinks to a point, the material is in the same state as at the beginning of the test and the stress path upon reloading is that typical of a virgin specimen, i.e. a straight line passing through the origin. In a sense, we can say that the specimen 'fails' during unloading in an oedometric test, since softening is activated. This fact has no dramatic consequences, however, since the state of the material is simply driven back to a virgin condition.

7   Conclusions

Failure of a material element is a primitive concept. When dealing with geomaterials such a concept becomes less neat, however. Although it is possible to define a convenient limit locus beyond which stress states are not feasible for a given material, in fact, unlimited strains may occur even for stress states within such locus, provided convenient loading conditions are imposed and controlling parameters appropriately chosen.

Such types of failures can occur in the hardening regime, provided the flow rule is non-associated and the stiffness matrix is not positive definite. An example is the loose sand specimen failure in a load controlled undrained test. Other types of loss of control can occur. Whenever a principal minor of the stiffness (or compliance) matrix is zero, there exists a special combination of stress and strain increments for which an infinity of solutions, in terms of the complementary variables of strain and stress, is possible. Homogeneous bifurcations may take place, therefore.

It is also possible to consider linear combinations of stresses and strains. In this case a generalized stiffness matrix can be defined and a similar conclusion holds in terms of generalized stresses and strains. In particular, it can be shown that even the shear band type of bifurcation can be described within this conceptual framework. In this latter case a non-homogeneous bifurcation is possible, in a specimen of finite size, since neither equilibrium nor compatibility is violated by two different, adjacent, stress-strain rate states. In situ, the non-homogeneous bifurcation associated to shear band condition can give rise to slope failures. If the behaviour within the band is undrained, the stability analysis of a slope coincides with the traditional limit analysis based on the total stress approach and the associated undrained strength.

All the aforementioned bifurcation states may occur even in the hardening regime. The homogeneous bifurcations can take place only if the appropriate control variables are chosen, however. The states for which such bifurcations can occur are 'unstable' (since infinite solutions to a single small perturbation are possible), but such an instability is highlighted only under special load conditions. The situation is similar to that of a mountaineer walking up a steep slope. His state is unstable along the entire path, since he can fall downhill if he stumbles. On the other hand, under appropriate control of his feet, he can quietly reach the mountain top.

Along certain load paths, geomaterials can experience softening. According to the theory of plasticity, this occurs when the hardening modulus is negative. Usually to the term 'softening' are associated the concepts of 'failure' and 'instability'. This is certainly the case when load controlled tests are considered. In general, however, softening is not necessarily associated to specimen collapse and loss of control. For instance, if an oedometric test on a cemented soil is considered, it was shown in the paper that, at variance with simple mind expectations, softening is associated to an increase in deviator stress. On the contrary, hardening is initially associated to a decrease in the deviator stress. The axial load increases in both cases, however, so that no loss of control occurs, even in partially load controlled tests.

If the soil bonds degrade with increasing plastic strains, as it can be the case with calcareous materials, for example, the interplay of hardening and softening is complex. The latter, linked to bond degradation, can prevail on the former, that is due to plastic volumetric compaction. Overall softening can take place together with volumetric compaction, therefore, and a peak in the axial stress-strain curve is recorded in fully kinematically controlled oedometric tests. As a consequence, in a test in which the axial load is controlled, sudden collapse of the specimen (i.e. loss of control) occurs at the stress level corresponding to the peak. Since the peak condition is associated to the nullity of the stiffness coefficient in the axial stress direction, a planar compaction band in the orthogonal direction can be formed. The stress strain state in the oedometric specimen can undergo a non-homogeneous bifurcation, with layers that are elastically unloaded and adjacent layers experiencing plastic strains.

Finally, it was shown that softening can occur in the oedometric test even for a non cemented geomaterial, when this is fully unloaded. The softening experienced erases from the specimen memory all the previous loading history and the material behaves as it would be fully virgin.

Failure and the related concepts of hardening and softening are by no means intuitive, therefore. Instabilities leading to large strains can occur in the hardening regime, while softening can be associated to perfectly controllable situations. Sudden collapses are possible in oedometric tests and softening can occur in unloading, regenerating virgin soil specimens. A conceptual framework, as the theory of strainhardening plasticity, is necessary in order to fully appreciate and control the complexity of the behavior of geomaterials.

References

Castellanza, R. 2002. Weathering effects on the mechanical behaviour of bonded geomaterials: an experimental, theoretical and numerical study. *PhD Thesis*, Politecnico di Milano.

Coop, M.R. & Atkinson, J.H. 1993. The mechanics of cemented carbonate sands. Géotechnique 43(1):53-67.

Crosta G., Imposimato S., Nova R. 1997 Instability analysis of submarine slopes *Proc. Progressive failure in geomechanics*, Adachi et al. eds., Nagoya, 689-694

Darve F. 1978 Une formulation incrementale des lois rhéologiques. Application aux sols , *Thèse docteur es- Sciences physiques, INPG Grenoble.*

Desrues J. (1984) La localisation de la déformation dans les matériaux granulaires. *Thèse de Docteur ès Sciences*, Grenoble.

Drescher, A., Vardoulakis, I. & Han, C. 1990. A biaxial apparatus for testing soils. *Geotechnical Testing Journal* ASTM 13(2): 226-234

di Prisco C., Matiotti R., Nova R. 1995 Theoretical investigation of the undrained stability of shallow submerged slopes *Géotechnique*, 45,3, 479-496

Gens, A. & Nova, R. 1993. Conceptual bases for a constitutive model for bonded soils and weak rocks. In A. Anagnostopoulos, F. Schlosser, N. Kalteziotis & R. Frank (eds) *Geotechnical Engineering of Hard Soils-Soft Rocks; Proc. Intern. Symp.* Athens: 485-494. Rotterdam: Balkema

Imposimato S. and Nova R. (1997) 'Instability of loose sand specimens in undrained tests' Proceed. of the 4th Int. Workshop on *Localization and Bifurcation Theory for Soils and Rocks*, 313-322, Gifu (Japan) 28 September-2 October

Imposimato, S. & Nova, R. 1998. An investigation on the uniqueness of the incremental response of elastoplastic models for virgin sand. *Mechanics of Cohesive Frictional Materials*, 3(1): 65-87

Imposimato, S. & Nova, R. 2001. On the value of the second order work in homogeneous tests on loose sand specimens. In H.-B. Muehlhaus, A.V. Dyskin & E. Pasternak (eds), *Bifurcation and localisation theory in Geomechanics*: 209-216. Lisse : Swets & Zeitlinger.

Lade, P.V. 1992. Static instability and liquefaction of loose fine sandy slopes. *J. Geotech. Engin. ASCE* 118: 51-71

Leddra, M.J. 1988. Deformation on chalk through compaction and flow. *PhD Thesis*, Univ. of London.

Nova, R. 1988. Sinfonietta classica: an exercise on classical soil modelling. In A. Saada & G. Bianchini (eds) *Constitutive Equations for Granular non-cohesive soils*; Proc. Intern. Symp., Cleveland: 501-519. Rotterdam: Balkema

Nova, R. 1992. Mathematical modelling of natural and engineered geomaterials. General lecture 1st E.C.S.M. Munchen, *Eur.J. Mech. A/Solids*, 11,Special issue: 135-154.

Nova, R., Castellanza, R. & Tamagnini, C. 2003. A constitutive model for bonded geomaterials subject to mechanical and/or chemical degradation. *International Journal Numerical and Analytical Methods in Geomechanics*. 27, 705-732.

Nova R., Wood D. M. 1979 A constitutive model for sand in triaxial compression *I. J. Num. Anal. Meth. Geomech.*, 3, 3, 255-278.

Olsson, W. A. 1999. Theoretical and experimental investigation of compaction bands in porous rocks. *J. Geophys. Res.* 104: 7219-7228.

Ostrowksy A. and Taussky O. (1951) 'On the variation of the determinant of a positive definite matrix', *Nederl. Akad. Wet. Proc.*, (A) 54, 333-351

Rudnicki J. W. 2002. Conditions for compaction and shear bands in a transversely isptropic material. *Int. J. Solids and Structures* 39: 3741-3756.

Rudnicki, J.W. & Rice, J.R. 1975. Conditions for the localisation of deformation in pressure sensitive dilatant materials. *J. Mech. Phys. Solids* 23: 371-394

Schofield, A.N. & Wroth, C. P. 1968. *Critical State Soil Mechanics*. Chichester: McGraw-Hill.

Ziegler, H. 1968. *Principles of structural stability*. Waltham: Blaisdell Publishing Company

Continuous and discrete modelling of failure in geomechanics

Félix Darve[*] and Cédric Lambert[*]

[*] Laboratoire Sols, Solides, Structures, RNVO, Alert Geomaterials,
INPG, UJF, CNRS, Grenoble, France

Abstract Practice shows that there are various modes of failure in geomaterials. Different criteria have been proposed to analyse these failures. Hill's Condition of Stability and diffuse modes of failure are particularly considered in this paper in a dual framework : continuum mechanics and discrete mechanics. With the assumption of continuous media and as an illustration, the experiments have shown that q-constant loading paths can exhibit non-localized failure modes and they are analyzed by the second order work criterion. More generally, the equations of the boundaries of the unstable domain and of the cones of unstable stress directions are established in axisymmetric and plane strain conditions. With the assumption of discrete media, grain avalanches are considered and, spatial and temporal correlations between bursts of kinetic energy and peaks of negative values of second order work are exhibited from discrete computations. It is concluded that the second order work criterion (under its double form : continuous and discrete) can be a proper tool to analyse diffuse modes of failure in geomaterials.

1 Introduction

One of the most basic question in geomechanics is how to define, analyse and simulate failure. The general way to introduce the phenomenon of failure is to note that in elementary laboratory experiments, some loading paths are leading to limit stress states. Along these paths, these stress states can not be passed : experimentally they constitute asymptotical limit states.

By restricting our study to elasto-plastic behaviors, the general expression of rate-independent constitutive equations is given by :

$$d\sigma_{ij} = L_{ijkl}d\varepsilon_{kl} \qquad (1.1)$$

where the elasto-plastic tangent operator $\underset{\sim}{L}$ depends on the previous stress-strain history (characterized by state variables and memory parameters χ) and on the direction of $\underset{\sim}{d\varepsilon}$ (characterized by unit matrix $\underset{\sim}{v} = \underset{\sim}{d\varepsilon}/\|\underset{\sim}{d\varepsilon}\|$).

In the six-dimensional associated spaces it comes :

$$d\sigma_\alpha = N_{\alpha\beta}(v_\gamma)d\varepsilon_\beta \qquad (\alpha, \beta = 1, \ldots, 6) \qquad (1.2)$$

$v_\gamma = d\varepsilon_\gamma / \|d\underset{\sim}{\varepsilon}\|$ and $\|d\underset{\sim}{\varepsilon}\| = \sqrt{d\varepsilon_\alpha d\varepsilon_\alpha}$.

From a mathematical point of view, limit stress points are characterized by :

$$d\underset{\sim}{\sigma} = \underset{\sim}{0} \qquad \text{with} \qquad \|d\underset{\sim}{\varepsilon}\| \neq 0 \qquad (1.3)$$

Condition (1.3) implies :

$$\det \underset{\sim}{N}_\chi (\underset{\sim}{v}) = 0 \qquad \text{and} \qquad \underset{\sim}{N}_\chi (\underset{\sim}{v}) d\underset{\sim}{\varepsilon} = 0 \qquad (1.4)$$

If the constitutive relation is incrementally linear, $\underset{\sim}{N}$ is independent on $\underset{\sim}{v}$. The first equation of (1.4) represents precisely the plastic limit criterion because this is the equation of an hyper-surface in the six-dimensional stress space. The second equation represents the material flow rule, since this equation gives the direction of $d\underset{\sim}{\varepsilon}$ when the plastic limit surface has been reached.

For incrementally piecewise linear constitutive relations, where $\underset{\sim}{N}$ depends in a discontinuous manner from $\underset{\sim}{v}$ (inside "tensorial zones", see the chapter by Darve and Servant), $\det \underset{\sim}{N}_\chi (\underset{\sim}{v}) = 0$ is still the equation of an hyper-surface in the stress space, possibly involving several plastic mechanisms. Besides, $\underset{\sim}{N}_\chi (\underset{\sim}{v}) d\underset{\sim}{\varepsilon} = 0$ is a singular generalized flow rule with vertices which are locally constituted by pyramids (Hill R. , 1967).

This classical elasto-plastic view of failure has given rise to the wide domain of the "limit analysis", where the whole body at a failure state is assumed to have reached the plastic limit condition.

However, in practice, various modes of failure are encountered : localized modes in shear bands (Rice J.R. , 1976; Vardoulakis I. & Sulem J. , 1995; Darve F. , 1984), in compaction bands or in dilation bands, diffuse modes due to geometric instabilities (buckling,...) or with a chaotic displacement field,... For non-associated materials (essentially because of the non-symmetry of the elasto-plastic matrix $\underset{\sim}{N}$), the elasto-plastic theory shows effectively that, by considering proper bifurcation criteria, these localized or diffuse modes of failure can appear before the plastic limit condition (Bigoni D. & Hueckel T. , 1991).

In relation with these various modes of failure, different bifurcation criteria are available. With respect to shear band formation by plastic strain localizations, Rice's criterion (1976) based on the description of such an incipient shear band of normal n corresponds to vanishing values of the so-called "acoustic tensor" :

$$\det(^t n \underset{\sim}{L} n) = 0 \qquad (1.5)$$

For non-associated materials, equation (1.5) can be satisfied before plasticity criterion (1.4). This has been verified experimentally for dense sand (Desrues J. , 1990). Strain localization corresponds to a bifurcation of the strain mode from a diffuse one to a strictly discontinuous one. This kind of bifurcation can be called "discontinuous bifurcation".

"Continuous bifurcations" (Darve F. & Roguiez X. , 1998) opposed to the previous one correspond also to a failure mode, but without strain localization. Such failure mode is called "diffuse failure" and this is the response path which is subjected to a bifurcation with a loss of constitutive uniqueness at the bifurcation point. According to the control mode (stress, strain or mixed control) of the loading path, different response paths are possible from the bifurcation point. For certain control modes, the loading is no more "controllable" in Nova's sense (Nova R. , 1994).

These continuous bifurcations and the related diffuse modes of failure can be detected by Hill's sufficient condition of stability (Hill R. , 1958) which corresponds, for unstable states, to vanishing values of second order work, that is to say vanishing values of the determinant of the symmetric part of matrix $\underset{\sim}{N}$ for incrementally linear constitutive relations. Section 2 of this chapter is devoted to recall briefly Lyapunov's definition of stability (Lyapunov A.M. , 1907) and Hill's condition, then to apply them to the well known case of the failure of undrained loose sands. The generalization to axisymmetric paths and to plane strain conditions allows to determine numerically and analytically the "unstable" domain (i.e. the failure domain) and, inside this domain, the "unstable" stress directions which form cones for both these cases.

In section 3, this analysis is applied to axisymmetric q-constant loading path, since some experiments (Chu J. & Leong W.K. (2003)) have shown failure modes without localization along these paths in the case of loose sands.

Section 2 and 3 have illustrated the application of Hill's condition in the framework of continuum mechanics. It is also possible to consider a discrete form of second order work (Mandel J. , 1966) and to apply it to boundary value problems described by discrete mechanics. In this perspective the most known examples of typical failures are the so-called "grain avalanches". This phenomenon has been extensively studied specially by physicists of granular media (see Hermann H.J. et al. (1998) for example). It has been also considered as a paradigm for the concept of "self-organized criticality" (Bak P. , 1996). In section 4 of this paper, grain avalanches are analyzed in the strict framework of discrete mechanics. The, more or less local, failures will be characterized by bursts of kinetic energy (computed by a discrete element method). Strong spatial and temporal correlations between these bursts and the peaks of negative values of second order work are exhibited. These results represent another illustration of the deep link which relates certain failure modes and second order work, which could constitute a novel indicator of failure.

2   Material instabilities

The most basic definition of stability was certainly proposed by Lyapunov A.M. (1907). Applied to the field of continuum mechanics it states that :

"A stress-strain state, for a given material after a given loading history, is called stable, if for every positive scalar ε, a positive number $\eta(\varepsilon)$ exists such that for all incremental loading bounded by η, the associated responses remain bounded by ε "

According to this definition, all limit stress states (as defined in the introduction) are unstable.

Indeed if a "small" incremental reverse stress from a limit stress state is considered, a "small" (more or less elastic) response will be induced, while a "small" incremental additional stress beyond the limit state produces a "large" strain response. Lyapunov's definition of stability is interesting because it states clearly that material instabilities can be conjectured in elasto-plastic media. Since some limit stress states are strictly inside Mohr-Coulomb plastic limit surface (because of non-associativity) - an example is given in section 2.1 - some instabilities can be expected before reaching Mohr-Coulomb plastic limit condition.

But, even if this first conclusion is essential, Lyapunov's definition is not a proper tool to investigate instabilities in geomaterials. For that, we are now considering Hill's sufficient condition of stability (Hill R. , 1958). According to Hill, a stress-strain state is unstable if it exists at least one loading direction which can be pursued in an infinitesimal manner without any input of energy from outside. In practice it means that the deformation process will continue "on one's own" under dead loads . Indeed, in some situations, failure needs some energy from outside of soil boby (for example, the weight of a structure on a foundation), while in other situations energy is not needed (landslides, rockfalls,...).

Finally Hill's condition of stability states that a stress-strain state is stable if, for all $(d\underset{\sim}{\sigma}, d\underset{\sim}{\varepsilon})$ linked by the constitutive relation, the second order work is strictly positive :

$$d^2W = d\underset{\sim}{\sigma}\, d\underset{\sim}{\varepsilon} > 0 \qquad (2.1)$$

For hyperelastic-plastic constitutive relations, Drucker's postulate ($d\underset{\sim}{\sigma}\, d\underset{\sim}{\varepsilon}^p > 0$, see Drucker D.C. (1959)) implies Hill's condition, since : $d\underset{\sim}{\sigma}\, d\underset{\sim}{\varepsilon}^e \geq 0$ in hyperelasticity, and : $d\underset{\sim}{\sigma}\, d\underset{\sim}{\varepsilon} = d\underset{\sim}{\sigma}\, d\underset{\sim}{\varepsilon}^e + d\underset{\sim}{\sigma}\, d\underset{\sim}{\varepsilon}^p$, with the usual decomposition of the incremental strain into an elastic part and a plastic part in classical elasto-plasticity.

The link between Lyapunov's definition and Hill's condition has not been established in a general manner. The equivalence has been proved by Koïter W.T. (1969) in the case of elasticity. If the equivalence between Lyapunov's definition and "non-controllable" paths in Nova's sense (Nova R. , 1994) is accepted, then non-controllability and Hill's condition are equivalent in classical elasto-plasticity (Nova R. , 1994). For hypoplastic relations (as defined by Chambon R. et al. (1994)) the positiveness of second order work assures the uniqueness (Chambon R. & Caillerie D. , 1999). In the general case of incrementally non-linear constitutive relations, counter-example might exist. In the next sub-section, all that is illustrated by the well known case of the undrained triaxial behavior of loose sands.

2.1   Undrained paths on loose sands

The undrained triaxial behavior of loose sands is typically represented in figure 1, where q is passing through a maximum which is situated strictly inside Mohr-Coulomb plastic limit surface. The experiments show that, if a "small" additional axial force is applied at q peak, there is a sudden failure of the whole sample without any localization pattern. Because of this sudden failure for a small perturbation at q peak, q peak is an unstable state according to Lyapunov's definition.

For axisymmetric conditions and coinciding principal axes, the second order work is equal to :

$$d^2W = d\sigma_1 d\varepsilon_1 + 2d\sigma_3 d\varepsilon_3 \qquad (2.2)$$

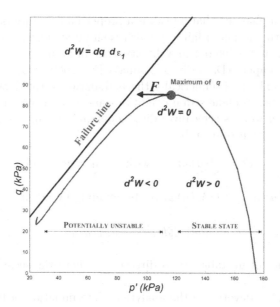

Figure 1. Example of an unstable state related to a diffuse mode of failure strictly inside Mohr-Coulomb plastic limit condition : typical undrained triaxial behavior of a loose sand.

and with the isochoric condition : $d\varepsilon_1 + 2d\varepsilon_3 = 0$, it comes :

$$d^2W = dq d\varepsilon_1 \qquad \text{with :} \qquad q = \sigma_1 - \sigma_3 \tag{2.3}$$

Thus according to Hill's condition , q peak has to be considered as an unstable state ((Darve F. & Chau B. , 1987)).

Let us see now that we can define at q peak a generalized plasticity criterion and a generalized flow rule (as defined in equation (1.4)).

Because $q - \varepsilon_1$ and $\varepsilon_v - \varepsilon_3$ (with : $\varepsilon_v = \varepsilon_1 + 2\varepsilon_3$) are conjugate variables with respect to energy : $\sigma_1\varepsilon_1 + 2\sigma_3\varepsilon_3 \equiv q\varepsilon_1 + \varepsilon_v\sigma_3$, the constitutive relation can be written under the following form :

$$\begin{bmatrix} dq \\ d\varepsilon_v \end{bmatrix} = \underset{\sim}{P} \begin{bmatrix} d\varepsilon_1 \\ d\sigma_3 \end{bmatrix} \tag{2.4}$$

For an undrained loading, the isochoric condition is fulfilled, thus :

$$d\varepsilon_v = 0 \tag{2.5}$$

As dq is vanishing at q peak, it comes :

$$\det \underset{\sim}{P} = 0 \qquad \text{and} \qquad \underset{\sim}{P} \begin{bmatrix} d\varepsilon_1 \\ d\sigma_3 \end{bmatrix} = \begin{bmatrix} 0 \\ 0 \end{bmatrix} \tag{2.6}$$

These relations generalize equations (1.4). The failure mode can be viewed as the eigen-mode associated to the vanishing eigenvalue of matrix $\underset{\sim}{P}$.

In conclusion it is emphasized that q peak is a proper failure state (satisfying all the criteria associated to the notion of failure) which is situated strictly inside Mohr-Coulomb plastic limit surface and is related to a diffuse mode of failure.

In several previous papers (Darve F. & Laouafa F. , 2000, 2001; Laouafa F. & Darve F. , 2001) we have exhibited numerically the boundaries of the unstable domain in axisymmetric conditions, in plane strain conditions and more generally for 3D conditions. Besides, second order work is typically a directional quantity in the stress space. Indeed, if the constitutive relation is given by :

$$d\underset{\sim}{\varepsilon} = \underset{\sim}{M}(\underset{\sim}{u})d\underset{\sim}{\sigma} \qquad \text{with} \qquad \underset{\sim}{u} = d\underset{\sim}{\sigma}/\|d\underset{\sim}{\sigma}\| \tag{2.7}$$

it comes : $d^2W = {}^t d\underset{\sim}{\sigma}\, d\underset{\sim}{\varepsilon} = {}^t d\underset{\sim}{\sigma}\, \underset{\sim}{M}(\underset{\sim}{u})d\underset{\sim}{\sigma}$ or, by dividing by $\|d\underset{\sim}{\sigma}\|^2$:

$$d^2W/\|d\underset{\sim}{\sigma}\|^2 = {}^t\underset{\sim}{u}\,\underset{\sim}{M}(\underset{\sim}{u})\underset{\sim}{u} \tag{2.8}$$

Thus there are cones of unstable stress directions, where the second order work takes negative values.

Next sub-section 2.2 is devoted to the analytical determination of the equations of the unstable domain and of the cones of unstable stress directions in the case of an incrementally piecewise linear constitutive relation for axisymmetric conditions.

2.2   Axisymmetric loading paths

Analytical explicit computations are feasible with our incrementally piecewise linear model,called "the octo-linear model" (see chapter 1 by Darve and Servant and (Darve F. & Labanieh S. , 1982)) because it is constituted by 8 linear relations between $d\underset{\sim}{\varepsilon}$ and $d\underset{\sim}{\sigma}$ in 8 "tensorial zones". Let us restrict the presentation to the axisymmetric case with 4 tensorial zones. In such a case, the expression of the model is given by :

$$\begin{bmatrix} d\varepsilon_1 \\ \sqrt{2}d\varepsilon_3 \end{bmatrix} = \frac{1}{2}\left(\underset{\sim}{Q}^+ + \underset{\sim}{Q}^-\right)\begin{bmatrix} d\sigma_1 \\ \sqrt{2}d\sigma_3 \end{bmatrix} + \frac{1}{2}\left(\underset{\sim}{Q}^+ - \underset{\sim}{Q}^-\right)\begin{bmatrix} |d\sigma_1| \\ \sqrt{2}|d\sigma_3| \end{bmatrix} \tag{2.9}$$

with :

$$\underset{\sim}{Q}^\pm = \begin{bmatrix} \dfrac{1}{E_1^\pm} & -\sqrt{2}\dfrac{V_3^{1\pm}}{E_3^\pm} \\ -\sqrt{2}\dfrac{V_1^{3\pm}}{E_1^\pm} & \dfrac{1-V_3^{3\pm}}{E_3^\pm} \end{bmatrix} \tag{2.10}$$

where index (+) means "compression" ($d\sigma_i > 0$) and (-) "extension" ($d\sigma_i < 0$) in the axial direction "i" of "generalized" triaxial paths (two lateral constant stresses σ_j and σ_k). E_i are "generalized" tangent Young moduli and V_i^j "generalized" tangent Poisson's ratios defined by :

$$E_i = \left(\frac{\partial\sigma_i}{\partial\varepsilon_i}\right)_{\sigma_j,\sigma_k} ; V_i^j = -\left(\frac{\partial\varepsilon_j}{\partial\varepsilon_i}\right)_{\sigma_j,\sigma_k} ; V_i^k = -\left(\frac{\partial\varepsilon_k}{\partial\varepsilon_i}\right)_{\sigma_j,\sigma_k}$$

According to the signs of $d\sigma_1$ and $d\sigma_3$ (4 possibilities), there are 4 linear relations between $d\underset{\sim}{\varepsilon}$ and $d\underset{\sim}{\sigma}$, (see details in chapter 1).

In order to investigate the sign of second order work at a given stress-strain state in axisymmetric conditions, it is necessary to consider all the directions for example in the axisymmetric strain space (i.e. Rendulic Plane). For that it is introduced axisymmetric proportional strain loading paths defined incrementally by:

$$\begin{cases} d\varepsilon_1 & = & \text{positive constant} \\ d\varepsilon_2 & = & d\varepsilon_3 \text{ (axisymmetric condition)} \\ d\varepsilon_1 + 2Rd\varepsilon_3 & = & 0 \ (R \text{ constant for a given path}) \end{cases}$$

$R = 1$ corresponds to the undrained path which is characterized by the isochoric condition (no volume variation).

For $0 < R < 1$, the imposed path is dilatant, while for $1 < R$ it is contractant.

For this class of paths, the proper conjugate variables are $(\sigma_1 - \sigma_3/R)$ versus ε_1 and $(\varepsilon_1 + 2R\varepsilon_3)$ versus (σ_3/R) since : $\sigma_1\varepsilon_1 + 2\sigma_3\varepsilon_3 \equiv (\sigma_1 - \sigma_3/R)\varepsilon_1 + \sigma_3(\varepsilon_1 + 2R\varepsilon_3)/R$.

We have verified (Darve F. & Laouafa F. , 2001) numerically that, for a range of small R values, curve $(\sigma_1 - \sigma_3/R)$ versus ε_1 is passing through a maximum. This range of R values depends, of course, on the initial density of the sand (R must be smaller for a higher density). These curves can be considered as generalizations of $q - \varepsilon_1$ diagrams for undrained triaxial paths.

It is easy to verify that, at $(\sigma_1 - \sigma_3/R)$ peak, Lyapunov's definition is verified.

As regards Hill's condition, the second order work is equal to :

$$d^2W = (d\sigma_1 - d\sigma_3/R)d\varepsilon_1 + d\sigma_3(d\varepsilon_1 + 2Rd\varepsilon_3)/R \qquad (2.11)$$

With the constraint : $d\varepsilon_1 + 2Rd\varepsilon_3 = 0$, it is clear that the second order work is vanishing at $(\sigma_1 - \sigma_3/R)$ peak and, after it, takes negative values. The constitutive equation can be written like that :

$$\begin{bmatrix} d\sigma_1 - d\sigma_3/R \\ d\varepsilon_1 + 2Rd\varepsilon_3 \end{bmatrix} = \begin{bmatrix} E_1^- & 2R\frac{E_1^-}{E_3^-}V_3^{1-} - 1 \\ 1 - 2RV_1^{3-} & \frac{2R^2(1-V_3^{3-}-2V_1^{3-}V_3^{1-})}{E_3^-} \end{bmatrix} \begin{bmatrix} d\varepsilon_1 \\ d\sigma_3/R \end{bmatrix} \qquad (2.12)$$

if we assume that $d\sigma_1$ and $d\sigma_3$ are negative (as given by the numerical results). At $(\sigma_1 - \sigma_3/R)$ peak, the bifurcation criterion is given by :

$$\det \underset{\sim}{S} = 0 \qquad (2.13)$$

if $\underset{\sim}{S}$ is the constitutive matrix of relation (2.12). (2.13) gives the following equation in R :

$$2\frac{E_1^-}{E_3^-}\left(1 - V_3^{3-}\right) R^2 - 2\left(V_1^{3-} + \frac{E_1^-}{E_3^-}V_3^{1-}\right)R + 1 = 0 \qquad (2.14)$$

which is an equation of second degree in R. It has two real solutions in R, if discriminant Δ is strictly positive :

$$\Delta = \left(V_1^{3-} + \frac{E_1^-}{E_3^-}V_3^{1-}\right)^2 - 2\frac{E_1^-}{E_3^-}\left(1 - V_3^{3-}\right) > 0 \qquad (2.15)$$

or equivalently :

$$\Delta = -2 \left(\frac{1 - V_3^{3-} - 2V_1^{3-}V_3^{1-}}{E_3^-} \right)^2 \det \underset{\sim}{Q^S} > 0$$

where $\underset{\sim}{Q^S}$ is the symmetric part of the constitutive matrix $\underset{\sim}{Q}$ given by :

$$\underset{\sim}{Q} = \begin{bmatrix} \frac{1}{E_1^-} & -\sqrt{2}\frac{V_3^{1-}}{E_3^-} \\ -\sqrt{2}\frac{V_1^{3-}}{E_1^-} & \frac{1-V_3^{3-}}{E_3^-} \end{bmatrix}$$

which relates $\begin{bmatrix} d\varepsilon_1 \\ \sqrt{2}d\varepsilon_3 \end{bmatrix}$ to $\begin{bmatrix} d\sigma_1 \\ \sqrt{2}d\sigma_3 \end{bmatrix}$ for $d\sigma_1 < 0$ and $d\sigma_3 < 0$ (see equation 2.10).

The unstable domain is thus characterized by : $\det \underset{\sim}{Q^S} < 0$ and its boundary by :

$$\det \underset{\sim}{Q^S} = 0 \tag{2.16}$$

Indeed, for incrementally linear constitutive relations,
$d^2W = \underset{\sim}{d\sigma}\, \underset{\sim}{d\varepsilon} = {}^t\underset{\sim}{d\sigma}\,\underset{\sim}{M}\,\underset{\sim}{d\sigma} \equiv {}^t\underset{\sim}{d\sigma}\,\underset{\sim}{M^S}\,\underset{\sim}{d\sigma}$, with : $\underset{\sim}{d\varepsilon} = \underset{\sim}{M}\,\underset{\sim}{d\sigma}$ and M^S symmetric part of $\underset{\sim}{M}$.
In such a simple case, d^2W is vanishing only if $\det \underset{\sim}{Q^S}$ is vanishing (if the eigenvalues
of $\underset{\sim}{M^S}$ are varying in a continuous manner from strictly positive initial values, what is
always verified for realistic constitutive matrices).
For an incrementally piecewise linear relation (as considered here with the octo-linear
model), we have 4 different constitutive matrices $\underset{\sim}{Q}$ following the signs of $d\sigma_1$ and $d\sigma_3$.

The discussion is thus more intricate, because the first unstable stress direction (which is
given by the unique solution of equation (2.14) when discriminant Δ is vanishing) must
belong to the tensorial zone defined by the signs of $d\sigma_1$ and $d\sigma_3$. Anyway the equation
of the boundary of the unstable domain (if it exists) is given by :

$$\det \underset{\sim}{Q^S} = 0$$

where $\underset{\sim}{Q^S}$ is the symmetric part of constitutive matrix $\underset{\sim}{Q}$ associated to the proper tensorial
zone.
At this boundary, the unique solution of (2.14) is equal to :

$$R_c = -\frac{\delta\varepsilon_1}{2\delta\varepsilon_3} = \frac{E_1^- V_3^{1-} + E_3^- V_1^{3-}}{2E_1^- \left(1 - V_3^{3-}\right)} \tag{2.17}$$

In Rendulic plane, the first unstable stress direction is then given by :

$$\frac{\delta\sigma_1}{\sqrt{2}\delta\sigma_3} = \frac{E_1^- \left(1 - V_3^{3-}\right) \left(E_1^- V_3^{1-} - E_3^- V_1^{3-}\right)}{\sqrt{2}E_3^- \left(E_1^- \left(1 - V_3^{3-} - V_1^{3-}V_3^{1-}\right) - E_3^- \left(V_1^{3-}\right)^2\right)} \tag{2.18}$$

with $\delta\sigma_1 < 0$ and $\delta\sigma_3 < 0$. These stress directions are plotted in figure 5.
The associated mode of rupture is characterized by the following rupture rule :

$$\underset{\sim}{S}\begin{bmatrix} d\varepsilon_1 \\ d\sigma_3/R \end{bmatrix} = \begin{bmatrix} 0 \\ 0 \end{bmatrix} \qquad \text{with} \qquad \det \underset{\sim}{S} = 0$$

where $\underset{\sim}{S}$ is the constitutive matrix defined by equation (2.12). It comes :

$$E_1^- d\varepsilon_1 + \left(2\frac{E_1^-}{E_3^-}V_3^{1-} - \frac{1}{R} \right) d\sigma_3 = 0 \tag{2.19}$$

At the bifurcation point (or equivalently at the limit stress-strain state) characterized by
:

$$\begin{cases} d\sigma_1 - d\sigma_3/R &= 0 \\ d\varepsilon_1 + 2R d\varepsilon_3 &= 0 \end{cases}$$

there are an infinite number of solutions (loss of uniqueness), but all these solutions must
fulfilled constraint (2.19).
Note that these conclusions are perfectly similar to the plastic limit conditions discussed
in the introduction (see general equations (1.3)and (1.4)), where :

$$\begin{cases} d\sigma_1 &= 0 \\ d\sigma_3 &= 0 \end{cases}$$

defines the plastic limit stress state, $\det \underset{\sim}{N} = 0$ is the plasticity criterion, and : $V_3^{1+}d\varepsilon_1 + d\varepsilon_3 = 0$ is the flow rule.
What is remarkable in equation (2.19) is the fact that it is a mixed condition, even if
essentially this is the generalization of the notion of flow rule.
All these results have been verified numerically with the octo-linear model and with
the thoroughly incrementally non-linear relation (for a presentation of this relation see
chapter 1 by Darve and Servant and Darve F. et al. (1995)).
The strain proportional loading paths have been simulated for various values of R. Some
maxima for $(\sigma_1 - \sigma_3/R)$ have been found for a certain range of R values (see figure 2).
Finally the cones of unstable stress directions have been plotted for a dense sand (figure
3) and a loose sand (figure 4). Note that for high stress levels there are two cones of
unstable stress directions. Indeed the previous analytical results have just shown that
there are possibly two solutions in R (what means one cone) inside one tensorial zone.
Thus several cones can exist, when there are several tensorial zones.
In the next section, the case of plain strain conditions is considered.

2.3 Plane strain conditions

For plane strain conditions ($\varepsilon_2 = 0$) and coinciding stress-strain principal axes, the
same reasoning can be developed.
The octo-linear constitutive model takes the following form :

$$\begin{bmatrix} d\varepsilon_1 \\ d\varepsilon_3 \end{bmatrix} = \begin{bmatrix} \frac{1-V_1^{2\pm}V_2^{1\pm}}{E_1^\pm} & -\frac{V_3^{1\pm}+V_3^{2\pm}V_2^{1\pm}}{E_3^\pm} \\ -\frac{V_1^{3\pm}+V_1^{2\pm}V_2^{3\pm}}{E_1^\pm} & \frac{1-V_2^{3\pm}V_3^{2\pm}}{E_3^\pm} \end{bmatrix} \begin{bmatrix} d\sigma_1 \\ d\sigma_3 \end{bmatrix} \tag{2.20}$$

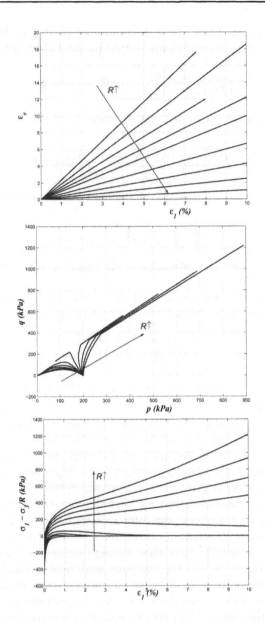

Figure 2. Proportional strain paths simulated by the incrementally non-linear model in the case of Hostun dense sand. $(\sigma_1 - \frac{\sigma_3}{R})$ possesses a maximum for low R values, $R \in \{0.3, 0.35, 0.4, 0.45, 0.5, 0.6, 0.7, 0.8, 0.9, 1.0\}$.

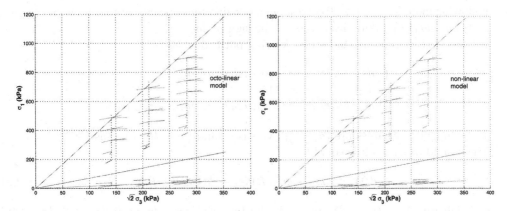

Figure 3. Cones of unstable stress directions and boundary of the unstable domain in Rendulic plane for dense Hostun sand in the cases of the octo-linear model (on the left) and of the non-linear one (on the right).

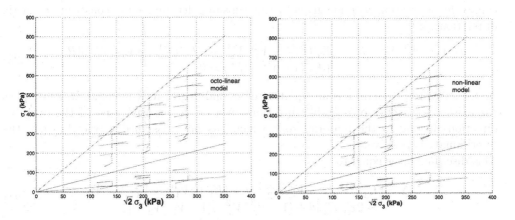

Figure 4. Cones of unstable stress directions and boundary of the unstable domain in Rendulic plane for loose Hostun sand in the cases of the octo-linear model (on the left) and of the non-linear one (on the right).

Following indices $(+,-)$, 2.20 corresponds to 8 different linear relations between $(d\varepsilon_1, d\varepsilon_3)$ and $(d\sigma_1, d\sigma_3)$ by taking into account the sign of $d\sigma_2$.

Let us consider proportional strain loading paths defined by :

$$\begin{cases} d\varepsilon_1 & = & \text{positive constant} \\ d\varepsilon_2 & = & 0 \\ d\varepsilon_1 + Rd\varepsilon_3 & = & 0 \, (R \text{ constant for a given path}) \end{cases}$$

The conjugate variables are $(\sigma_1 - \frac{\sigma_3}{R})$ versus ε_1 and $(\varepsilon_1 + R\varepsilon_3)$ versus $(\frac{\sigma_3}{R})$.

The second order work is equal to :

$$d^2W = (d\sigma_1 - d\sigma_3/R)d\varepsilon_1 + d\sigma_3(d\varepsilon_1 + Rd\varepsilon_3)/R \tag{2.21}$$

and will vanish at $(\sigma_1 - \frac{\sigma_3}{R})$ peak. If the tensorial zone $d\sigma_1 < 0, d\sigma_2 < 0, d\sigma_3 < 0$ is considered (this assumption has to be verified a posteriori), the octo-linear model can be written under the following form :

$$\begin{bmatrix} d\sigma_1 - d\sigma_3/R \\ d\varepsilon_1 + Rd\varepsilon_3 \end{bmatrix} = \begin{bmatrix} \frac{E_1^-}{1-V_1^{2-}V_2^{1-}} & \frac{E_1^-(V_3^{1-}+V_3^{2-}V_2^{1-})}{E_3^-(1-V_1^{2-}V_2^{1-})}R - 1 \\ 1 - \frac{V_1^{3-}+V_1^{2-}V_2^{3-}}{1-V_1^{2-}V_2^{1-}}R & \frac{dR^2}{E_3^-(1-V_1^{2-}V_2^{1-})} \end{bmatrix} \begin{bmatrix} d\varepsilon_1 \\ \frac{d\sigma_3}{R} \end{bmatrix} \tag{2.22}$$

with $d = 1 - V_1^{2-}V_2^{1-} - V_1^{3-}V_3^{1-} - V_2^{3-}V_3^{2-} - V_1^{2-}V_2^{3-}V_3^{1-} - V_2^{1-}V_1^{3-}V_3^{2-}$.

Taking into account the loading constraint $d\varepsilon_1 + Rd\varepsilon_3 = 0$, and, at $(\sigma_1 - \frac{\sigma_3}{R})$ peak, it comes the bifurcation criterion given by the vanishing determinant of constitutive matrix of relation 2.22 :

$$E_1^-(1 - V_2^{3-}V_3^{2-})R^2 - [E_3^-(V_1^{3-} + V_1^{2-}V_2^{3-}) + E_1^-(V_3^{1-} + V_3^{2-}V_2^{1-})]R \\ + E_3^-(1 - V_1^{2-}V_2^{1-}) = 0 \tag{2.23}$$

As for axisymetric conditions an equation of second degree in R is found, which implies the existence of cones of unstable directions in a given tensorial zone (if the unstable directions belong to the tensorial zone which has been assumed a priori), when the discriminant is positive.

This discriminant is equal to :

$$\Delta = (E_3^-)^2(V_1^{3-} + V_1^{2-}V_2^{3-})^2 + (E_1^-)^2(V_3^{1-} + V_3^{2-}V_2^{1-})^2 \\ - 2E_1^- E_3^- [d + (1 - V_2^{3-}V_3^{2-})(1 - V_1^{2-}V_2^{1-})] \tag{2.24}$$

which can be put under the following form :

$$\Delta = -4(E_1^-)^2(E_3^-)^2 \det \underset{\sim}{P^S} \tag{2.25}$$

where $\underset{\sim}{P^S}$ is the symmetric part of the constitutive matrix of relation 2.20 (with indices "_").

Once more, the boundary of the unstable domain is given by :

$$\det \underset{\sim}{P^S} = 0 \tag{2.26}$$

(if the unstable directions correspond to $d\sigma_1 < 0$ and $d\sigma_3 < 0$)
The first unstable direction corresponds to :

$$R_c = -\left(\frac{d\varepsilon_1}{d\varepsilon_3}\right)_c = \frac{E_3^-\left(V_1^{3-} + V_1^{2-}V_2^{3-}\right) + E_1^-\left(V_3^{1-} + V_3^{2-}V_2^{1-}\right)}{2E_1^-\left(1 - V_2^{3-}V_3^{2-}\right)} \qquad (2.27)$$

or

$$\left(\frac{d\sigma_1}{d\sigma_3}\right)_c = \frac{E_1^-\left(1 - V_2^{3-}V_3^{2-}\right)\left[E_1^-\left(V_3^{1-} + V_3^{2-}V_2^{1-}\right) - E_3^-\left(V_1^{3-} + V_1^{2-}V_2^{3-}\right)\right]}{E_3^-\left[E_1^-\left(d + \left(1 - V_1^{2-}V_2^{1-}\right)\left(1 - V_2^{3-}V_3^{2-}\right)\right) - E_3^-\left(V_1^{3-} + V_1^{2-}V_2^{3-}\right)^2\right]} \qquad (2.28)$$

Figures 5, 6 and 7 present the "first" unstable stress states (according to the level of stress deviator) respectively for the axisymmetric conditions (figure 5), for plane strain conditions (figure 6) and for 3D conditions in a deviatoric stress plane (figure 7). The stress curve or surface, which is obtained by linking these stress states, corresponds to the boundary of the "unstable" domain. Inside this stress unstable domain one can find various kinds of bifurcations and losses of uniqueness, which means various failure modes by material instabilities (localized failure, diffuse failure...) and geometric instabilities. These failure modes depend on the previous stress-strain history, on the current incremental stress direction and on the loading mode (stress controlled, strain controlled...).

Hill's condition of stability is essentially a directional quantity (see equation 2.8) since

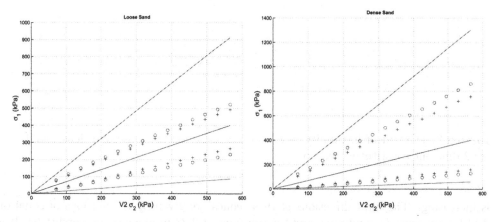

Figure 5. Domains of instability in Rendulic Plane and axisymmetric conditions for the loose sand on the left, for the dense sand on the right and for 2 different constitutive relations (symbols "o" and "+" for the incrementally non-linear and the octolinear models respectively).

:

$$\frac{d^2W}{\|d\sigma\|^2} =^t \underset{\sim}{u}\,M(\underset{\sim}{u})\,\underset{\sim}{u} \qquad (2.29)$$

It follows that there are cones of unstable stress directions, as plotted in figures 3 and 4. The first unstable direction is progressively opening into a cone when the stress deviator

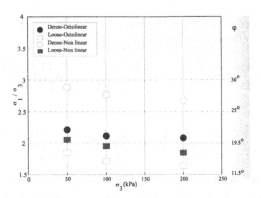

Figure 6. Lowest unstable stress levels in plane strain conditions for the loose and dense sand, for 2 different constitutive relations.

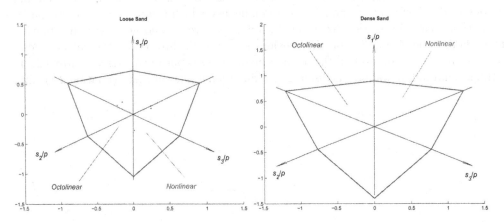

Figure 7. Domain of instability in deviatoric plane for the loose and dense sand, and for 2 different constitutive relations.

is increasing. The limit directions of these cones are given by th solutions of equation 2.14 for axisymetric conditions and the solutions of 2.23 in plane strain conditions. It is worth to note in figures 3 and 4 that the vertical stress direction (which corresponds to drained triaxial tests) is never inside the cones. Along drained triaxial compressions or extensions it is impossible to obtain diffuse failure, which is a well known experimental fact.

3 q-constant loading paths : an example

3.1   Analysis

At a given stress-strain state, q-constant axisymmetric loading paths consist in maintaining q constant by an incrementally isotropic unloading defined by : $d\sigma_1 = d\sigma_2 =$

$d\sigma_3 =$ negative constant.

The proper conjugate variables to analyse this path are : $\varepsilon_v - \sigma_3$ and $q - \varepsilon_1$, where : $\varepsilon_v = \varepsilon_1 + 2\varepsilon_3$, because we have :

$$W = \sigma_1\varepsilon_1 + 2\sigma_3\varepsilon_3 = \varepsilon_v\sigma_3 + q\varepsilon_1 \tag{3.1}$$

The second order work can thus be written as follows : $d^2W = d\varepsilon_v d\sigma_3 + dq d\varepsilon_1$, and with the constraint : $dq = 0$, it comes :

$$d^2W = d\varepsilon_v d\sigma_3 \tag{3.2}$$

If ε_v passes through a minimum value, this minimum will be an unstable state according to Hill's condition. On the contrary σ_3 is monotonously decreasing.

As the relative volume variations $\frac{\Delta V}{V_0}$ is equal to $(-\varepsilon_v)$: $\frac{\Delta V}{V_0} = -\varepsilon_v$, the minimum for ε_v is a maximum for $\frac{\Delta V}{V_0}$.

Thus a "small" additional positive volume variation (for a volume variation controlled loading) induces a sudden failure of the sample. It can be concluded that this maximum of volume variations is unstable following Lyapunov's definition of stability.

Now if the constitutive relation is written under the following form (for axisymmetric conditions) :

$$\begin{bmatrix} d\varepsilon_v \\ dq \end{bmatrix} = \underset{\sim}{T} \begin{bmatrix} d\sigma_3 \\ d\varepsilon_1 \end{bmatrix} \tag{3.3}$$

the analysis can be carried out as previously.

Along the path the constraint : $dq = 0$ is fulfilled and, at ε_v minimum, $d\varepsilon_v$ is zero. Thus, the bifurcation criterion is given by :

$$\det \underset{\sim}{T} = 0 \tag{3.4}$$

and the rupture rule by :

$$\underset{\sim}{T} \begin{bmatrix} d\sigma_3 \\ d\varepsilon_1 \end{bmatrix} = \begin{bmatrix} 0 \\ 0 \end{bmatrix} \tag{3.5}$$

Note that equations (3.4) and (3.5) are exactly equivalent to equations (2.6), what means that bifurcation criterion and rupture rule are the same for q-constant loading path and for undrained loading. Thus the results of the subsection 2.2 apply with constant R equal to 1.

With our octo-linear model, the bifurcation criterion is given by :

$$2\frac{E_1^-}{E_3^-}\left(1 - V_3^{3-} - V_3^{1-}\right) + 1 - 2V_1^{3-} = 0 \tag{3.6}$$

The rupture rule corresponds to :

$$E_1^- d\varepsilon_1 + \left(2\frac{E_1^-}{E_3^-}V_3^{1-} - 1\right)d\sigma_3 = 0 \tag{3.7}$$

The unstable stress direction is obviously given by : $d\sigma_1 = d\sigma_3$ and it is parallel to the hydrostatic line.

Finally one question is remaining open : are the volume variations possessing a maximum for a certain range of initial densities and a certain range of stress levels ? To answer to this question in the next subsection, we have simulated q-constant loading paths with our constitutive relations.

3.2   Numerical modelling

q-constant paths have been simulated with our octo-linear model and with the non-linear one. Figures 8 and 9 present the results for loose Hostun sand and for initial isotropic pressures respectively equal to 100 kPa and 500 kPa. When the second order work is negative, the computed points are replaced by different symbols (triangles, squares, circles, depending on the value of q, see figure 8). The diagrams are composed of two parts. A first part is constituted by a triaxial compression until reaching various q values. Then q is maintained constant.

These results allow to validate our previous assumptions. For example, lateral pressure is decreasing monotonously. On the contrary the volume variations are indeed possessing a maximum. Of course it is verified that from this maximum, the second order work is negative.

Figure 10 shows the results obtained with dense Hostun sand. The same comments as previously are pertinent, except the fact that the volume variations do not reach any maximum. Note that the computations are stopped when Mohr-Coulomb plastic limit condition is reached. The second order work is always strictly positive.

These results could have been drawn from figures 3 and 4. q-constant loading paths are parallel to the hydrostatic axis. We see that this direction is included inside the cone of unstable directions for low q/p values for loose Hostun sand and not at all for any q/p value for the dense one.

Modelling has been also performed with the incrementally non-linear model with similar conclusions (see figure 11 for such a comparison).

Several conclusions can be drawn.

The most important one is certainly the fact that a diffuse mode of failure, which was repeatedly noted in experiments (Chu J. & Leong W.K., 2003), can be properly analyzed by Hill's condition of stability.

It has been also shown that the bifurcation criterion and the failure rule are the same for undrained (i.e. isochoric) paths and q-constant paths. Finally a generalization of the notion of limit state has been brought out. For q-constant paths the stresses are monotonously decreasing ($d\sigma_1 = d\sigma_2 = d\sigma_3 < 0$) while the dilatancy is reaching a maximum value. If one injects at this maximum a small amount of water inside the sample, a sudden failure is occurring without any localization pattern. Thus we can expect the existence of a limit surface in the strain space (dual of the classical limit surface in the stress space). Indeed the strains are limited on the hydrostatic axis or for one-dimensional compressions.

Next section is now devoted to investigate a possible application of Hill's condition in discrete mechanics.

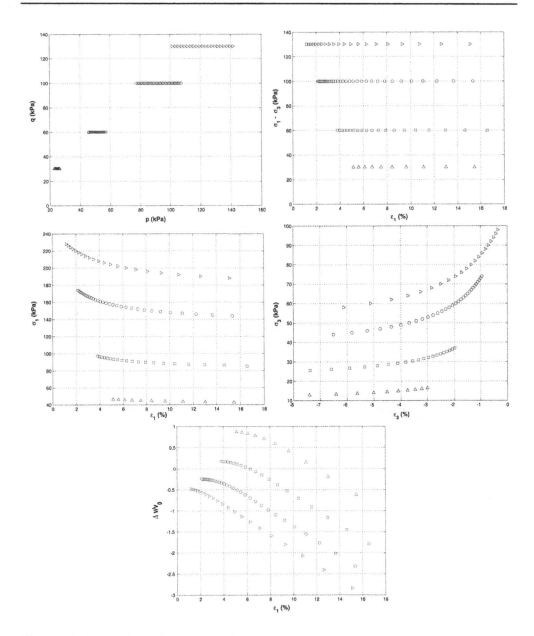

Figure 8. Simulation of q-constant loading paths for loose Hostun sand by the octo-linear constitutive model. Points are replaced by other symbols when the second order work takes negative values from the maximum of volume variations. The initial isotropic pressure is equal to 100 kPa.

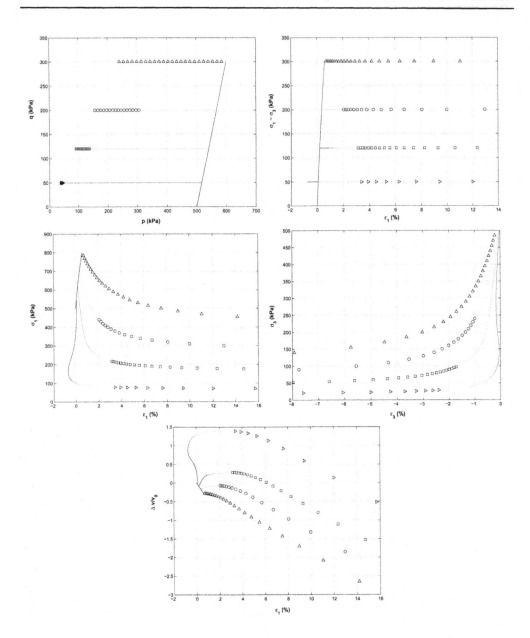

Figure 9. Simulation of q-constant loading paths for loose Hostun sand by the octo-linear constitutive model. Points are replaced by other symbols when the second order work takes negative values from the maximum of volume variations. The initial isotropic pressure is equal to 500 kPa.

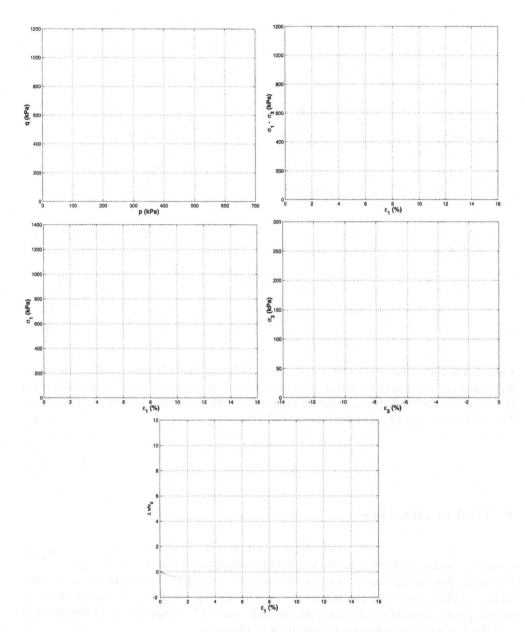

Figure 10. Simulation of q-constant loading paths for dense Hostun sand by the octo-linear constitutive model. The second order work is never vanishing.

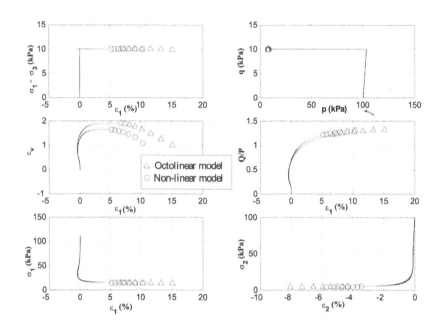

Figure 11. Simulation of q-constant loading paths for loose Hostun sand by the octo-linear and the non-linear constitutive model. Points are replaced by other symbols when the second order work takes negative values from the maximum of volume variations. The initial isotropic pressure is equal to 100 kPa.

4 Grain avalanches

Granular media are characterized by a double nature : continuous and discrete.
For some phenomena (experiments on homogeneous samples, in situ behavior at large scale,...) they behave like continuous media , the assumption of matter continuity represents a proper description, stresses, strains and constitutive relations are pertinent tools and the framework of continuum mechanics is adequate.
For other phenomena (granular segregation, mixing of two granular populations,...) they behave like discrete media, each grain has to be considered as a specific individual, forces, displacements, rotations and local granular interactions are pertinent tools, while the framework of discrete mechanics is convenient. Grain avalanches along a slope are certainly a consequence of discrete nature of granular media. Thus, in this section, we have to introduce discrete notions, discrete tools and finally to perform discrete computations.

4.1 Discrete second order work and kinetic energy

In this section we consider only two-dimensional granular materials which are constituted by piled cylinders (the so-called Schneebeli material). All the variables are thus defined in a two-dimensional space.

A given grain is subjected to several forces due to the other grains in contact and due to gravity. The resultant force is called $\overrightarrow{F}$. Under the force increment $\overrightarrow{dF}$ the grain is moving of $\overrightarrow{dl}$. All these forces are also developing torques along the cylinder axes. The resultant torque is noted C, whose consequence is a rotation Ω around the axis. Thus the discrete second order work is defined by (Mandel J. , 1966) :

$$d^2W = \overrightarrow{dF}.\overrightarrow{dl} + dCd\Omega \qquad (4.1)$$

The objective is to investigate a possible link between discrete second order work and failure in discrete media, as an extension of the previous results which have shown such a link between second order work and diffuse failure.

In discrete media, failure at free surface is generally characterized by granular avalanches. During these avalanches the kinetic energy is varying a lot from zero values to maxima, then again to zero values. Thus kinetic energy has been chosen as a proper variable to characterize a grain avalanche from a mechanical point of view.

For a given grain its expression is :

$$e_c = \frac{1}{2}mv^2 + \frac{1}{2}I\omega^2 \qquad (4.2)$$

where m is the grain mass, v its speed, I the inertia moment around the cylinder axis and ω the rotation speed.

Second order work and kinetic energy have been defined for a given grain. It is also interesting to consider their values for the whole assembly of grains as follows :

$$\begin{cases} D^2W &= \sum_{k=1}^{N} d^2W_k \\ E_c &= \sum_{k=1}^{N} e_c^k \end{cases} \qquad (4.3)$$

where N is the total number of grains.

4.2 Discrete material and discrete numerical method

The granular medium is a two-dimensional material constituted by wood piled cylinders. The diameters are equal to 13, 18 and 28 mm. There are around 350 particles for a given experiment simulated by a discrete numerical method. The large dimensions of the grains allow an automatic image acquisition and data processing. From this photographic technique it is possible to extract incremental displacement and rotation fields (or the corresponding velocity fields) for the whole granular assembly.

The experiments have been performed in a plane shear box (Joer H. et al. , 1992) which allows to apply loading programs at the boundary of the slope. In this paper two cases are presented in a more detailed manner, even if the conclusions are also valid for the other cases not considered here. The first case is constituted by an active failure of the slope by pulling the retaining vertical wall outside of the granular assembly (see figure 12(a)), the second case is a slope "loaded at his top", where avalanches are induced by a vertical loading on the flat top of the slope (see figure 12(b)). The configurations of the grain assemblies, as they are appearing on figure 12, are the real experimental ones. These initial configurations are exactly reproduced numerically in order to have a realistic granular assembly. Indeed the numerical experiments have shown the great influence of the initial configuration on the following deformations and failures. With this procedure (i.e. the simulation of experimental initial configurations), a good agreement has been found between the experimental displacement/rotation fields and the numerical ones (Darve F. et al. , 2003). This question is not discussed here, where it is chosen to focus on the mechanical analysis of the numerical results.   The numerical discrete

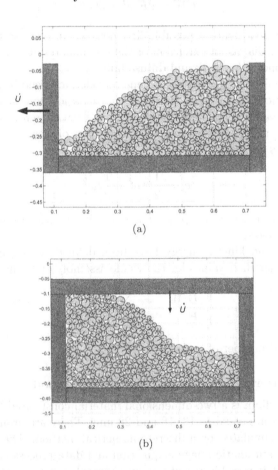

(a)

(b)

Figure 12. Configurations of the slope and loading programmes to induce granular avalanches in a two-dimensional medium.

method which has been utilized is based on the "Contact Dynamics" theory as developed by Moreau J.J. (1994). The numerical code has been achieved by Lanier J. & Jean M.

(1994).

This method is essentially different from the one proposed by Cundall P.A. & Strack O.D.L. (1979), because the numerical algorithm for time integration is implicit and also because the interactions between grains are not described by springs, slides and dashpots. In the Contact Dynamics method, the grains are assumed perfectly rigid. Thus there is a kinematic constraint constituted by the non-penetrability condition (the so-called Signorini condition, see figure 13(a)). The granular interactions are described by a simple coulombian friction, as represented on figure 13(b). If μ is the friction coefficient between grains, a tangential displacement U_t is allowed only if the tangential component R_t of the interaction force is equal to : $\pm \mu R_n$, where R_n is the normal component (figure 13(b)). For these simulations the friction coefficient between grains has been estimated to 0.5317 and between grains and walls to 0.25. Note that both these coefficients are the only mechanical parameters introduced in the computation. The other data have a geometric nature and are constituted by the initial configuration of the grain assembly, as discussed before.

(a)   (b)

Figure 13. Illustrations of the non-penetrability condition 13(a) and of the coulombian friction 13(b), where R_n and R_t are respectively the normal and tangential components of the interaction force and μ the friction coefficient. δ is the "distance" (roughly) between two grains and U_t the tangential inter-granular displacement.

4.3 Grain avalanches by active failure of the slope

In this case the number of grains is equal to : $N = 335$, the displacement speed (see figure 12(a)) to 10^{-2} m/s (the physical time has however not any influence on the computation). There are 1080 increments of computation (the time increment is : 5.610^{-3}s).

Figure 14 presents a comparison between the measured displacement fields (on the left) and the computed values (on the right) for 4 values of the plate displacement. A first conclusion is obvious : the utilized discrete element method is able to capture the main characteristics of the successive grain avalanches thanks to the taking into account of the initial configuration of the grain assembly in the experiments. A second aspect is the fact that various failure modes are visible on figure 14. Some modes are deep and seem to be localized because of parallel displacement vectors (as for rigid body motions), while others are surficial and correspond to classical grain avalanches. These different failure modes (local or more global) will be characterized later on by bursts of kinetic

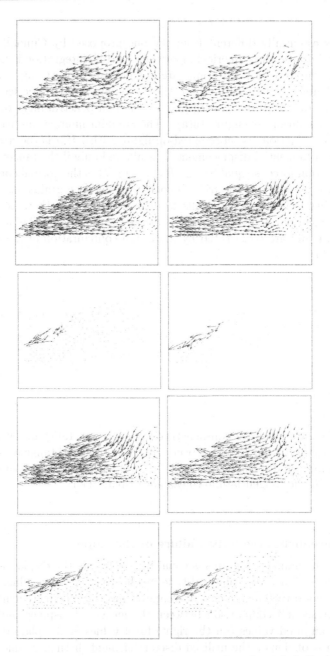

Figure 14. Experimental (on the left) and numerical (on the right) displacements fields for various plate displacements (see figure 12(a))

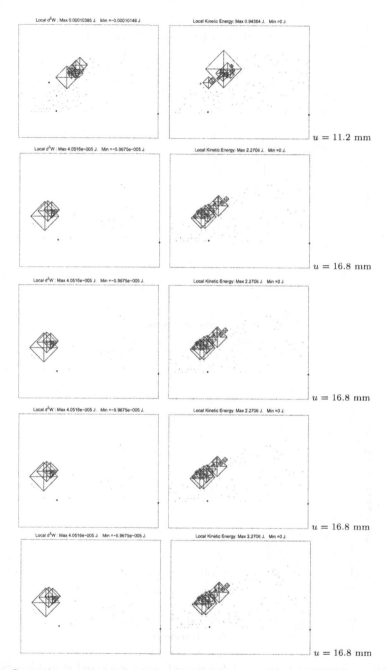

$u = 11.2$ mm

$u = 16.8$ mm

$u = 16.8$ mm

$u = 16.8$ mm

$u = 16.8$ mm

Figure 15. Spatial comparison between negative values of second order work d^2W (on the left) and the values of kinetic energy e_c (on the right).

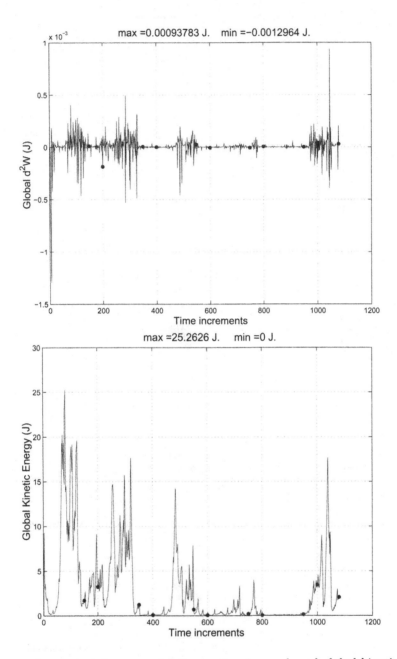

Figure 16. Comparison between global second order work and global kinetic energy as functions of plate displacement (equivalently number of time increments).

energy of various amplitudes.

Figure 15 shows a spatial comparison between the negative values of discrete second order work and the kinetic energy of the grains. The dimensions of the square symbols are proportional to the intensity of the scalar variables d^2W and e_c. The main failure mode is surficial with erratic fields of kinetic energy confirming the essential features of grain avalanches. However the most important point to be noticed is the spatial correlation between both the area where grains have negative second order work and where grains have bursts of kinetic energy. It is worthy to note also that the correlation is not attached to grains (a grain with or without negative second order work may or not move) but to area which can be characterized as "unstable".

This correlation is also a temporal one. Let us consider now the global quantities D^2W and E_c determined by adding the local values of each grain. Figure 16 presents such a comparison. Each point on both graphs corresponds to one computation, which means that there are exactly 1080 points.

All the values of D^2W have been plotted : the positive as the negative ones. Some conclusions are obvious from these results like the fact that the global second order work is varying suddenly from positive values to negative ones. Moreover, large positive values are associated to strong negative ones (the diagram is more or less symmetric with respect to the axis $D^2W = 0$). Indeed, the second order work is linked to energy, which can not be extracted in a finite manner from the medium. As regards the global kinetic energy variations, figure 16 shows successive 5 large failures for numbers of increments equal to 0, 90, 310, 490 and 1040. Smaller events (i.e. more surficial grain avalanches) occur for increment numbers of 200, 520, 710 and 760. Now let us go to the global second order work variations on same figure. We find 5 large minima for the negative values of D^2W and also 4 smaller relative minima for respectively the same 9 increment numbers. For example the small grain avalanches at increment 710 and 760 are clearly noticeable on D^2W graph by relative minima. This strong correlation is the manifestation of the deep link between failure and second order work.

Another example is presented in the next sub section.

4.4 Grain avalanches by loading at the top

Here the number of grains is equal to : $N = 377$. The vertical displacement rate of the plate is : 10^{-3} m/s (see figure 12(b)). The loading programme is still divided into 1000 increments. The friction coefficients are the same as before.

Figure 17 shows the incremental displacements fields for 8 values of the vertical displacement of the plate. Two different failure modes appear here. For plate displacements equal to 2.1, 25.2, 51.1 mm it seems that there is a deep localized failure mechanism exhibiting a rotation of a rather solid body. On the contrary, for plate displacements equal to 5.6, 23.1, 31.5, 35 mm the failure mechanism is rather a grain avalanche with a chaotic displacement field for the involved grains. From this rough analysis it appears difficult to conclude for the last case (plate displacement equal to 57.4 mm).

On figure 18 the spatial comparison between the area where the local second order work takes negative values and the one where the local kinetic energy is finite is presented. As before (see figure 15), a correlation between both areas seems to exist.

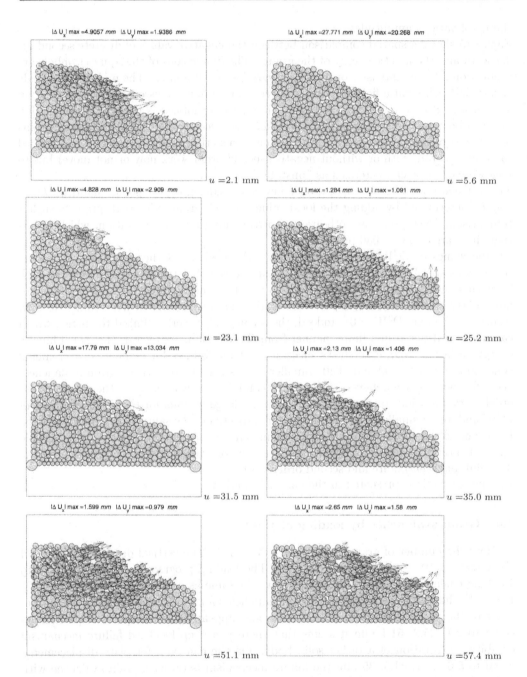

Figure 17. Field of incremental displacements at different values of vertical plate displacement u.

Figure 19 shows the temporal correlation between the bursts of global kinetic energy and the minima of negative values for the global second order work. From the kinetic energy variations there are 5 visible events for plate displacements of about 7, 14, 32, 46 and 65 mm. 5 events are also clearly distinguishable on the second order work diagram for 5, 14, 31, 46 and 65 mm.

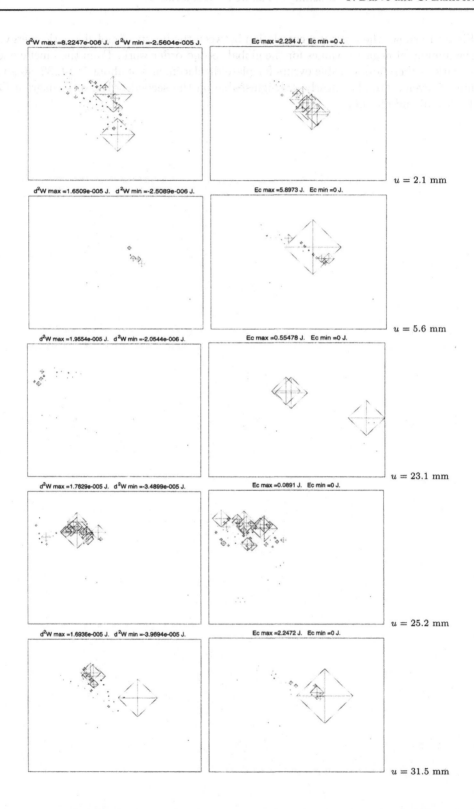

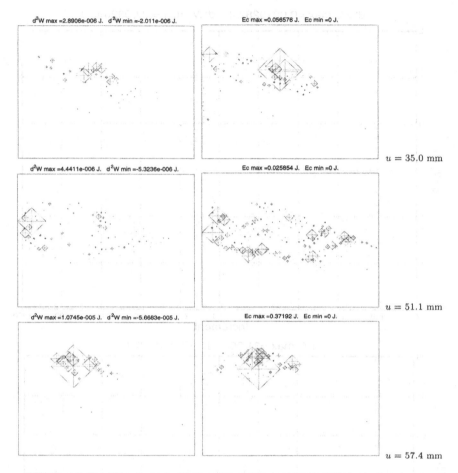

Figure 18. Spatial comparison between the negative values of second order work d^2W (on the left) and the values of kinetic energy e_c (on the right).

5   Conclusion

From an experimental point of view first, it has been shown that the question of failure in geomaterials has to be revisited. The peak of q in undrained conditions for a loose sand does satisfy neither the Mohr-Coulomb's plastic limit condition nor a strain localization criterion, but the failure is predicted by Hill's condition. Secondly this kind of failure states (located strictly inside the plastic limit surface and a localization condition) can be expected from the theory of non-associated hardening elasto-plasticity. Thirdly the numerical analysis carried out with an incrementally piece-wise linear and an incrementally non-linear constitutive relation have shown the existence of a whole stress domain where Hill's condition of stability is lost. In this domain, failure is characterized by bifurcation phenomena, losses of uniqueness and instabilities in Lyapunov's sense. Various modes

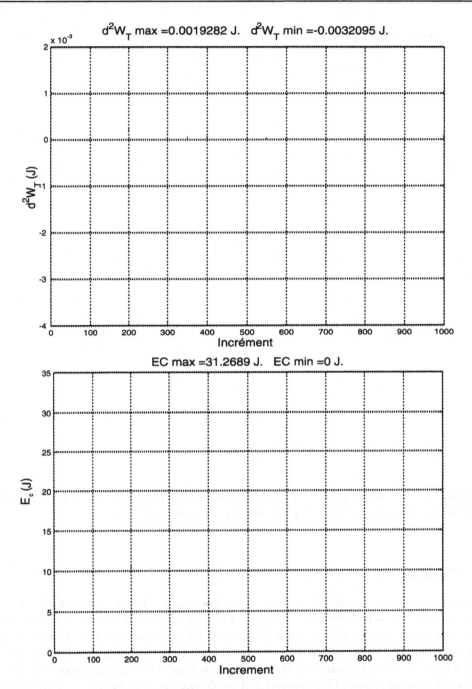

Figure 19. Comparison between global second order work and global kinetic energy as functions of the vertical plate displacement(or equivalently the number of time increments).

of failure can develop (localized, diffuse, due to geometric instabilities...). It seems that diffuse modes of failure can be analyzed by Hill's condition of stability.

The equation of the boundaries of the failure domain has been given and discussed, as the equations of the cones of unstable stress directions. Diffuse modes of failure can appear only if the stress state lies inside the failure domain, if the loading direction is included inside the cone of unstable directions and if proper loading modes (stress controlled) are considered. For boundary value problems (cases not considered in this paper), the global second order work (i.e. integrated on the whole body) and the symmetric part of the matrix of rigidity have to be analyzed.

The specific case of q-constant loading paths has been considered. It has been shown that the unexplained diffuse modes, exhibited in triaxial experiments on a loose sans along these paths, can be predicted by Hill's condition of stability.

In a second part, the phenomenon (of discrete nature) of grain avalanches has been tackled. The grain motions have been characterized by local and global kinetic energies. Surprisingly strong correlations between the bursts of kinetic energy and the negative values of second order work have been derived.

As a general conclusion, it appears that the sign of second order work (under both its continuous and its discrete forms) could constitute a proper criterion to detect certain modes of failure, which appear before the localized failure but for specific loading conditions given in the paper in the framework of continuum mechanics.

The application to natural hazards driven by gravity (landslides, debris flows, rockfalls, snow avalanches) is of great interest since some of them can not be explained in a classical plasticity framework.

Acknowledgements

The supports of European projects DIGA (5th PCRD), LESSLOSS (6th PCRD) and french national projects PIR (RGCU) and ACI CatNat (2002) are gratefully acknowledged.

Figure 1 Example of an unstable state related to a diffuse mode of failure strictly inside
Mohr-Coulomb plastic limit condition : typical undrained triaxial behavior of a
loose sand.

Figure 2 Proportional strain paths simulated by the incrementally non-linear model in
the case of Hostun dense sand. $(\sigma_1 - \frac{\sigma_3}{R})$ possesses a maximum for low R values,
$R \in \{0.3, 0.35, 0.4, 0.45, 0.5, 0.6, 0.7, 0.8, 0.9, 1.0\}$.

Figure 3 Cones of unstable stress directions and boundary of the unstable domain in
Rendulic plane for dense Hostun sand in the cases of the octo-linear model (on the
left) and of the non-linear one (on the right).

Figure 4 Cones of unstable stress directions and boundary of the unstable domain in
Rendulic plane for loose Hostun sand in the cases of the octo-linear model (on the
left) and of the non-linear one (on the right).

Figure 5 Domains of instability in Rendulic Plane and axisymmetric conditions for the
loose sand on the left, for the dense sand on the right and for 2 different constitutive
relations (symbols "o" and "+" for the incrementally non-linear and the octolinear
models respectively).

Figure 6 Lowest unstable stress levels in plane strain conditions for the loose and dense
sand, for 2 different constitutive relations.

Figure 7 Domain of instability in deviatoric plane for the loose and dense sand, and for
2 different constitutive relations.

Figure 8 Simulation of q-constant loading paths for loose Hostun sand by the octo-
linear constitutive model. Points are replaced by other symbols when the second
order work takes negative values from the maximum of volume variations. The
initial isotropic pressure is equal to 100 kPa.

Figure 9 Simulation of q-constant loading paths for loose Hostun sand by the octo-
linear constitutive model. Points are replaced by other symbols when the second
order work takes negative values from the maximum of volume variations. The
initial isotropic pressure is equal to 500 kPa.

Figure 10 Simulation of q-constant loading paths for dense Hostun sand by the octo-
linear constitutive model. The second order work is never vanishing.

Figure 11 Simulation of q-constant loading paths for loose Hostun sand by the octo-
linear and the non-linear constitutive model. Points are placed by other symbols
when the second order work takes negative values from the maximum of volume
variations. The initial isotropic pressure is equal to 100 kPa.

Figure 12 Configurations of the slope and loading programmes to induce granular
avalanches in a bi-dimensional medium.

Figure 13 Illustrations of the non-penetrability condition 13(a) and of the coulombian
friction 13(b), where R_n and R_t are respectively the normal and tangential components of the interaction force and μ the friction coefficient. δ is the "distance"
(roughly) between two grains and U_t the tangential inter-granular displacement.

Figure 14 Field of numerical incremental displacements at different values of rotation
angle α.

Figure 15 Spatial comparison between negative values of second order work d^2W (on
the left) and the values of kinetic energy e_c (on the right).

Figure 16 Comparison between global second order work and global kinetic energy as functions of rotation angle α.

Figure 17 Field of incremental displacements at different values of vertical plate displacement u.

Figure 18 Spatial comparison between the negative values of second order work d^2W (on the left) and the values of kinetic energy e_c (on the right).

Figure 19 Comparison between global second order work and global kinetic energy as functions of the vertical plate displacement.

Bibliography

Darve F., Labanieh S., 1982 "Incremental law for sands and clays. Simulations of monotonic and cyclic tests", Int. J. Num. Anal. Meth. in Geomech., 6 : 243-275

Hill R., 1967 "Eigenmodal deformations in elastic-plastic continua", J. Mech. Phys.
Solids, 15 : 371-386

Rice J.R., 1976 "The localization of plastic deformation", Proceedings IUTAM Congress,
1 : 207-220

Vardoulakis I., Sulem J., 1995 "Bifurcation analysis in Geomechanics", Chapman & Hall
publisher

Darve F., 1984 "An incrementally non-linear constitutive law of second order and its application to localization", in Mechanics of Engineering Materials, Desai and Gallagher
eds, 179-196

Bigoni D., Hueckel T., 1991 "Uniqueness and Localization- I. Associative and Nonasso-
ciative Elastoplasticity",Int. J. Solids Structures, 28(2) : 197-213

Desrues J., 1990 "Shear band initiation in granular materials : experimentation and
theory", in Geomaterials Constitutive Equations and Modelling, Darve ed., Elsevier,
Taylor and Francis Books, 283-310

Darve F., Roguiez, X., 1998 "Homogeneous bifurcation in soils", in Localization and
Bifurcation Theory for Soils and Rocks, Adachi et al. eds, Balkema publ., 43-50

Nova R., 1994 "Controllability of the incremental response of soils specimens subjected
to arbitrary loading programmes", J. Mech. of Behav. Mater., 5(2) : 193-201

Hill R., 1958 "A general theory of uniqueness and stability in elastic-plastic solids", J.
of the Mech. and Phys. of Solids, 6 : 239-249

Lyapunov A.M., 1907 "Problème général de la stabilité des mouvements", Annales de la
faculté des sciences de Toulouse,9 : ,203-274

Chu J., Leong W.K., 2003 "Recent progress in experimental studies on instability of granular soil", Int. Workshop on Bifurcations and Instabilities in Geomechanics, Labuz et
al. eds, Zwets and Zeitlinger publ., 175-192

Mandel J., 1966 "Conditions de stabilité et postulat de Drucker", Rheology and Soil
Mechanics, J. Kravtchenko and P. M. Sirieys eds, Springer, Berlin, 58-68

Hermann H.J., Hovi J.P., Luding S., 1998 "Physics of Dry Granular Media", Kluwer
Academic publ.

Bak P., 1996 "How nature works", Springer Verlag publ.

Drucker D.C., 1959 "A definition of stable inelastic material", J. Applied Mech., 26 :
101-186

Koïter W.T., 1969 "On the thermodynamic background of elastic stability theory",in
Problems of Hydrodynamics and continuum Mechanics, SIAM, Philadelphia, 423-433

Chambon R., Desrues J., Hammad W., Charlier R., 1994 "CLoE, a new rate-type constitutive model for geomaterials. Theoretical basis and implementation", Int. J. Numer.
Anal. Meth. Geomech., 18 : 253-278

Chambon R., Caillerie D., 1999 "Existence and uniqueness theorems for boundary value
problems involving incrementally non-linear models", Int. J. Solids Structures, 5089-
5099

Darve F., Chau B., 1987 "Constitutive instabilities in incrementally non-linear modelling", in Constitutive Laws for Engineering Materials, ed. C.S. Desai, Elsevier publ., 301-310

Darve F., Laouafa F., 2000 "Instabilities in Granular Materials and Application to Landslides", Mech. Cohes. Frict. Mater., 5(8) : 627-652

Darve F., Laoufa F., 2001 "Modelling of granular avalanches as material instabilities", in Bifurcation and Localization in Geomechanics, Muehlhaus et al. eds, Zwets and Zeitlinger publ., 29-36

Laouafa F., Darve F., 2001 "Modelling of slope failure by a material instability mechanism", Computers and Geotechn., 29 : 301-325

Darve F., Flavigny E., Méghachou M., 1995 "Yield surfaces and principle of superposition revisited by incrementally non-linear constitutive relations", Int. J. Plasticity, 11(8) : 927-948

Joer H., Lanier J., Desrues J., Flavigny E., 1992 "$1\gamma2\varepsilon$: A new shear apparatus to study the behavior of granular media", Geotechn. Testing J., 15(2) : 129-137

Darve F., Laouafa F., Servant G., 2003 "Discrete instabilities in granular materials", Italian Geotechn. J., 37(3) : 57-67

Moreau J.J., 1994 "Some numerical methods in multibody dynamics, application to granular materials", Eur. J. Mech A., 13(4) : 93-114

Lanier J., Jean M., 1994 "Experiments and numerical simulations with 2D-disks assembly",Powder Technology, 109 : 206-221

Cundall P.A., Strack O.D.L., 1979 " A discrete numerical model for granular assemblies", Geotechnique, 29(1) : 47-65

Micromechanical approach to slope stability analysis

Francesco Calvetti and Roberto Nova

Department of Structural Engineering, Politecnico (University of Technology) di Milano
Piazza Leonardo da Vinci, 32 - 20133 Milano, Italy

Abstract This paper deals with the micromechanical approach to the study of unstable slopes, by means of Distinct Element simulations.

In the first part general issues related to this kind of approach are discussed, by highlighting advantages and difficulties with respect to other (more common) methods.

A particular attention is devoted to define procedures which are necessary for calibrating parameters. This point is addressed by introducing a micro–macro correlation among the contact parameters and the friction angle of the assembly. This relationship is "empirically" obtained by inspection of the results of a series of simulations of a small scale retaining wall.

Then, the obtained correlation is verified by studying a reference small scale model of an infinite slope. The possibility of introducing the effects of seepage, although in a simplified manner, is discussed. The stability of the numerical slope is analyzed for different values of (simulated) water-table level, and the obtained results are compared with those from classical limit analysis.

In the final part of the paper the previously acquired expertise is applied to the study of a representative real–scale landslide. Two main points are investigated: evaluation of the factor of safety, and influence of seepage conditions.

1 Introduction

The Distinct Element Method (DEM) is a numerical tool devoted to the modelling of assemblies of particles, and it has been used since late seventies to study the (micro)mechanical behavior of granular materials, mainly in the field of soil mechanics. Since the pioneering work by Cundall and Strack (1979) DEM based studies provided interesting, but only qualitative, highlights on the basic mechanisms that are responsible of the complex behavior of granular materials. The amount of information obtained from DEM (and from other similar approaches) is not easily transferrable into a constitutive model for the equivalent continuum.

On the other hand, DEM models composed by a relatively small number of idealized particles (in general spheres, or disks in 2D) are sufficient to qualitatively and quantitatively reproduce the behavior of granular soils in element tests (Kruyt, 1993; Calvetti et al., 2003b). Starting from this consideration, and taking advantage of the continuous increase of computing power, it is possible to envisage a different use of numerical simulations: that of analyzing geotechnical and rock engineering problems by modelling the material as a large assembly of distinct elements. Some particular phenomena have already been successfully studied following this approach: rockfall (Calvetti, 1998), debris flows (Calvetti et al., 2000) and granular flows in silos (Masson and Martinez, 2000), landslide interaction with a buried pipeline (Calvetti et al., 2003a).

These problems are particularly suited for a DEM simulation, because they are characterized by finite displacements, large rearrangements of the soil fabric, and dynamic effects. All these features are embedded in the DEM, and make the discrete approach attractive with respect to classical FEM analysis.

The main advantages of the discrete approach can be summarized as follows:

1. the need for a constitutive model for the equivalent continuum is bypassed, not to mention the computational difficulties related to some particular features of the mechanical behavior of granular soils (softening, non-associativeness of the flow rule);
2. in principle, the same DEM model can be used to study in a comprehensive way the problem, including triggering, propagation, run-out, interaction with sheltering structures.

As a counterpart, a DEM model requires the evaluation of micromechanical parameters (namely interparticle friction and particle stiffness). This point is critical: in fact, too many differences exist between a real soil and the assembly of numerical particles employed in DEM simulations, which makes not realistic the direct use of the stiffness and surface friction of actual soil particles (in most cases these measurements are not available, or impossible to make). The mentioned differences are mainly due to the shape of the particles and to the relatively small number of distinct elements that can be used in order to have sufficiently rapid calculations. As a consequence, in most cases the micromechanical parameters must be obtained by back–analysis (Calvetti, 2003b).

Besides these general considerations, the use of DEM models to solve practice oriented (*i.e.*, not purely academical) slope stability problems has to face the implementation of the grain-fluid interaction. At least, the seepage effects in a saturated material must be accounted for: a simplified modelling approach will be presented in this paper (see Section 3.2). The reproduction of partially saturated materials represents a further degree of refinement (Gili and Alonso (2002) and Liu and Sun (2002)), not to mention the difficulties related to the modelling of some of the mechanisms which are invoked by several authors to explain the high mobility of rock avalanches and debris flows, such as hovercraft or air layer effects, rock melting or water vaporization due to frictional heating.

1.1   On the need for micro-macro relationships

In order to perform a quantitative DEM slope stability analysis, the prerequisite is that the "numerical soil" is representative of the in-situ material not only qualitatively (which is easy to obtain with DEM models) but even quantitatively. If the numerical simulations aim at the evaluation of the factor of safety, this apparently trivial consideration implies that at least the strength (*i.e.*, friction angle for granular materials) is properly reproduced. It is well known in soil mechanics that the friction angle is determined by a series of factors, in part related to the geometrical properties of the grains and of the assembly (relative density, grain-size curve, shape and angularity of grains), in part depending on the stress conditions at failure (confining pressure, intermediate principal stress).

Several strategies are suitable to calibrate the interparticle parameters so that the strength of the reference material is properly matched, as widely discussed by various authors (see, *e.g.*, Calvetti (2003b)); some possible approaches are here in the following briefly mentioned:

1. element tests: for a given numerical material, characterized by its porosity and grain-size curve, the influence of the numerical parameters which determine strength can be in principle numerically investigated by performing DEM simulations of (true)triaxial tests in which different sets of contact parameters (interparticle friction and stiffness of grains) are considered, and the corresponding continuum equivalent parameters (in particular friction angles, but also "elastic" modulus) are evaluated.

2. in-situ tests: alternatively, the friction angle can be evaluated indirectly (provided that the previously discussed method can be labelled as "direct") by performing DEM simulations of classical in-situ tests: for instance, Monti (2003) studied numerically the influence of contact parameters on the stiffness and strength of a numerical soil under load-plate.

3. engineering problems: in this case, simulations of selected problems can be used to assess "empirical" correlations between micromechanical and constitutive parameters.

2 Assessment of a micro-macro relationship: the retaining wall problem

In this paper, the last method is adopted: micro-macro relationships are obtained from the simulations of a small scale model of earth retaining structure. The simulation of a reference slope stability problem (infinite slope) is used as verification of the obtained correlation (see Section 3.1).

Numerical simulations were performed by means of the DEM code PFC-2D, which numerically models the behavior of an assembly of rigid disks with compliant contacts (Itasca, 1995). The following points apply to the simulations that will be presented in the following:

- no numerical damping is employed;
- simulations are performed either with free or inhibited rotations: this point will be thoroughly discussed in the following;
- the influence of contact stiffness on the strength of the numerical material is negligible, as it was observed during a preliminary series of simulations.

The retaining wall is represented by a smooth plate element ($H = 20$ cm, see Figure 1), that can translate horizontally. During the simulations, the displacement of the wall, s, is controlled, and the supporting reaction, S, is recorded. By analogy with an equivalent continuum, the wall reaction can be written as:

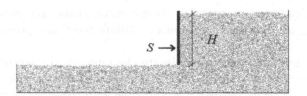

Figure 1. 2D model of retaining wall

$$S = \frac{1}{2}\gamma_{dry}H^2K \qquad (2.1)$$

From the numerically determined wall support reaction, the corresponding earth pressure coefficient (that can be interpreted as a normalized wall reaction) can be easily computed as:

$$K = \frac{2S}{\gamma_{dry}H^2} \qquad (2.2)$$

The backfill is composed by an assembly of about 4000 disks, according to a given grain size curve (average diameter, $D_{ave} = 1$ cm, diameter ratio, $d = D_{max}/D_{min} = 2.5$). In a first phase of the simulations, the wall is kept fixed and gravity force is applied, until an equilibrium configuration is reached. This state is attained for a porosity $n \simeq 0.23$, with small influence of contact friction; the corresponding unit weight is $\gamma_{dry} = 20$ kN/m^3. It is worth noting that the 2D nature of the numerical model makes impossible a direct comparison with porosity and unit weight of a real granular soil. For the same reason, the diameter ratio, d, is used instead of the analogous uniformity coefficient (D_{60}/D_{10}) of the grain size curve.

2.1 Active earth pressure

If the wall is progressively horizontally moved away from the supported soil, a decrease of the supporting reaction is observed, until an active condition is met. In Figure 2 the earth pressure coefficient (Equation 2.2) is plotted as a function of the normalized wall displacement ($\delta = s/H$) for different values of the interparticle friction angle, Φ_μ.

The plotted data refer to the inhibited rotations condition, but qualitatively similar trends are observed with free rotations. A comparison with the passive pressure conditions is shown in Section 2.2.

The numerically determined K_A values are presented in Table 1, where Φ_μ is the interparticle friction angle, and $\mu = \tan \Phi_\mu$ is the interparticle friction coefficient.

As a first comment, for a given interparticle friction angle K_A is smaller with inhibited rotations, and the difference with the corresponding "free rotation" values becomes significant for large Φ_μ. From the obtained K_A values it is immediate to calculate the corresponding friction angle of the equivalent continuum, according to the usual K_A definition:

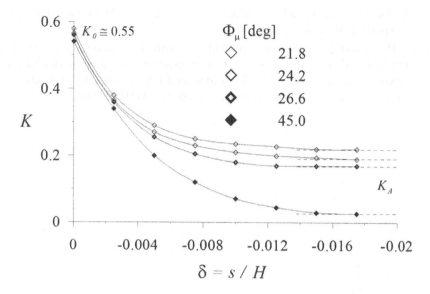

Figure 2. K coefficient as a function of normalized wall displacement (inhibited rotations)

Table 1. numerically determined K_A values

μ	Φ_μ	K_A, free rotations	K_A, fixed rotations
0.1	5.7	0.65	0.54
0.25	14.0	0.58	-
0.4	21.8	-	0.22
0.45	24.2	-	0.19
0.5	26.6	0.475	0.17
0.75	36.9	0.425	-
1.0	45.0	0.4	0.025

$$K_A = \frac{1 - \sin \Phi}{1 + \sin \Phi} \Rightarrow \Phi = \sin^{-1}(\frac{1 - K_A}{1 + K_A}) \tag{2.3}$$

In Figure 3 the friction angle of the numerical material as computed according to Equation 2.3, is plotted as a function of the interparticle friction, for the two aforementioned conditions regarding particle rotations. As a synthetic comment the following points are underlined:

1. with free rotations, the friction angle keeps well under 30°, irrespectively of the interparticle friction;

2. with inhibited rotations, the correlation between Φ_μ and Φ is linear (see Equation 2.4);

3. the influence of particle rotations is particularly evident for large values of the interparticle friction;

4. on the contrary, both trends indicate a common, non zero, friction angle for Φ_μ that tends to zero. In this case, shear resistance is completely due to interlocking; moreover, the behavior of such an idealized material is non dissipative: a specific research is currently devoted to this topics (Calvetti, 2003a).

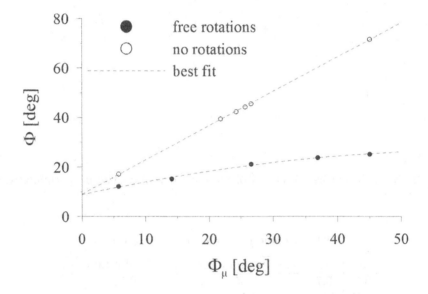

Figure 3. Friction angle as a function of interparticle friction angle

The influence of particle shape on the shear resistance of granular materials is commonly acknowledged in soil mechanics. The use of assemblies of spheres (disks) emphasizes these effects, as it was first observed by Skinner (1969), who performed tests on lead beads; in the same line are the results presented by Rothenburg and Bathurst (1992), who performed DEM simulations with disks and elliptical particles. Assemblies of spherical particles are characterized by very low friction angles ($\Phi < 30°$), irrespectively of the value of the intergranular friction angle. As a consequence, models composed of round particles are not suitable for quantitative oriented calculations in geotechnical engineering, where friction angles larger than 30° are common. The constraint imposed on particle rotations should be intended as an artificial way to compensate the shape effects. Note that preventing particle rotation does not affect the capability of the numerical model to reproduce the experimental behavior of granular materials in element tests, provided that the interparticle friction angle is properly calibrated, as it was shown by Calvetti et al. (2002) and Calvetti et al. (2003b). Other strategies are proposed by various authors, who suggest the use of non-circular elements (polygons or ellipses, in 2D; agglomerates of spheres or polyhedra in 3D), or the modification of the interparticle force-displacement relationship, which can be improved by allowing contacts to transfer couples, *i.e.*, by introducing the so called "rolling resistance" (see, *e.g.*, Iwashita and Oda (1998) with reference to shear band formation, and Ferellec et al. (2001) with application to silo flow). These approaches make the model more similar to the real material, but additional parameters and computational drawbacks are introduced. Note also that the fixed rotations constraint is equivalent to the introduction of an infinite rolling resistance at contacts.

With inhibited rotations, the following correlation is used to interpolate the results:

$$\Phi = \Phi(\Phi_\mu) = a \cdot \Phi_\mu + b \qquad (2.4)$$

Equation 2.4 is in principle valid for any DEM model with inhibited rotations, but it is likely that the parameters a and b depend on grain size curve (d and D_{ave}) and porosity of the material. For the employed assembly, the values:

$$a = 1.38 \qquad b = 10° \qquad (2.5)$$

provide the best fit with the numerically computed friction angles.

In order to investigate this point thoroughly a specific research effort is further required. At the moment, a first verification is given by a series of analogous simulations performed with a numerical specimen characterized by the same porosity, n, and diameter ratio, d, of the reference one, but with 10 times larger D_{ave}: the results of these simulations perfectly agree with the linear $\Phi_\mu - \Phi$ correlation, and the same values of a and b are obtained. This result shows that the friction angle of this idealized granular material does not depend on the size of the grains, the opposite of what is commonly retained. In fact, it is observed in laboratory tests that friction angle decreases with grain size, *i.e.*, passing from gravel to silt. This discrepancy however could be due to the fact that in a real granular material the variation in size is accompanied by the change of other "hidden" parameters, for instance grain shape. These issues deserve further investigation, in particular as far as the influence of the coefficient of uniformity of the grain-size curve (here represented by the diameter ratio) is concerned.

As a first confirmation of the obtained results, the active thrust simulations were continued until the wall was completely removed: in this case a slope is obtained, with an angle β which is close to Φ for all the investigated values of Φ_μ. On the contrary, a fluid-like behavior is observed in the case of free rotations (see Figure 4).

2.2   Passive earth pressure

Passive thrust state is attained by pushing the wall towards the supported soil: starting from the K_0 condition, a progressive increase of the earth pressure is observed, until a passive condition is met. As a representative example, the simulation with $\Phi_\mu = 26°$ and free rotations is presented in Figure 5. Analogous results are obtained for all the remaining investigated conditions.

From a qualitative point of view, it is observed that passive conditions are attained for a normalized wall displacement far larger than the displacement necessary to reach the active limit state ($\delta = 15\%$ and 1.5 %, respectively). This result is in agreement with experimental evidence, though the displacement at active conditions is overestimated.

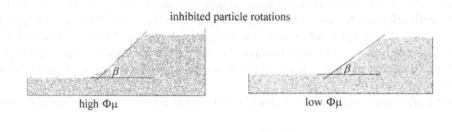

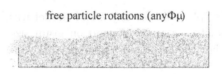

Figure 4. Slope configurations. Top: inhibited rotations (two values of Φ_μ), bottom: free rotations

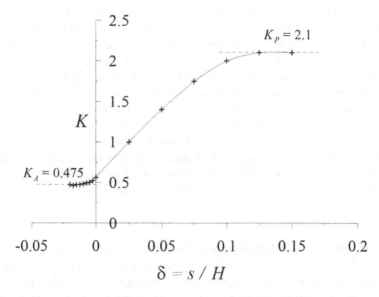

Figure 5. K coefficient as a function of wall displacement

From a quantitative point of view, the numerically determined values of the active and passive earth pressure coefficients respect the relationship $K_A = 1/K_P$, as expected for the equivalent continuum. This result is intended as a confirmation of the $\Phi_\mu - \Phi$ correlation.

3 Application to slope stability analysis

In the following, the micro-macro relationship that was obtained in the previous simulations is transferred to the study of slope stability. In a first phase (Section 3.1), the numerical model is employed to reproduce a simple problem (small scale infinite slope) for which the analytical solution is well known. Then, an original technique is proposed for modelling the effects of seepage (Section 3.2).

3.1 Infinite slope

As a first verification of the $\Phi_\mu - \Phi$ correlation (Equations 2.4 and 2.5) a simple infinite slope problem is studied. The employed model is shown in Figure 6. In order that a verification is possible the assembly is characterized by the same grain-size curve and porosity of the specimen employed for the retaining wall simulations. Also note that particle rotations are inhibited. The thickness of the slope is far smaller than its longitudinal extension, in order to avoid side effects. The bedrock is represented by a linear wall element, with given inclination, α. The simulation procedure is described in the following:

1. disks are randomly generated to form the slope, and gravity force is applied. The elements in contact with the wall are glued to it, in order to minimize boundary effects;
2. the interparticle friction angle is progressively decreased until failure is reached. The corresponding value, Φ_μ^F, is recorded;
3. the described steps are repeated for different α values.

As expected, the numerical results indicate that Φ_μ^F increases with α. More important, the corresponding value of the friction angle (Equations 2.4 and 2.5) is such that: $\alpha \simeq \Phi(\Phi_\mu^F)$ for all the investigated inclinations. This result is a further verification of the $\Phi_\mu - \Phi$ correlation that was obtained with reference to the earth retaining wall problem. A more detailed discussion about this point will be given in Section 3.3.

3.2 Seepage effects

In its original formulation DEM is devoted to the modelling of dry assemblies of particles. A relevant research effort is currently devoted to implement in the Distinct Element Method the coupling with water. However, several issues related with this task are of relevant difficulty, because of the different nature (a discrete structure of grains on one side, and a continuum fluid on the other) of the involved species. A recent tendency is that of using two numerical codes in parallel, in order to uncouple the problem: the first one (based on a Finite Element or Finite Differences discretization) is used to solve seepage through a fictitious porous material; the second code (typically a DEM based

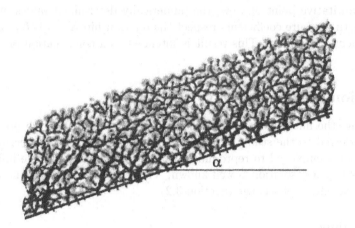

Figure 6. Small scale model of an infinite slope (representative slice).

one) is used to model the behavior of the granular material, taking into account the seepage forces previously calculated. It is worth noting that an averaging procedure is necessary in order to obtain volumetric strains from the displacements of particles. All in all, this modelling strategy is time consuming and to some extent artificial.

In this paper a simplified approach is proposed, in which simple seepage conditions are analyzed, and it is assumed that the grain displacements do not affect the water pressure distribution, *i.e.*, only drained conditions are taken into account. Moreover, it is assumed that the material above the water-table is dry, *i.e.*, capillary effects are neglected.

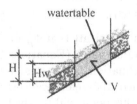

Figure 7. Infinite slope: seepage conditions.

The effects of seepage are simulated by applying to each particle within the seepage region equivalent forces, namely a vertical uplift (F'), and a drag force (J_P) parallel to the seepage direction. With reference to Figure 7, where an infinite slope is considered, the seepage force acting on volume V is:

$$J = jV = \gamma_w iV = \gamma_w \sin\alpha V \qquad (3.1)$$

The seepage-equivalent drag force (J_P) that pertains to each particle, is evaluated as:

$$J_P = jV_P\chi = \gamma_w \sin\alpha V_P\chi \tag{3.2}$$

where V_P is the volume of each individual particle, and

$$\chi = V / \sum V_P = 1/(1-n) \tag{3.3}$$

is introduced in order that $J = \sum J_P$. Note that the summations are extended to all the particles within the reference volume V.

The uplift force is simply computed as:

$$F' = \gamma_w V_P \tag{3.4}$$

With reference to Figure 7, the factor of safety of an infinite slope composed by a purely frictional material is:

$$F_S = \frac{\gamma' H_W + \gamma_{dry}(H - H_W)}{\gamma_{sat} H_W + \gamma_{dry}(H - H_W)} \frac{\tan\Phi}{\tan\alpha} \tag{3.5}$$

For given inclination of the slope the critical water-table level H_{Wc}, which can be easily computed by setting $F_S = 1$ in Equation 3.5, depends on the friction angle Φ.

In order to verify the reliability of the proposed modelling, an infinite slope with $\alpha = 30$ is built, according to Section 3.1. Then, (simulated) water-table is progressively raised until failure occurs (see Figure 8) and the corresponding level is recorded as H_{Wc}. Note that during the simulations each particle is monitored in order to assess whether its center lies above or below the current water-table level: in the latter case, the seepage-equivalent forces (J_P and F') are automatically applied to that particle. The slip line indicated in Figure 8 is traced on the basis of the numerically determined displacements, and should be considered part of the (interpreted) solution provided by the code. The observed pattern (failure line is close to the base of the layer) is in good agreement with the fact that F_S decreases with depth according to Equation 3.5.

The described simulation procedure is repeated for different values of Φ ($\Phi \geq \alpha$), i.e., different values of Φ_μ, according to Equations 2.4 and 2.5.

The numerically determined values of H_{Wc} are close to those analytically calculated (Equation 3.5), as it is shown in Figure 9, though the numerical solution slightly overestimates or underestimates the critical water-table level for low and high friction angles, respectively.

3.3 Preliminary discussion of results

On the basis of the results of the numerical simulations presented in Sections 2 and 3, the following conclusions are drawn:

1. the correlation among the interparticle friction angle and the friction angle demonstrated a solid reliability; originally determined by analyzing active pressure conditions (retaining wall problem), it holds when passive conditions are considered, or when a different problem is analyzed (infinite slope). However, it is necessary to point out that the numerical soils employed in the different problems must share

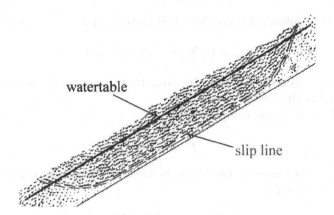

Figure 8. Displacement increment field at failure (infinite slope, $\alpha = 30°$).

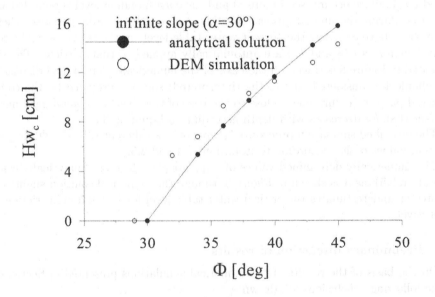

Figure 9. Critical water-table level; comparison between analytical solution and DEM results (infinite slope, $\alpha = 30°$)

the same grading and porosity; on the contrary, it is not necessary for the size of the particles to be the same. This latter point allows for scaling the particles according to the size of the problem, and permits to perform simulations of real scale which are not too much time consuming;

2. the simplified approach to the solid–fluid interaction gives results that are in very good agreement with the analytical solution, in terms of critical water-table level, at least for the simple case of an infinite slope; however, it is necessary to note that with more complex geometries the approach requires the evaluation (through simplifying assumptions) hydraulic gradient (see Section 4.3).

4 A practical application: Bema landslide

As an example of application of the proposed modelling techniques a landslide close to the village of Bema (Sondrio, Italy) is considered (Figure 10). This landslide undergoes periodic reactivation, the most recent dating back to the Summer of 1987, during the same heavy rains that provoked diffuse floods and landslides (among them the Valpola rock-avalanche) in Valtellina (the valley of river Adda). The Bema landslide is of particular concern because of the risk of occlusion of the Bitto torrent, an Adda river tributary, that flows at the bottom of the valley. This event (that do occurred in 1987) represents a threat for the town of Morbegno that lies along Adda river a few kilometers downstream.

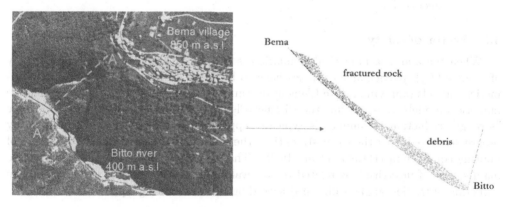

Figure 10. Plan view and schematic section of the Bema landslide.

The slope can be schematically divided into two parts (Figure 10): the superior half is composed of a fractured rock mass, while the bottom is formed of debris, resulting from previous landslides, that lie over an intact bedrock. In view of the numerical simulations (that are limited to study the stability of the debris) a campaign of laboratory geotechnical tests were performed on reconstituted samples of the debris. In particular, the friction angle was evaluated from a series of triaxial and direct shear tests: the retained value was $\Phi = 42°$. The cohesion value was negligible (Bongio, 1999).

4.1 Description of the model

The debris is represented by an assembly of about 4000 particles with porosity $n = 0.23$, and diameter between $D_{min} = 1.2$ m and $D_{max} = 3$ m. Note that both the diameter ratio ($d = D_{max}/D_{min} = 2.5$) and the porosity are equal to those employed in all the previously performed simulations (retaining wall and infinite slope). Also note that the average diameter of particles is far larger than that used in the small scale simulations (Sections 2 and 3), in order to have sufficiently rapid calculations: this does not affect the $\Phi - \Phi_\mu$ correlation, as it was previously discussed. Therefore, Equation 2.4 can be used, with the same set of interpolating parameters of Equation 2.5.

The simulation procedure is as follows:

1. 4000 disks are randomly generated in the area corresponding to the debris of Figure 10, and gravity force is applied;
2. the relevant numerical parameter, Φ_μ, is chosen according to Equations 2.4 and 2.5 in order that the friction angle of the numerical debris is equal to that of the in-situ material ($\Phi = 42° \Rightarrow \Phi_\mu = 22°$, see Figure 3);
3. two failure conditions are considered:
   - dry analysis: interparticle friction is progressively reduced until failure, and factor of safety, F_S, is computed accordingly (see Section 4.2). These simulations are similar to those presented in Section 3.1;
   - interaction with water: water-table level is progressively increased until failure occurs (see Section 4.3). These simulations are similar to those presented in Section 3.2.

4.2 Factor of safety

When the aim of a numerical simulation is that of determining the factor of safety of a slope (F_S), particular techniques have to be used. In fact, neither continuum numerical models (applying Finite Element or Finite Difference discretization) nor discrete numerical models (mostly using the Distinct Element Method) provide directly F_S. The "strength reduction technique" is typically applied in factor of safety calculations by progressively reducing the shear strength of the material to bring the slope to a state of limiting equilibrium (Dawson et al., 1999). This technique implicitly assumes that the main source of incertitude is related to the evaluation of the strength parameters. The reduction of friction angle is obtained according to the following equation:

$$\Phi_R = \arctan\left(\frac{\tan \Phi}{F_{S_{trial}}}\right) \tag{4.1}$$

by progressively increasing the trial factor of safety, $F_{S_{trial}}$. Following this approach, the factor of safety of the slope is simply taken as $F_{S_{trial}}$ at failure, *i.e.*:

$$F_S = \frac{\tan \Phi}{\tan \Phi_F} \tag{4.2}$$

where Φ_F is the value attained by the (reduced) friction angle at failure. In addition, note that F_S is written in the same form of that corresponding to an infinite slope, for which failure occurs with $\Phi_F = \alpha$. This might be confusing and could induce to conclude that such a definition of F_S is oversimplified: in reality, the complexity of the problem is transferred to the evaluation of Φ_F, that pertains to the numerical model.

In Figure 11 the progressive evolution of the debris profile under decreasing friction angle is shown.

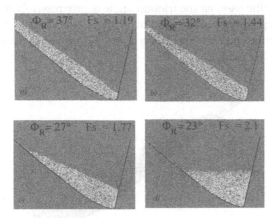

Figure 11. Post–failure configurations for decreasing Φ.

During the reduction of the friction angle, the following situations are attained:

1. $37° < \Phi_R \le 42°$: the slope is stable;
2. $\Phi_R = 37°$: first failure occurs, the debris flow until a new stable configuration is reached (Figure 11a). The corresponding factor of safety, is 1.19 (Equation 4.2);
3. $32° < \Phi_R \le 37°$: the slope is stable;
4. $\Phi_R = 32°$: second failure occurs (Figure 11b). The factor of safety with respect to this failure is 1.44;
5. $27° < \Phi_R \le 32°$: the slope is stable;
6. $\Phi_R = 27°$: third failure occurs (Figure 11c). $F_S = 1.77$;
7. $23° < \Phi_R \le 27°$: the slope is stable;
8. $\Phi_R = 23°$: fourth failure occurs (Figure 11d). $F_S = 2.1$.

As a preliminary comment, it is reported that the porosity of the (stable) debris does not vary significantly with respect to the initial one: this allows to estimate Φ by using Equations 2.4 and 2.5 all over the simulation.

The factor of safety corresponding to the first failure ($F_S = 1.19$) is close to that determined with a classical slice method (Janbu's method, $F_S = 1.21$).

Each failure is followed by the reprofiling of the slope, towards a more stable configuration. The dynamic nature of this phase is particularly evident if Figure 11d is considered: in this case, failure is triggered with $\Phi = 23°$, but the post–failure profile is nearly horizontal.

Remarkably, an unique DEM model is able to follow the evolution of the slope, including: triggering, flow and post–failure configuration. However, the modelling of the flow phase is here intended as qualitative, for the lack of comparison with in-situ data, and because of the possible influence of numerical damping during this phase (see Calvetti et al. (2000)).

It is worth noting that each post–failure configuration is associated to a specific factor of safety (Figure 11), which is of great importance for risk assessment and territorial planning. In this specific case, as the friction angle is progressively reduced, the thickness of the debris in correspondence of the bottom of the valley increases, which means possible occlusion of the regular flow of the Bitto river. However, a quantitative evaluation requires the use of a 3D model.

Figure 12. Post–failure contact force network ($\Phi = 27°$).

As a concluding comment, DEM analysis provide interesting insight at the microme-chanical scale, which can be useful for interpreting the results. As an example, a repre-sentative post–failure distribution of contact forces, which is characterized by a strong non-homogeneity with a concentration of locked-in forces parallel to the slip line at the bottom of the debris, is shown (Figure 12). The influence of the distribution of contact forces (the analogous of the initial state of stress in a continuum numerical analysis) will be discussed in Section 4.3.

4.3 Seepage effects

The simulations presented in this Section are similar to the corresponding ones per-formed on the infinite slope model (Section 3.2). The model of the debris as defined according to Section 4.1 is employed in a series of simulations with increasing water-table level until failure occurs. Note that the evaluation of this critical water-table level is of relevant practical importance, as it can be used to guide monitoring of a landslide by means of piezometric measurements. The simulations are repeated for different values of Φ in order that a comparison with the solution given by limit equilibrium is made (with $\Phi \geq 37°$, see Section 4.2).

With respect to the analogous simulations with the infinite slope model two main differences are highlighted:

1. water-table is assumed to be parallel to the free surface of the debris, and it is vertically raised from the bottom; the seepage forces are computed under these

simplifying assumptions: each point within the seepage area is mapped vertically onto the water-table; the inclination of the water-table in correspondence to the target point is used to compute the hydraulic gradient by analogy with the infinite slope;

2. the effects of the initial distribution of contact forces are analyzed, by considering the two configurations shown in Figure 13a, b. The corresponding numerical analysis will be labelled as "DEM simulation 1" and "DEM simulation 2", respectively.

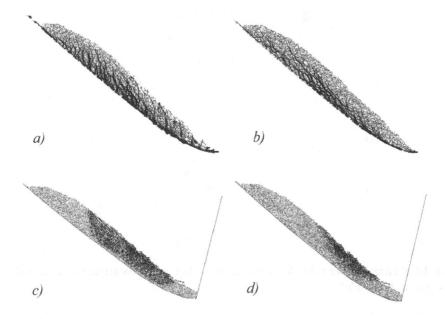

Figure 13. Initial distribution of contact forces (top), with the corresponding failure mechanisms (bottom). ($\Phi = 40°$)

The study of the influence of the contact forces distribution was inspired by the previously described post–failure pattern (Figure 12) which is characterized by non negligible non-uniformity. The contact force distribution of Figure 13a is that corresponding to the preparation procedure described in Section 4.1. Note that in this case the forces are organized in nearly vertical columns. The contact force distribution of Figure 13b, is more similar to the post–failure one, and it was obtained from the reference one by artificially triggering and stopping a failure mechanism. This was obtained by decreasing (for a limited time period) the interparticle friction angle. Note that, with the exception of the micromechanical information provided by the numerical model, no direct information regarding the in-situ stress state is available to guide the choice between the two considered conditions.

Figure 13c, d shows the failure mechanisms that correspond to the two distributions of contact forces. With the reference contact force pattern a deep failure is observed, in the same line of the one that occurs with the infinite slope model (see Figure 8). In the other case, the displacement increment field defines a superficial slip line: a comparison with the corresponding pattern of contact forces indicates that failure is limited to the less confined region. It is worth noting that also the water-table levels at failure are different (see Figure 14). In particular, during DEM simulation 2, failure does not occur as long as the water-table is within the more confined region.

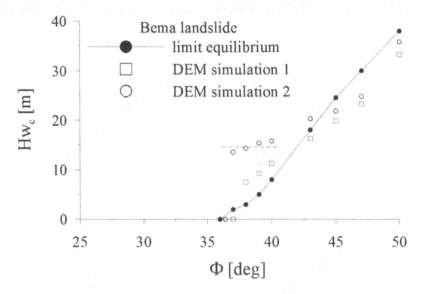

Figure 14. Critical water-table level; comparison between analytical solution and DEM results (Bema landslide)

The comparison between the numerically determined critical water-table levels (for the two mentioned conditions) and the corresponding values obtained from the application of the Janbu's slice method is shown in Figure 14.

The following comments are suggested by the inspection of the results:

1. failure occurs nearly for the same friction angle in dry conditions;
2. the results of DEM simulation 1 are in good agreement with the solution provided by Janbu's method, though the differences are slightly larger than those observed between the DEM results and the analytical solution for the infinite slope (See Figure 9;
3. the results of DEM simulation 2 are affected by the non uniformity of the contact force distribution; within the range $37° \leq \Phi \leq 40°$, H_{Wc} is almost constant, and it corresponds to the thickness of the region where contact forces are larger.

As a general comment, it is worth noting that the presented results cannot be used to conclude that DEM simulation 1 is more reliable than DEM simulation 2. DEM simulations are sensible to aspects that cannot be incorporated in a classical limit equilibrium analysis. It is certainly true that the data provided by the numerical model are affected by the preparation of the specimen, but the way in which results are influenced by the "initial state of stress" is mechanically sound. This capability of the numerical model requires to be "handled with care", but it can turn out to be useful when the (recent) history of an unstable slope is known (*e.g.*, in the case of a possible reactivation of a landslide).

5 Conclusions

A micromechanical way to the study of slope stability is proposed in this paper.

This approach, is particularly suited for problems where finite displacements and large rearrangements of the soil fabric occur, and dynamic effects play a relevant role. Note that in a more traditional FEM analysis, these features require the adoption of particular numerical techniques, and the use of an appropriate constitutive model: these points are completely by–passed using a discrete model.

Another important point is that in principle, the same DEM model can be used to study in a comprehensive way the problem, including triggering, propagation, run-out and interaction with sheltering structures. This latter issue was not addressed in the text, but it can be straightforwardly included in the presented analysis.

The simplified way to account for the solid–fluid interaction gives results that are in very good agreement with the analytical solution, in terms of critical water-table level, at least for the simple case of an infinite slope. However, this approach is completely uncoupled: it refers to drained conditions only, with the additional requirement that the hydraulic gradient has to be estimated with simplifying assumptions.

Despite the satisfactory performances of the numerical model, some points are critical to this kind of approach: micromechanical parameters should be chosen according to a general procedure, if predictive analysis is envisaged. The proposed micro–macro correlation represents a possible solution, but several points (influence of the porosity, and of the grading) need to be further analyzed. Also note that scale effects should be carefully checked, with particular reference to the influence of the confining stress on the friction angle. However, as a preliminary comment, these effects are in general not relevant if contact stiffness is sufficiently high, so that particle overlapping is negligible.

Acknowledgements

This paper is written within the framework of DIGA network (supported by the European Commission under contract n. HPRN-CT2002-00220) and ALERT Geomaterials. The support of Itasca and Ce.A.S. Milano is also gratefully acknowledged.

Bibliography

P. Bongio. *Analisi della frana di Bema*. Politecnico di Milano, 1999. Tesi di Laurea (in Italian).

F. Calvetti. Distinct element evaluation of the rock-fall design load for shelters. *Rivista Italiana di Geotecnica*, 32(3):63–83, 1998.

F. Calvetti. Energy balance in non-dissipative materials. 2003a. in preparation.

F. Calvetti. Limitations and perspectives of the micromechanical modelling of granular materials. *Mathematical and Computer Modelling*, 37(5/6):485–495, 2003b.

F. Calvetti, G.B. Crosta, and M. Tatarella. Numerical simulation of dry granular flows: from the reproduction of small-scale experiments to the prediction of rock avalanches. *Rivista Italiana di Geotecnica*, 34(2):21–38, 2000.

F. Calvetti, C. di Prisco, and R. Nova. Experimental and numerical analysis of soil–pipe interaction. *J. Geotech. Engng., ASCE*, 2003a. accepted.

F. Calvetti, C. Tamagnini, and G. Viggiani. On the incremental behaviour of granular soils. In G.N. Pande and S. Pietruszczak, editors, *NUMOG VIII*, pages 3–9. Balkema, Lisse, 2002.

F. Calvetti, G. Viggiani, and C. Tamagnini. A numerical investigation of the incremental behavior of granular soils. *Rivista Italiana di Geotecnica*, 37(3):11–29, 2003b.

P.A. Cundall and O.D.L. Strack. A discrete numerical model for granular assemblies. *Géotechnique*, 29(1):47–65, 1979.

E.M. Dawson, W.H. Roth, and A. Drescher. Slope stability analysis by strength reduction. *Géotechnique*, 49(6):835–840, 1999.

J.F. Ferellec, J. Martinez, S. Masson, and K. Iwashita. Influence of particle rolling resistance on silo flow in DEM simulations. In Y. Kishino, editor, *Powders & Grains 2001*, pages 409–412. Balkema, Rotterdam, 2001.

J.A. Gili and E.E. Alonso. Microstructural deformation mechanisms of unsaturated granular soils. *Int. J. Num. Anal. Meth. Geomech.*, 26:433–468, 2002.

Itasca. *PFC-2D, User's manual.* Minneapolis, 1995.

K. Iwashita and M. Oda. Rolling resistance at contacts in simulation of shear band development by DEM. *J. Engng. Mech., ASCE*, 124(3):285–292, 1998.

N.P. Kruyt. Towards micro-mechanical constitutive relations for granular materials. In D. Kolymbas, editor, *Modern Approaches to Plasticity*, pages 147–178. Elsevier, 1993.

S.H. Liu and D.A. Sun. Simulating the collapse of unsaturated soil by DEM. *Int. J. Num. Anal. Meth. Geomech.*, 26:633–646, 2002.

S. Masson and J. Martinez. Multiscale simulations of the mechanical behaviour of an ensiled granular material. *Mech. Cohesive-Frictional Materials*, 5:425–442, 2000.

M. Monti. *Analisi numerica delle proprieta' meccaniche di uno strato ammortizzante.* Politecnico di Milano, 2003. Tesi di Laurea (in Italian).

L. Rothenburg and R.J. Bathurst. Micromechanical features of idealised of granular assemblies with planar elliptical particles. *Géotechnique*, 42(1):75–95, 1992.

A.E. Skinner. A note on the influence of interparticle friction on the shearing strength of a random assembly of spherical particles. *Géotechnique*, 19:150–157, 1969.

Ageing and durability of concrete structures

Gilles Pijaudier-Cabot, Khalil Haidar, Ahmed Loukili, and Mirvat Omar

R&DO – GeM, Ecole Centrale de Nantes, France

Abstract: The serviceability of concrete structures is a coupled problem in which fracture and damage are coupled with several environmental attacks. In this chapter, we start with the investigation of chemo-mechanical damage due to calcium leaching. We show that the fracture characteristics, namely the internal length in continuum models evolve with ageing. This observation is confirmed with experiments on model materials. Acoustic emission data and size effect tests data are strongly correlated with the evolution of the microstructure of the material. This chapter concludes on some creep tests designed to investigate the coupled effect between creep and fracture.

9.1.   Introduction

The serviceability concrete structures used for instance in waste containment or electric production facilities is a coupled problem. These concrete structures are subjected to various mechanical loads, soil structure interactions, and at the same time to chemical aggressions. In order to determine the service life of such structures, prediction of the material response and structural failure with time are required. Typically, this kind of problem involves two sources of damage: one due to mechanical loads, and another one due to environmental actions. In the following, we are going to give examples of such coupled mechanisms. The first one deals with leaching of concrete. It is related to the safety of nuclear waste disposal systems. Then, we will discuss results obtained with model materials aimed at describing this type of chemical damage and we will be interested with the variation of the internal length in the course of the ageing process. Finally, recent experiments on creep – fracture interaction will be presented. These tests aim at evaluating the capacity of prestressed concrete structures and its variation during ageing.

9.2.   Leaching of cementitious materials

The long-term behaviour of concrete structures subjected to environmental damage is a topic of the utmost importance for waste containment disposals. In the case of long-life wastes (such as radioactive wastes), the models cannot rely on experimental data over the entire service life of the disposal, which spans over several hundred years. Assessing such structures and comparing several possible designs prior to their construction must rely on robust and accurate computational tools, in which the knowledge of the various degradation mechanisms at the material level must be incorporated. Here, the mechanism of degradation considered is leaching due to water, and particularly calcium leaching as calcium is the most important element in the hydrated cement paste. The consequences of calcium leaching are an increase of porosity, a decrease of the elastic stiffness and a decrease of strength (Carde 1996, Gérard 1996, Carde and Francois 1997). It is this decrease of the mechanical properties that is addressed more specifically. The main reason is that for lifetime assessments, chemical degradation mechanisms must be superimposed to mechanical loads so that cracking may be predicted.

As leaching is a very slow process, accelerated experimental procedures are needed in order to calibrate models, and to validate them subsequently. Goncalves and Rodrigues (1991) and Schneider and Chen (1998, 1999) among others, have studied the influence of calcium leaching on the residual strength of beams for different cements subjected to different aggressive solutions. Carde (1996) used ammonium nitrate as an aggressive solution for accelerated leaching. Le Bellégo et al. (2000) have carried out size effect tests on leached specimens for different degradation rates, and more recently Ulm and co-workers (2001) used a similar method in order to accelerate leaching.

Various models have been developed to describe calcium leaching, its effect on the mechanical properties of the leached material, and mechanical damage as well. Ulm and co-workers (1999, 2001) used a chemo-plastic approach based on the mechanics of porous materials. Kuhl et al. (2000) devised a continuum damage model on the same basis. Gérard (1996) proposed a non-local damage model coupled with leaching effects. It is based on the definition of two damage variables: one is due to mechanical loads and the other one, which could be called a chemical ageing variable, accounts for chemical damage. These models could provide predictions of the residual strength and of the lifetime of concrete structures subjected to aggressive water but their range of validity in structural computations needs to be explored still. Such an assessment is the prime objective of this section, in which the experiments obtained by Le Bellégo et al. (2000) will serve as reference data. The constitutive relations due to Gérard will be considered as an illustration. Among the models mentioned above, it is the only one that contains an internal length. Examined with the help of the size effect law, which contains a definition of brittleness as the transition of a strength of material criterion to a linear elastic fracture mechanics one in design (Bazant and Planas, 1998), the model fails to describe the increase of brittleness of the specimens during leaching. Such an increase may be represented by a decrease of the internal length and by an increase of the damage threshold essentially.

9.2.1 Constitutive relation

We are going to describe chemical damage very briefly only, and see how it can be coupled with a scalar mechanical damage model. Full details on the chemical damage model can be found in Gérard (1996).

9.2.1.1 Chemical damage

Concrete is a reactive porous material. Hydrates and water are the main components. Solid phases are in thermodynamic equilibrium with the surrounding pore solution chemistry and calcium is the main chemical component of the binder. The hydrates leaching process is caused by the difference of composition and chemical activity between water in contact with concrete and the pore solution inside concrete. The difference between the pore solution and the external solution causes the motion of ions out of the cement paste. This overall motion of ions breaks the thermodynamic equilibrium established between the pore solution and the hydration products, and favours the dissolution of solid phases. The present simplified approach avoids considering all the elementary chemical mechanisms and focuses on the evolution of one variable only, the amount of calcium ions in the liquid phase of the hardened cement paste (Gérard et al. 1999). It is obtained by solving a non-linear diffusion equation:

$$\frac{\partial C_{solid}}{\partial C}\frac{\partial C}{\partial t} = div\left[D(C)\mathrm{gr\vec{a}d}(C)\right]$$ (1)

where C_{solid} denotes the amount of calcium in the solid phase and C denotes the amount of calcium ions in the liquid phase. A mathematical relationship for the calcium solid-liquid equilibrium can be found in Gérard (1996). It is a function of the cement chemistry and of the water to cement ratio. The diffusion coefficient D is not constant. It is related to the volume fraction of the different solid phases initially. During the leaching process, it varies between that of the virgin material and the diffusion coefficient of calcium ions in water (when the material is totally cracked).

9.2.1.2 Mechanical damage

Micro cracking due to mechanical loads is described with a scalar damage model due to Mazars (1984). It is briefly recalled in the following. The stress – strain relation is:

$$\sigma_{ij} = (1-d)\Lambda_{ijkl}\varepsilon_{kl}$$ (2)

where σ_{ij} and ε_{ij} are the components of the stress and strain tensors, Λ_{ijkl} are the initial stiffness moduli, and d is the damage variable. E and v are the initial Young's modulus and Poisson's ratio respectively. For the purpose of defining damage growth, the equivalent strain is introduced first:

$$\tilde{\varepsilon} = \sqrt{\sum (<\varepsilon_i>)^2}$$ (3)

where $\langle.\rangle_+$ is the Macauley bracket and ε_i are the principal strains. In order to avoid ill-posedness due to strain softening, the mechanical model is enriched with an internal length (Pijaudier-Cabot and Bazant 1987). The non-local variable $\bar{\varepsilon}$ is defined:

$$\bar{\varepsilon}(x) = \frac{1}{V_r(x)}\int_\Omega \psi(x-s)\tilde{\varepsilon}(s)ds \text{ with } V_r(x) = \int_\Omega \psi(x-s)ds$$ (4)

where Ω is the volume of the structure, $V_r(x)$ is the representative volume at point x, and $\psi(x-s)$ is the weight function:

$$\psi(x-s) = \exp\left(-\frac{4\|x-s\|^2}{l_c^2}\right)$$ (5)

l_c is the internal length of the non local continuum. $\bar{\varepsilon}$ is the variable that controls the growth of damage according to the following conditions:

$$F(\bar{\varepsilon}) = \bar{\varepsilon} - \kappa$$ (6a)

if $F(\bar{\varepsilon}) = 0$ and $\dot{\bar{\varepsilon}} > 0$ then $\begin{cases} \dot{d} = h(\kappa) \\ \dot{\kappa} = \dot{\bar{\varepsilon}} \end{cases}$ with the condition $\dot{d} \geq 0$

(6b)

else $\begin{cases} \dot{d} = 0 \\ \dot{\kappa} = 0 \end{cases}$

κ is the softening parameter and takes the largest value of $\bar{\varepsilon}$ ever reached during the previous loading history at a given time and at the considered point in the medium. Initially $\kappa = \kappa_0$, where κ_0 is the threshold of damage. The damage evolution law $h(\kappa)$ in (6b), is defined in an integrated form. In particular, the tensile damage growth is:

$$d_t = 1 - \frac{\kappa_0(1-A_t)}{\kappa} - \frac{A_t}{\exp(B_t(\kappa-\kappa_0))}$$ (7)

where constants A_t, B_t are model parameters which must be obtained from a fit of experimental data (e.g. bending beams).

9.2.1.3   Chemo-mechanical coupling

The influence of chemical damage on the mechanical properties of the material may be introduced in the stress – strain relations following at least two methods. In the first one, which is inspired from the works of Ulm and Coussy (1996), the chemical degradation enters in the plastic yield function and, at the same time, changes the elastic stiffness of the solid skeleton of the material. Accordingly and within the context of continuum damage mechanics, the evolution of damage can be rewritten in a generic form as:

$$\dot{d} = F(\bar{\varepsilon}) + G(C)$$ (8)

For the sake of simplicity a single phase material is considered only. The contribution to the growth of damage of each separate mechanism appears explicitly: damage growth due to mechanical loads is $F(\bar{\varepsilon})$ (e.g. as defined by Eqs. 6a,b) and damage growth due to leaching is defined by the function $G(C)$. One has to check, however, that the damage variable ranges between 0 and 1, which means that the functions $F(\bar{\varepsilon})$ and $G(C)$ cannot be independent.

A second method consists in adding a new chemical damage variable V on top of the mechanical one. This chemically induced damage variable is an independent contribution to the weakening of the material. It is a new internal variable. This method as been chosen by Gérard (1996), Gérard et al. (1998), Saetta et al. (1999), Kuhl et al. (2000), Meftah et al. (2000), and Stabler and Baker (2000) for environmental (chemical and thermal) damage modelling. The corresponding stress – strain relation reads:

$$\sigma_{ij} = (1-d)(1-V)\Lambda_{ijkl}\varepsilon_{kl} \qquad\qquad (9)$$

V is the chemical damage, a function of the concentration C of calcium ions in the pore solution: $V = g(C)$. g is an experimentally determined function shown in Fig. 1.

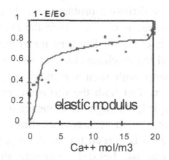

Figure 1. Consequence of calcium leaching on the cement paste on elasticity (after Gérard (1996)).

It may be also related to the porosity of the material in the case of a cement paste (Ulm et al. 1999). Figure 2 shows the material responses in tension and in compression at different levels of chemical damage.

The mechanical evolution of damage will affect the diffusion of the ions in the material too. In the absence of experimental data, the variation of the diffusion coefficient due to damage can be neglected. In more advanced computations, it can be arbitrarily fixed as a S-shaped function ranging in between D_0 which is the value of the diffusion coefficient in the absence of chemical attack and $D_{\max}$ which is the value for a cracked material which cannot exceed the diffusion coefficient in free water.

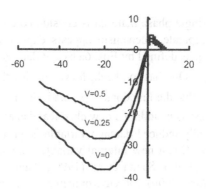

Figure 2. Constitutive law of a concrete as a function of the ageing variable V (stresses on the vertical axis are in MPa, strains on the horizontal axis are m/m 10^{-4}).

9.2.2   Sensitivity analysis

The chemo-mechanical damage model has been implemented in a finite element code according to a staggered scheme. The diffusion problem (Eq. 1) is solved first over a time step, considering that mechanical damage is constant, then the mechanical problem is solved assuming that the chemical damage is constant. The time integration scheme in the diffusion problem is unconditionally stable and the mechanical solver uses a secant stiffness algorithm that provides sufficiently robust results with such a damage model (Rodriguez-Ferran and Huerta 2000). It should be stated here that both the diffusion problem and the mechanical problem are non linear. They are coupled, although the influence of mechanical damage on the diffusion problem is quite small. For a good accuracy, the finite element grid and the time steps are rather small (smaller than what is needed for the damage problem in the absence of coupled effects). These are two reasons, besides simplicity of implementation, why a staggered scheme has been preferred to a fully coupled computational approach as it provides a reasonable accuracy. Note also that it is possible to control the accuracy of the time integration in the diffusion process (Diez et al. 2002).

Let us perform first a sensitivity analysis and consider a three point bending beam subjected to an external constant deflection and a chemical attack by pure water on the tensile bottom face (Fig. 3). This is a plane stress calculation. Vertical displacements are prevented at the bottom supports, and fixed at mid span on the upper part of the beam under the load. Horizontal displacement is prevented at the left-side support. The central half of the beam is treated with the damage model while the rest is assumed to remain elastic. The diffusion coefficient is assumed to remain constant (Table 1). All the computations have been performed with the same mesh of 4 nodded square elements of size 2.5 mm. Initially, the tensile stress at the bottom is 80 % of the tensile strength. Due to leaching, chemical damage propagates from the bottom to the upper faces of the beam. The strain increases and mechanical damage is triggered in the central part of the beam. As a consequence, the beam carries less load and failure occurs eventually. The time to failure is the lifetime of the beam.

E	At	Bt	κ_0	l_c	v	D
35000 MPa	1	6000	1.510-4	1cm	0.2	$7\ 10^{-10}\ m^2/s$

Table 1. Model parameters for the evolution of damage.

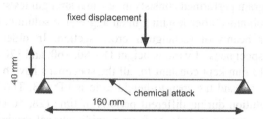

Figure 3. Three point bend beam subjected to a chemical attack and a constant deflection.

For a mode I crack propagating in an infinite body, it is possible to show that the fracture energy is roughly proportional to the area under the stress strain curve and to the internal length (Mazars and Pijaudier-Cabot, 1996). Because the internal length is proportional to the fracture energy, it is sufficient to change this parameter in order to change the fracture energy without changing the tensile strength. We can observe in Fig. 4 that the time to failure is extremely sensitive to the fracture energy. A 100% increase of the fracture energy G_f results in a 600% increase of the time to failure.

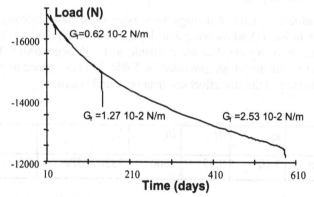

Figure 4. Variation of life time predictions with the fracture energy.

These results are also quite sensitive on the tensile strength (Le Bellégo et al. 2003b).

9.2.3   Residual strength of leached beams

We are going now to compare the model predictions to experimental results. The experimental data have been taken from Le Bellégo et al. (2000) and Le Bellégo (2001).

9.2.3.1 Size effect tests on leached specimens

The experimental program performed consists in residual strength tests: the residual behaviour of mortar beams was obtained after leaching in an aggressive solution (amonium nitrate). The specimens are mortar beams of rectangular cross section. In order to study size effect, geometrically similar specimens of various height D = 80, 160, and 320 mm, of length L = 4D, and of thickness b = 40 mm kept constant for all the specimens have been tested. The length-to-height ratio is L/D = 4 and the span-to-height ratio is l/D = 3. The beams were immersed into the aggressive solution during different periods of time (28, 56 and 98 days). Only the two lateral sides were exposed in order to have a unidirectional leaching front. The leaching depth X_f was measured at the end of each time period. It helped at defining the leaching rate $L_R = 2X_f / b$ which is the ratio of the leached cross section to the initial cross section of the beam. For 28, 56 and 98 days of leaching, the leaching rates are 45%, 62%, and 84% respectively. After the chemical degradation tests, the residual mechanical behaviour of the beams were obtained from three point bending tests. A notch of depth D/10 and thickness 3 mm was sawed just before the mechanical test. The notch thickness was kept the same for all specimens. Its size is approximately that of the sand particles and the notch could be seen as a local defect in the beam. The response of bending beams entirely leached was also extrapolated from these data following the method described in Le Bellégo et al. (2000).

9.2.3.2 Response of leached beams

The parameters controlling chemical damage have been fitted independently, including the diffusion coefficient in Eq. (1) whose variation depends on the type of accelerated leaching used in the experiments (complete data are available in Le Bellégo 2001). The parameters entering in the mechanical model are provided in Table 2. These parameters result from a manual (trial and error) fit of the size effect test data on sound beams.

l_c	κ_0	At	Bt	E	ν
40 mm	$3\ 10^{-5}$	0,95	9000	38500 MPa	0.24

Table 2. Set of model parameter in the size effect computations.

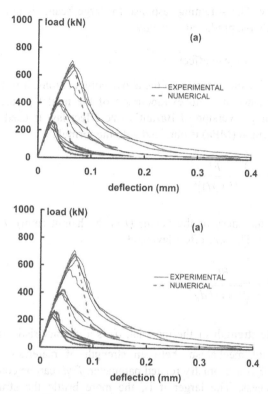

Figure 5. Prediction of the response of leached beams for two leaching rates 45% (a) and 64% (b). The thin lines correspond to the experimental results for each size and the thick lines correspond to the computations.

It is now possible to compute the responses of beams at different leaching rates and to compare with the experiments. A one dimensional diffusion problem is solved first within the beam thickness in order to compute the calcium concentration. Then, the distribution of calcium ions is converted into a distribution of chemical damage according to the function plotted in Fig. 1. It is averaged over the thickness of the beam. Then, the average chemical damage is inserted into the mechanical model (decrease of Young's modulus) and the load deflection curves of the beams are computed. Within the fracture process zone, the finite element should be smaller than the internal length. The finite element meshes of geometrically similar beams should not be geometrically similar (unless the above requirement is met for each size). In the present computations, the finite elements placed in the fracture process zone (constant strain triangles) have a size of 1/3 of the internal length approximately.

Figure 5 shows the numerical predictions for leaching rates of 45 % and 64%. Due to chemical damage, the average Young's moduli of the beams have been decreased by 25% and 40% in the mechanical computations respectively. These comparisons can be considered to be quite satisfactory. The softening response for large beams is not very well described but the peak loads are almost predicted correctly.

9.2.3.3 Prediction of size effect

In order to examine the accuracy of the above computations, it is helpful to consider them on a size effect plot since it is an extrapolation of the model predictions for different sizes of specimens. A simple version of Bazant's size effect law is used (Bazant and Planas, 1998). The nominal strength (MPa) is obtained with the formula:

$$\sigma = \frac{3}{2} \frac{Fl}{b(0.9D)^2} \tag{10}$$

where b is the thickness of the beam, D is the height (mm), l the span (mm), and F the maximal load (N). The size effect law reads:

$$\sigma = \frac{Bf_t}{\sqrt{1+D/d_0}} \tag{11}$$

f_t is the tensile strength of the material, d_0 is a characteristic size which corresponds to a change of failure mechanism between strength of materials and linear elastic fracture mechanics, and B is a geometry related parameter. D/d_0 can be considered as a measure of the structural brittleness. The larger it is, the more brittle the structure, which tends to fail according to linear elastic fracture mechanics. Bf_t and d_0 are obtained from a linear regression. The fracture energy G_f is also obtained with this regression. Table 3 shows these parameters for sound and entirely leached beams.

	Bf_t (MPa)	G_f (N/mm)	d_0 (mm)
Sound	5.25	89	416
Leached	3,04	26,9	123

Table 3. Material characteristics from size effect regression on experimental data.

Figure 6a shows the size effect plots for the computations of sound beams and of leached beams at different leaching rates. Figure 6b shows the experimental results, which are quite different from the computations. These plots have been obtained assuming that the beams are made of a homogeneous material equivalent to the leached and sound materials in the cross section. For each stage of leaching, new values of Bf_t and of d_0 are obtained from the regression and used to normalise the data in Figs. 6a,b. The tensile strength f_t, the parameter d_0 and the fracture energy which result from these fits of Eq. (11) should not be regarded as intrinsic material properties as they depend on the leaching rate in the cross section of the beam. It is their initial values (sound material) and their asymptotic values (leaching rate of 100%) which have a sense in terms of constitutive modelling and therefore are intrinsic, since in these cases the beams are indeed made of a homogeneous material.

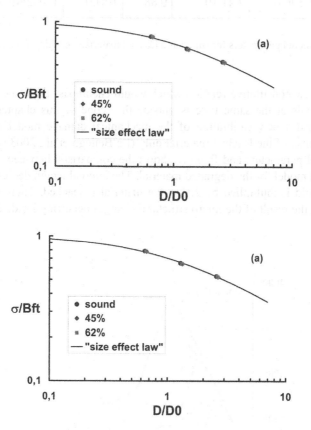

Figure 6. Prediction of size effect at several leaching rates (a), and experimental size effect at several leaching rates (b).

Figure 6a shows that the model cannot reproduce the shift of the experimental results to the right on the horizontal axis of the size effect plot. This shift, observed on Fig. 6b, corresponds to an increase of brittleness of the beams upon leaching. For such a shift to be captured, d_0 should change with the leaching rate at least. Computations performed by Le Bellégo et al. (2003a) have shown that d_0 is proportional to the internal length l_c which does not depend on chemical damage here. Hence d_0 does not depend on the leaching rate and a shift to the right or left of the data points on the horizontal axis of the size effect plot cannot be described.

	l_c	κ_0	At	Bt	E	ν
Sound	45.9 mm	$2.92\ 10^{-5}$	0.72	10000	38500 MPa	0.24
Leached	35.3 mm	$7.41\ 10^{-5}$	0.88	10500	13052 MPa	0.24

Table 4. Set of model parameters for sound and leached materials resulting from optimised fits.

Developing a new constitutive relation which would encompass the response of the sound and leached materials at the same time is outside the scope of this chapter. It is possible, however, to perform a new calibration of the mechanical damage model assuming that it describes the response of the leached material only (Le Bellégo et al., 2003b). Table 4 shows the resulting model parameters and figure 7 shows the comparison between the experimental data and numerical model for the degraded material. The internal length decreases by 24% and the damage threshold is multiplied by 2.5 as the material is leached. There is no doubt that such variations are the result of the micro structural changes occurring inside the cement paste upon leaching.

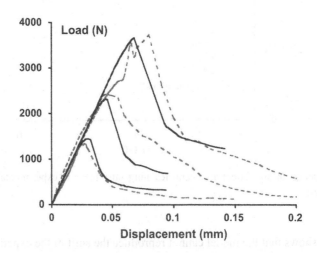

Figure 7. Size effect experiments – comparison between the experimental results (dotted curves) and the numerical results (plain curves) obtained after calibration of the model for degraded material.

   The computations considered in this section could be performed with any other constitutive model aimed at describing chemo-mechanical damage due to calcium leaching. It is found here that the mechanical consequences of leaching are better described if the model parameters in the damage evolution laws are changed. The most important variations deal with the internal length and the damage threshold. These variations remain to be explained. In order to investigate them more easily, experiments on model materials can be performed. Such data are going to be reviewed next.

9.3.   Experiments on model materials

In quasi-brittle materials, fracture exhibits a finite size process zone. Macro-cracking is the result of progressive material damage in which micro-cracks appear first in a rather diffuse way and then coalescence occurs in order to form the crack. The size of the resulting fracture process zone (FPZ) is not dependent on the structural size, provided it does not interfere with the boundaries of the considered body. It is controlled by local heterogeneity, and by the state of stress as well. From the modelling point of view, the description of the FPZ has to involve the introduction of an internal length in the governing equations. It can be in the form of a characteristic length which is related to the length of the process zone, or in the form of an internal length in non local constitutive relations.

Figure 8. Distribution of polystyrene beads in 1.4 density mixture.

   In most existing proposals, the internal length is a constant parameter. Still, there are some theoretical indications that suggest that the internal length should change in the course of the fracture process: with the help of micromechanics, Bazant (1994) arrived at the conclusion that when the interactions between cracks and voids develops in the course of failure, the weight function entering in the calculation of the non local variable controlling damage should change. The aim of this section is to examine, with the help of model materials, the correlation between the width of the fracture process zone observed with acoustic emission (AE), the parameters in the size effect law, and the internal length in non local constitutive relations. In addition, tests on model materials designed to investigate ageing effects are presented in order to exhibit variations of the internal length due to the evolution of the microstructure.

All test specimens were made with a mix which consists of ordinary Portland cement CPA-CEMI 52.5, polystyrene beads, normal density fine sand with a maximum size of 2 mm, a superplasticizing agent (Glenium 51) and water. Expanded polystyrene spheres of 3 – 7 mm mean diameter were used as aggregate in the mix design. Expanded polystyrene consists essentially of air. Four different mixes of densities 2.0 – 1.8 – 1.6 and 1.4 (see Fig. 8), having polystyrene content g of 13 - 22 - 31 and 39 % respectively, were achieved in addition to the reference material (mortar without inclusions). All mixes have a cement/sand ratio of 0.46 and a water/cement ratio of 0.4. For the bending tests, four different sizes of geometrically notched concrete specimens were used. The depths were $D = 40$, 80 , 160 and 320 mm while the thickness was kept constant for all the specimens $b = 40$ mm. The length to depth ratio was $L/D = 8{:}3$ and the span to depth ratio was $l/D = 2.5$ for all specimens. One notch of depth $D/6$ and thickness 1.5 mm (same for all dimensions) was placed in each bending specimen by putting steel plates in the moulds before casting.

9.3.1.1 Size effect tests

The size effect tests followed the guidelines established by RILEM (1990). The tests were notch opening controlled with a constant CMOD rate of 0.1 μm/s for $D = 40 – 80$ mm, 0.20 μm/s for $D = 160$ mm, and 0.25 μm/s for $D = 320$ mm. Figure 9 shows the test set up and figure 10 shows the specimens tested.

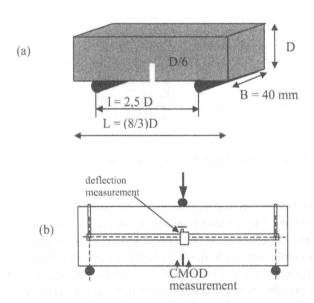

Figure 9. (a) Description of mortar samples; (b) description of instrumentation for mechanical test.

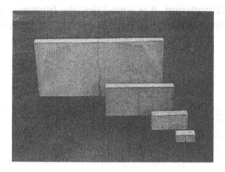

Figure 10. Specimens for the size effect tests.

Figure 11 shows the response of medium size specimens (D=40mm) for each material density. Note that the deflection at peak is almost independent from the density of the material. The material density influences still the mechanical behaviour of beams; the lower the density, the lower the stiffness and the peak load.

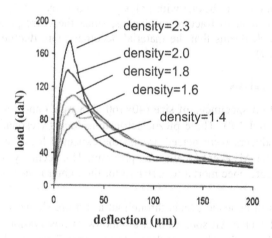

Figure 11. Influence of density on structural behavior, average load-deflection curves, for different material densities on 40 × 40 × 107 mm3 beams.

From size effect interpretation, it is found that the fracture energy G_f shifts from the value 88.5 N/m for the reference material to 52.30 N/m for the material density 1.4, a decrease of 41%.

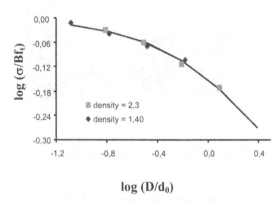

$$\log (D/d_0)$$

Figure 12. *Bazant's size effect curve: calibration for material density 2.3 and 1.4.*

In figure 12, size effect results are presented in a log-log diagram only for the two extreme densities (1.4 and 2.3) for more clarity. The larger the beam, the lower the relative strength. It can be noticed that the failure of the beams with polystyrene (density = 1.4) tends to adhere more to strength of material than to fracture mechanics, since the corresponding data shift right on the size effect plot. It means that the material becomes more ductile as the density decreases (increasing porosity).

9.3.1.2 Acoustic emission analysis

AE analysis was performed on specimens of size ($40\times160\times428$ mm^3) and for three different material densities (2.3, 2.0 and 1.8). Three piezoelectric transducers (resonant frequency of 150 kHz) were used. Transducers were placed around the expected location of the process zone to minimise errors in the AE event localisation program. The accuracy of the technique ranged from ±4 mm for the reference mortar to ±10 mm for the lighter material tested (density 1.8).

The cumulated locations of acoustic events throughout a test are shown in Figure 13: the plotted points indicate the detected AE sources over a window of observation, centered at the notch, of width 130 mm and covering the beam depth. In the same figure, we have plotted the observed crack path that appeared after the test on the lateral surface of the specimen.

The major aim of the AE analysis is to obtain an experimental characterisation of the FPZ. More specifically, it is the width of the FPZ which is the quantity of interest since it is proportional to the internal length in continuum models (Mazars and Pijaudier-Cabot 1996) and to the parameter d_0 obtained from Bazant's size effect analysis (Le Bellégo et al. 2003a). An useful approach, well suited to measure the crack band width, is to divide the specimen into an array of rectangular elements and to count the AE events located within each element. A grid of size 1×1 cm is used here. The cumulative number of events, i.e. the sum over the entire record during the experiment, is plotted as a function of its horizontal position x for various vertical positions y over the depth of the specimen. Then, the crack band width is defined as the length of the segment of a horizontal straight line placed at 20% of the maximum number of counts (Nb_{max}), which intersects the average distribution of AE events (see Haidar et al. 2003 for more details). These values are about 50, 67 and 85 mm for the material densities 2.3, 2.0 and 1.8 respectively.

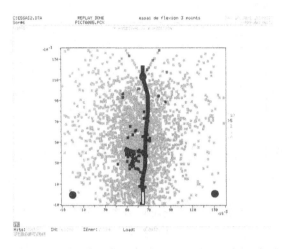

Figure 13. Cumulative location of AE events (material density 2.3).

We have also studied the case where the horizontal straight line intersects the vertical axis at the value = 10 % of Nb_{max}, and we have observed that the evolution of the width of the FPZ, as a function of the polystyrene content g, is the same.

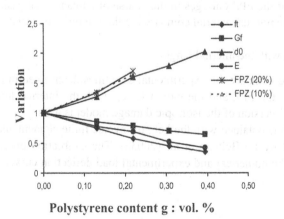

Polystyrene content g : vol. %

Figure 14. Evolution of fracture properties with polystyrene content.

Figure 14 shows the evolution of the width of the FPZ defined according to this criterion with the mass density of the specimens. On the same graph, the evolution of d_0, of the fracture energy G_f, of the tensile strength f_t, and of the Young's modulus E are also plotted. Note that there is a very good agreement between the evolution of d_0 and the evolution of the width of the FPZ.

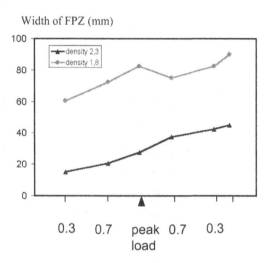

Figure 15. Evolution of the width of the fracture process zone v.s. the applied load for the two material densities. The load on the horizontal axis is increasing and then decreasing in the post peak regime.

This technique allows the determination of the shape of fracture process zone in its final stage, but incremental counts (in between two loading stages) provide also the evolution of the FPZ as the failure process develops. Such an evolution of the width of the FPZ is shown in Fig. 15. The width of the FPZ enlarges in the course of damage progression. Furthermore, the width is larger for the porous material compared to the reference one (without inclusions).

9.3.1.3 Correlation with the internal length

We are going now to compare the experimental quantities determined in the previous section, namely the width of the FPZ and the parameter d_0, with the internal length in a continuum model. The non local version of the isotropic damage model is used

The internal length is obtained with the help of inverse finite element analysis, following the procedure described by Le Bellégo et al. (2003a). The calibration procedure is based on a simultaneous fit of the numerical and experimental load deflection curves for the four sizes of specimens.

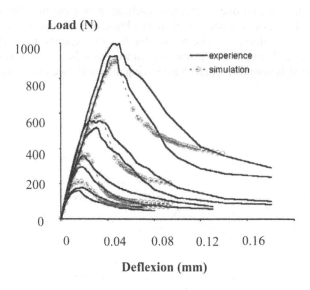

Figure 16. Prediction of the response of mortar beam with the material density 2.3.

Figure 16 shows the fits obtained for the reference material. The internal lengths and the other model parameters obtained as a result of these fits are reported in Table 5 for all the mass densities.

Density	l_c (mm)	κ_0	At	Bt
2.3	34.4	$4.61\ 10^{-5}$	0.79	9836
1.8	51.5	$4.66\ 10^{-5}$	0.65	9220
1.4	64.2	$4.79\ 10^{-5}$	0.52	7893

Table 5. Set of model parameters resulting from optimised fits.

Same as for the acoustic emission tests we have also measured the width of the FPZ obtained in the numerical analyses. It is 75 mm for the reference material, 120 mm for material density 1.8, and 139 mm for the lowest material density. These values are 40 percent greater than the experimental ones but the ratio between the computed and measured widths of the FPZ is constant whatever the mass density of the material as shown in Fig. 17. In the same figure, we have plotted also the evolution of Bazant's size effect parameter d_0 and of the internal length with the mass density of the specimens. The respective variations of these parameters are very similar, which is quite remarkable. This is also in good agreement with the numerical result obtained by Bazant and Pijaudier-Cabot (1988), where the width of the zone of localised damage was shown to be proportional to the internal length of the material. These results yield two conclusions: firstly there are evidences taht the internal length should evolve during the fracture process; secondly if the variation of the mass density of the material is intended to simulate ageing due to calcium leaching, it is clear that the observed increase of the internal length is not consistent with data obtained from accelerated ageing where this parameter is shown to decrease. The reason for this discrepancy is probably that the model material simulates a decrease of the material viewed as a homogeneous one while in reality material dissolution is selective, each phase having different mechanical properties.

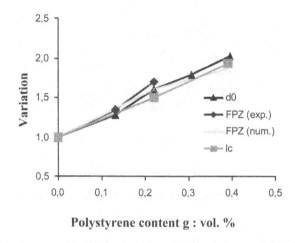

Figure 17. Evolution of d_0, FPZ (experimental), FPZ (numerical) and l_c with polystyrene content.

9.4. Damage and non linear creep

In common practice, it is usually assumed that concrete exhibits a linear visco-elastic response for low load levels and that the instantaneous mechanical response of concrete is elastic. For high load levels, deviation from linearity of the creep behaviour of concrete is expected. Under high sustained loads, cracks grow and interact with visco-elasticity. Some experimental and analytical results concerning non-linear creep can be found in the literature (see among others Bazant, 1988, Gettu and Bazant 1992, Mazzotti and Savoia 2003, Rüsch et al. 1957, Rüsch 1958) but much remains to be learned.

Modelling non linear creep of concrete is of fundamental importance for severely loaded structures as creep may decrease the material strength with increasing time. The evaluation of the residual carrying capacity of structural components is a problem of growing importance for civil engineering structures such as nuclear power plants. Actually, these structures are often exposed to high stresses for a long time, due to what damage develops. Therefore, the results arising from short time tests cannot be considered a sufficient basis for judging the safety of such structures and it is of great interest to devise methods for evaluating their residual capacity. This problem has drawn the attention of some authors, among them Rüsch and co-workers (1957, 1958), who have carried out the first experiments in order to determine the effect of continuous loading duration on the resistance and deflections of concrete specimens under compression tests. They found that the capacity of a structure subjected to creep loads seems to be 70 to 80% of that observed in short time compression tests.

In this section, the main results of a set of experimental creep tests on concrete three points bend specimens at different load levels are presented. The objective is to investigate the ranges of variation of the time response under constant load due to variations of the load level. For each creep test, a different load level has been applied, starting from low levels (36% of peak load) where linear visco-elasticity applies, to high levels (80% of peak load) where tertiary creep causing failure at a finite time can be observed. Concrete B11 used for the construction of the Civaux power plant in France and tested by Granger (1995) is being used in this experimental program. The specimens were made with this mix which consisted of ordinary Portland cement CPA-CEMII 42.5, fine sand with a maximum size of 5 mm, crushed gravel of size 5 to 25 mm, a superplasticizer a gent (Glenium 21) and water. This mixture is characterised by a water-cement ratio of 0.56 and a slump of 4 cm.

9.4.1.1 Fracture tests

Three different types of tests were performed in order to determine the mechanical characteristics of concrete. Six cylindrical concrete specimens of diameter 16 cm and length 32 cm were used for compressive and splitting tests. The compression tests were performed using 300 KN capacity hydraulic testing machine at a loading rate of 0.5 MPa/s until failure.

The third kind of tests consisted on three points bend size effect experiments on prismatic geometrically similar notched beams. Three different sizes were used, the depths were D1 = 10cm, D2= 20cm, and D3= 40 cm with respective lengths 35, 70, and 140 cm while the thickness was kept constant for all the specimens b = 10 cm. The span to depth ratio was l/D = 3 for all specimens. One notch of depth 0.15D and thickness 3 mm (same for all dimensions) was performed in each specimen by placing a plastic plate at the midpoint perpendicular to the long direction of the mould before casting. Nine prismatic specimens were cast, three for each size.

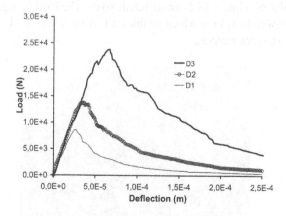

Figure 18. average load-deflection curves for small (D1), medium (D2) and large (D3) sizes

Figure 18 shows the average responses for all dimensions. Curing conditions for all specimens were 28 days at 50% RH and controlled temperature of 20° C. Mean values of physical and mechanical properties are summarised in table 6.

Another set of three points bending tests was carried out on beams of size D2:10x20x70 cm without notches. These tests were aimed at determining the maximum load of unotched beams in order to calibrate the applied load in the creep tests, where beams having the same geometry have been used. The average maximum load determined from flexural tests on three specimens is about 2409 daN.

Properties	Mean values
f_c (MPa)	41.25
f_t (MPa)	3.48
E_{dyn} (MPa)	39000
D1 : F_{max} (daN)	863.42
σ_N (MPa)	5.61
D2 : F_{max} (daN)	1401.13
σ_N (MPa)	4.42
D3 : F_{max} (daN)	2377.31
σ_N (MPa)	4.81
G_f (N/m)	180

Table 6. Mechanical properties of concrete B11.

9.4.1.2 Creep tests

The creep tests were performed on frames designed at R&DO which apply a constant load on flexural beams. The frames have a capacity ranging from 5 to 50 KN and can accommodate geometrically similar specimens of three different sizes. The load is applied by gravity with a weight and counter-weight system which enables a fine tuning of the load. Figure 19 shows a general view of these creep frames.

Figure 19. General view of creep frames.

The specimens tested have the same thickness of 10 cm. The smallest D1 is 10 cm high and 35 cm long. The medium D2 is 20 cm high and 70 cm long and the largest D3 is 40 cm high and 140 cm long. Two kinds of creep tests have been performed. The first one is intended to study the influence of the load level on creep. In the second kind of test (which is still in progress), the interaction between creep and fracture of concrete is considered via the associated size effect in structures and the decrease of the fracture energy due to creep. The applied loads in the creep tests have been determined from the previous fracture tests on the same material and the same geometry.

Twelve prismatic beams were used. For each creep test, a different load level has been applied, starting from low levels to very high levels. The applied loads are given in table 7.

Applied load	notched beams	notched beams	unotched beams
	D1 - daN	D2 - daN	D2 - daN
36% F_{peak}	313	-	-
50% F_{peak}	-		1198
60% F_{peak}	510	746	1444
70% F_{peak}	-	-	1678
80% F_{peak}	672	1033	1940

Table 7. Applied loads in creep tests.

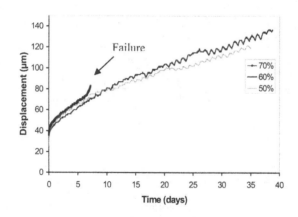

Figure 20. Deflection of unotched beams D2 for different load levels.

The first set consists of 4 unotched beams of dimension D2 loaded at 50%, 60%, 70% and 80% of the maximum load. These tests are performed at 20°C and 50% RH and the specimens are not protected. Therefore, basic creep and drying creep occur at the same time. The age of concrete at the beginning of the creep tests is 28 days. Figure 20 presents the total displacement (creep displacement plus instantaneous displacement) measured on the specimens versus time for the first three loading levels. These curves indicate that for all load levels, creep kinetics are comparable with the exception of the specimen loaded at 70% from peak load, where failure due to tertiary creep occurs after 8 days. In this test, we observed an important acceleration of creep and then failure. Concerning the specimen loaded at 80% from peak load, total failure occurred 2 minutes after applying the load.

The second test series consists of two notched beams of dimension D1 loaded at 60% and 36% of the maximum load and one notched beam of dimension D2 loaded at 60% of the maximum load. These tests were followed by a third set consisting of two notched beams of dimension D2 and three notched beams of dimension D1 loaded at 80% of the peak load. One of these beams failed after 4 days from loading, and results obtained for this test are not reported here. Specimens in the second and third test series were protected from desiccation by a double layer of self-adhesive aluminium paper. Hence basic creep was considered only. The experimental measurement of basic creep of concrete requires also that drying shrinkage be prevented. The curing conditions (3 months at 100% RH and 20°C) guaranteed to avoid early age autogeneous shrinkage so that basic creep could be measured only. The basic creep displacement was determined by subtracting the instantaneous elastic displacement from the total displacement. Figure 21 compare the displacements due to basic creep for the smaller specimens (size D1). It shows the influence of the load level on the basis creep evolution versus the elapsed time.

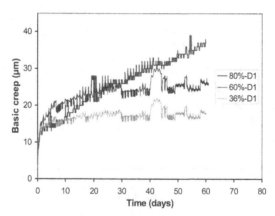

Figure 21. Basic creep displacement of notched beams D1 for 36% , 60% , and 80% of peak load.

As expected, increasing the applied load increases the basic creep magnitude. Moreover, creep develops very fast in the first days of loading and stabilises after a few weeks for the two lower loading levels. The specimens loaded at 36% and 60% of the maximum load were cast from the same batch. The results show a variation of the amplitude of basic creep of the 36% and 60% loaded specimens, which increases and decreases again between 40 days and 50 days of loading. This variation is due to a temperature regulation problem, which occurred on the climate control system and which lead to this elevation of the magnitude of deflections measured on loaded specimens. The specimen loaded at 80% is part of the third set of creep tests. It was made from another batch but followed the same curing conditions. For this specimen, basic creep increases rapidly.

Figure 22 presents a comparison between basic creep displacements obtained on two specimens of size D1 and D2, both of them loaded at 60% from their corresponding maximum load. These results show that the basic creep kinetic for dimension D2 is greater than that of dimension D1. This difference is more obvious in the case of a 80% loading level as seen in figure 23. On this plot, we show also the results obtained for the two specimens of each size subjected to the same creep load. The reproducibility of the test is quite satisfactory.

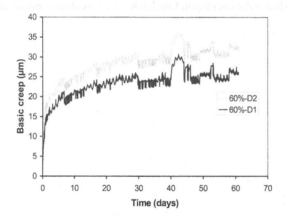

Figure 22. Basic creep displacement of notched beams D1 and D2 loaded at 60% of peak load.

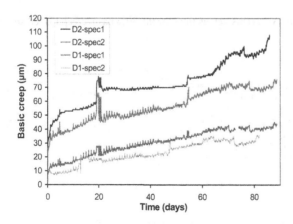

Figure 23. Basic creep displacement of notched beams D1 and D2 loaded at 80% of peak load.

9.4.1.3 Acoustic Emission in basic creep tests

We applied the AE technique to the creep tests presented above. Transducers were placed around the expected location of the process zone, on one side of the specimen, and in a linear array.

The primary aim is to correlate the evolution of basic creep with the total number of acoustic events. In other terms, this study is intended to clarify the qualitative relationship between basic creep strain and damage occurring inside the material. More accurate analyses might be performed which include acoustic event localisation in order to visualise the locations of damage. Such tests which require a large number of transducers placed on the same beam will be performed in future experiments. Figure 24 presents the total number of acoustic events versus time and figure 25 shows the corresponding basic creep evolution occurring at the same time.

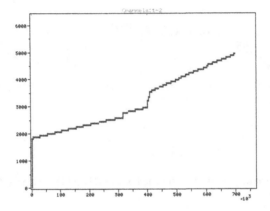

Figure 24. Total number of AE events versus time for the 80% basic creep loaded specimen D2.

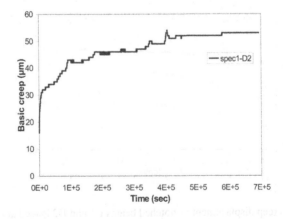

Figure 25. Deflection due to basic creep of a 80% loaded specimen D2.

We may observe that the number of AE events increases as basic creep develops. The evolution of the AE number is almost linear. There is a change of slope of the curve, which seems to correspond to a small peak in Fig. 25. This peak is due again to a slight defect in the room temperature control. If one admits that AE events are due to microcracking, these results show clearly that there is a coupling between creep and damage. It follows that the mechanical responses of beams subjected to creep compared to those not subjected to creep, ought to exhibit a difference due to initial damage accumulated during creep loads. A reduction of the residual carrying capacity due to basic creep has to be expected in particular.

9.4.1.4   Residual capacity tests

In order to investigate the variation of the residual capacity due to creep, comparison specimens cast at the same time than those subjected to creep were kept under the same conditions of temperature and relative humidity. Three points bending experiments till failure were carried out on the creep loaded and on the comparison specimens. Results deals with unotched beams under combined drying and basic creep. After 40 days of loading, beams of dimension D2 subjected to loads of 50% and 60% of the maximum load were removed from the creep frames and then immediately subjected to three point bending loading up to failure with a constant loading rate.

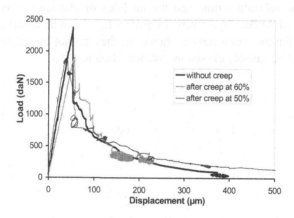

Figure 26. Residual capacity of unotched beams of dimension D2.

The chosen rate of loading is 0,1μm/s, which is the same rate used as in the flexural experiments intended to determine the average maximum load for this kind of specimens. Figure 26 displays the load-displacement curves obtained for the specimens after creep load and for the unloaded comparison specimens. The maximum carrying force of the specimens subjected to creep initially is reduced of about 20% in comparison to those of the unloaded specimens. Furthermore, it is found that there is no significant impact of the loading level on the response of beams. The 50% and 60% loaded specimens have approximately the same peak load. The results obtained on notched specimens of dimension D1 loaded at 36% and

60% of the maximum load during 60 days (basic creep tests) did not exhibit such a decrease of the maximum load. This underlines the importance of drying creep, at least for low load levels.

9.4.1.5 Damage coupled to creep: preliminary constitutive modelling

We consider here basic creep only. It is modeled with visco-elasticity, in the form of a generalized Maxwell chain (Fig. 27). The constitutive law is written with the help of a relaxation function R:

$$\sigma'(t) = R(t_0,t).\varepsilon\ (t_0) + \int_{t_0}^{t} R(\tau,t) \cdot \dot{\varepsilon}\ (\tau) \cdot d\tau \tag{12}$$

σ' is the effective stress applied to the Maxwell chain, t_0 is the age of loading. Note that the formulation is purely one-dimensional here. The relaxation function is:

$$R(\tau,t) = E_0(\tau) + \sum_{\mu=1}^{n} E_\mu(\tau).\exp(-\tau_\mu(t-\tau)) \tag{13}$$

τ_μ and E_μ are the relaxation time and the modulus of elasticity of each branch μ in the generalized Maxwell model respectively. In general $E_\mu(t_0)$ can be calculated by the method of least squares. As for the parameters τ_μ, however, they cannot be calculated from measured creep data but must be suitably chosen in advance (Bazant, 1988).

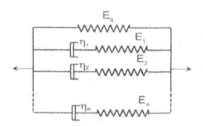

Figure 27. Generalized Maxwel model.

In order to incorporate damage, which is a degradation of the overall elastic stiffness, into the time dependent creep functions, we use the effective stress approach (Omar et al. 2003).

$$\sigma = \sigma'(1-d) \tag{14}$$

It is a classical approach, in which we consider a time independent growth of damage. Mechanical damage is controlled by the total strain. We have used this formulation to compute two examples. The first one is a creep test of a tensile bar. Figure 28 shows the results. Note that tertiary creep can be described, including failure after some load duration. Such a result cannot be obtained in a series arrangement between visco-elasticity and damage or plasticity. The second case corresponds to experiments in which beams are subjected to a constant displacement at mid-span for a given period of time, and then loaded up to complete failure. This test is aimed at obtaining the influence of creep on the capacity of bending beams. Figure 29 shows the model predictions for several values of the internal length.

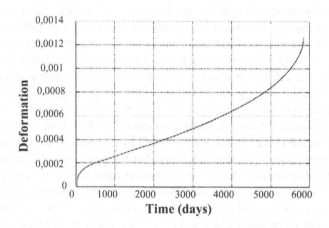

Figure 28. Creep loading of a tensile bar.

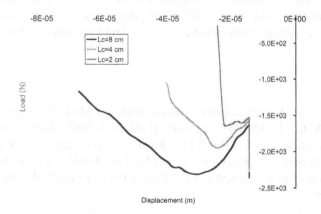

Figure 29. Influence of the relaxation on the residual capacities of beams for several internal lengths.

Again, the internal length has a great influence on the model predictions, which underlines the necessity of an accurate determination of this model parameter, and a good knowledge of its variation due to creep. According to experiments by Gettu and Bazant (1992), the size of the fracture process zone decreases as the loading is slowed down. If this decrease results in a decrease of the internal length, it would mean that due to creep, the capacity of structural elements might be overestimated if the internal length is assumed to remain constant.

9.5.   Discussion and closure

The test data shown for calcium leaching and for model materials show that there is a strong correlation between the internal length in non local damage models and the evolution of microstructure due to ageing. The evolution of this parameter is pretty difficult to measure experimentally and it requires combination of several tests, including size effect test with inverse analysis based on non linear computations. Variations of this length should not be neglected. As seen in the calcium leaching case, a decrease of the internal length induces a material that is more brittle, more prone to sudden failure. Neglecting such a decrease places the design on the unsafe side as the capacity of the structure is over estimated.

Similar effect are expected for creep fracture interaction. For high load level, tertiary creep i.e. a visco-elastic response of the material coupled with an evolution of damage, occurs. Basic creep on bending beams seems to be size dependent. For two geometrically similar specimens (sizes D1 and D2), the displacement due to creep is neither equal nor multiplied by two (which is the ratio between the sizes of the two beams). Residual capacity tests show that drying creep has a great influence on the residual carrying capacity of the beams. For the tests presented, which are limited to low creep levels (50% and 60% of the maximum load), basic creep does not influence the structural strength of the beam.

New constitutive relations with evolving internal length, and coupled creep damage effects, ougth to be devised in order to capture these phenomena. They can be inspired from simplified micromechanics (Pijaudier-Cabot et al. 2003) but they remain to be compared with experiments. For such a purpose, model materials such as those presented in this chapter may be of great help.

Acknowledgments:

Financial support from the European Commission through the MAECENAS project (FIS5-2001-00100), through the DIGA RTN network (HPRN-CT-2002-00220), and from the partnership between Electricité de France and the R&DO group through the MECEN project are gratefully acknowledged. The study on calcium leaching would not have been possible without the cooperation of Bruno Gérard and Caroline le Bellégo from EDF. The first author is very grateful to them.

9.6.   References

Bazant Z.P. 1988, Mathematical modeling of creep and shrinkage of concrete, John Wiley & Sons Ltd.

Bazant Z.P., 1994, Nonlocal Damage Theory based on Micromechanics of Crack Interactions, ASCE Journal of Engineering Mechanics, 120, 593 – 617.

Bazant Z.P. and Planas J., 1998, Fracture and size effect in concrete and other quasibrittle materials ", CRC press, Boca Raton and London, 1998.

Bazant Z.P. and Pijaudier-Cabot G., 1988, Nonlocal Continuum Damage, Localization Instability and Convergence, Journal of Applied Mechanics, ASME, 55, 287-294.

Carde C., 1996, Characterization and modeling of the alteration of material properties due to leaching of cement-based materials, Ph.D. thesis, Université Paul Sabatier, Toulouse, France, 218p. (in French).

Carde C. and Francois R., 1997, Effect of Leaching of Calcium Hydroxyde from Cement Paste on Mechanical and Physical Properties, Cement and Concrete Res., 27, 539-550.

Díez, P., Arroyo, M., Huerta, A., 2002, Adaptive Simulation of the Coupled Chemo-Mechanical Concrete Degradation, Proceedings of the Fifth World Congress on Computational Mechanics (WCCM V), July 7-12, 2002, Vienna, Austria, Editors: Mang, H.A.; Rammerstorfer, F.G.; Eberhardsteiner, J., Publisher: Vienna University of Technology, Austria, ISBN 3-9501554-0-6, http://wccm.tuwien.ac.at.

Gettu, R., Bazant, Z.P., 1992, Rate effects and load relaxation: static fracture of concrete. ACI Materials J. 89 (5), 456-468.

Gérard, B., 1996, Contribution of the mechanical, chemical, and transport couplings in the long-term behavior of radioactive waste repository structures, Ph.D. Thesis, Département de Génie Civil, Université Laval, Québec, Canada / École Normale Supérieure de Cachan, France, 278p. (in French).

Gérard B., Pijaudier-Cabot G. and La Borderie C., 1998, Coupled Diffusion-Damage Modelling and the Implications on Failure due to Strain Localisation, Int. J. Solids & Structures, 35, 4105-4120.

Gérard B., Pijaudier-Cabot G., and Le Bellégo C., 1999, Calcium Leaching of Cement Based Materials: a Chemo-Mechanics Application, Construction Materials – Theory and Application, Hans – Wolf Reinhardt Zum 6 Geburstag, R. Eligehausen Ed., ibidem, 313-329.

Goncalves A., Rodrigues X., 1991, The Resistance of Cement to Ammonium Nitrate Attack, Durability of concrete, 2nd International Conference, Montreal, Canada.

Granger L.; 1995, Comportement différé du béton dans les enceintes de centrales nucléaires (analyse et modélisation) . Thèse de doctorat de l'ENPC.

Haidar K., Pijaudier-Cabot G., Dubé J.F. and Loukili A., 2003, Correlation Between the Internal Length, the Fracture Process Zone and Size Effect in Mortar and Model Materials. Submitted for publication to Concrete Science and Engineering.

Kuhl D., Bangert F., and Meschke G.,2000, An Extension of Damage Theory to Coupled Chemo-Mechanical Processes, Proc. ECCOMAS 2000, Barcelona, Sept.

Le Bellégo C., Gérard B., and Pijaudier-Cabot G., 2000, Chemomechanical Effects in Mortar Beams Subjected to Water Hydrolysis, J. Engrg. Mech. ASCE, 126, 266-272.

Le Bellégo C., 2001, Couplages chimie-mécanique dans les structures en béton attaquées par l'eau : Etude expérimentale et analyse numérique, Ph.D. Dissertation, École Normale Supérieure de Cachan, France, 236p.

Le Bellégo C., Dubé J.F., Pijaudier-Cabot G., and Gérard B., 2003a, Calibration of Non Local Damage Model from Size Effect Tests, Eur. J. of Mechanics A/Solids, 22, 33-46.

Le Bellégo C., Pijaudier-Cabot G., Gérard B., Dubé, J.F. and Molez L., 2003b, Coupled Chemical and Mechanical Damage in Calcium Leached Cementitious Structures, ASCE J. Engrg. Mech., 129, 333-341.

Mazars J., 1984, Application de la mécanique de l'endommagement au comportement non linéaire et à la rupture de béton de structure, thèse de Doctorat d'Etat, Université Paris VI, France.

Mazars J., Pijaudier-Cabot G., 1996, From Damage to Fracture Mechanics and Conversely: a Combined Approach, Int. J. Solids & Structures, 33, 3327-3342.

Mazzotti C. and Savoia M., 2003, Nonlinear creep damage model for concrete under uniaxial compression, J. Engrg. Mech. ASCE, 129, 1065-1075.

Meftah, F., Nechnech, W., and Reynouard, J.M., 2000, An Elasto-Plastic Damage Model for Plain Concrete Subjected to Combined Mechanical and High Temperature Loads, Proc. EM2000, edited by J. L. Tassoulas, University of Austin, Texas.

Omar M., Pijaudier-Cabot G., and Loukili A., 2003, Etude du couplage endommagement – fissuration, Revue Française de Génie Civil, in press.

Pijaudier-Cabot G. and Bazant Z.P., 1987, Nonlocal Damage Theory, J. of Engrg. Mech., ASCE, 113, 1512-1533.

Pijaudier-Cabot G., Haidar K., and Dubé J.F., 2003, Non Local Damage Model with Evolving Internal Length, Int. J. Num. Anal. Meths. Geomech., in press.

RILEM Draft Recommendations, 1990, Size effect method for determining fracture energy and process zone size of concrete, 23, 461-465.

Rodriguez-Ferran, A. and Huerta A. , 2000, Error Estimation and Adaptivity for Non Local Models, Int. J. Solids & Struct., 37,7501-7528.

Rüsch H., 1957, Versuche zur Bestimmung des Einflusses der Zeit auf Festigkeit and Verformung, (Experimental Determination of the effect of the Duration of Loading on Strength and Deformation), *Final Report*, Fifth Congress, International Association for Bridge and Structural Engineering, Portugal, 237-244.

Rüsch H., Sell R., Rasch C., Stöckl S., 1958, Investigations on the Strength of Concrete under Sustained Load, RILEM Symposium on the Influence of Time on the Strength and Deformation of concrete, Munich.

Saetta, A., Scotta, R. and Vitaliani, R., 1999, Coupled Environmental-Mechanical Damage Model of RC Structures, J. Engrg. Mech. ASCE, 125, 930-940.

Schneider U. and Chen S.W., 1998, The Chemomechanical Effect and the Mechanochemical Effect on High-Performance Concrete Subjected to Stress Corrosion, Cement and Concrete Research, 28, 509-522.

Schneider U. and Chen S. W., 1999, Behavior of High-Performance Concrete under Ammonium Nitrate Solution and Sustained Load, ACI Materials Journal, 96, 47-51.

Stabler J. and Baker G., 2000, On the Form of Free Energy and Specific Heat in Coupled Thermo-Elasticity with Isotropic Damage, Int. J. Solids Struct., 37, 4691-4713.

Ulm F.J. and Coussy O., 1996, Strength Growth as Chemo-Plastic Hardening in Early Age Concrete, J. of Enrg. Mech. ASCE, 122, 1123-1132.

Ulm F.J., Torrenti J.M., and Adenot F., 1999, Chemoporoplasticity of Calcium Leaching in Concrete, J. of Engrg. Mech. ASCE, 125, 1200-1211.

Ulm F.J, Heukamp F.H. and Germaine J.T., 2001, Durability Mechanics of Calcium Leaching of Concrete and Beyond, Proc. of Framcos 4, R. de Borst et al. Eds, Balkema Pubs., 133-143.

Modelling of Landslides: (I) Failure Mechanisms*

M.Pastor [*†] , J.A.Fernandez Merodo [*†] , E.Gonzalez [*†] , P.Mira [*†] , T.Li [‡] and X.Liu [‡]

[*] Centro de Estudios y Experimentación de Obras Públicas, Madrid, Spain
[†] M2i (Math.Model.Eng.Group), Department of Applied Mathematics,
ETS de Ingenieros de Caminos, UPM Madrid, Spain
[‡] Hohai University, China

Abstract This paper presents a theoretical and numerical framework to model the initiation mechanisms of catastrophic landslides. The equations describing the coupling between the solid skeleton and the pore fluids are presented following an eulerian approach based on the mixture theory that can provide a unified formulation for both initiation and propagation phases. The system of Partial Differential Equations is then discretized using the classical Galerkin Finite Element Method and neglecting the convective terms. Some applications to localized and diffuse failure will be presented.

1 Introduction

Landslides are one of natural catastrophes causing important losses of human lives and damage to property.

There is a wide variety of types of landslides, depending on the materials involved and triggering mechanism. The time scale involved can range from minutes to years. For a classification, it is worth consulting the reference Dikau (1996) where a detailed description is provided.

Landslides and failure of slopes are caused by changes in the effective stresses, variation of material properties or changes in the geometry. Changes of effective stresses can be induced either directly, as consequence of variation of the external forces (earthquakes, human action), or indirectly through pore pressures (rainfall effects). Variations in material properties can be caused by processes of degradation (weathering and chemical attack). Finally, geometry can change because of natural causes (erosion) or human action (excavation, construction, reshaping...).

The study of landslides and their consequences has become a multidisciplinary subject, where geographical, pedological and urban planning aspects are important. Here we will deal only with engineering aspects and will focus on the **prediction** of the initiation and propagation of landslides. The first aspect is important not only to know what has

---

*The authors gratefully acknowledge the financial support provided by the Spanish Agency for International Cooperation (AECI), by the Spanish Ministry of Science and Technology (Projects Andes and Ramón y Cajal contract) and by the European Union (Projects Diga and Lamé).

really happened but also to avoid possible landslides. Prediction tools can be applied, for instance, to tailing dams, to propose the rate at which they can be built, their slopes, and their maximum allowable height.

Concerning the propagation phase, once the landslide has taken place, it is important to know the velocity of the flow, how long will it reach, and what will be the path followed by them. In this way, it is possible to propose strategies based on channeling and protection structures. Another interesting example is that of landslide affecting a reservoir. Here, we will need to know both the speed and the mass of soil involved in the problem in order to feed the hydrodynamic model with proper data.

The prediction tools which will be presented in this work are based on mathematical and constitutive models for which there are extremely few analytical solutions. Therefore, numerical models such as the finite element model are required to produce suitable numerical models.

It is possible to describe the whole process (initiation and propagation) using a single mathematical model. However, there exist a difficulty in the cases where the landslide evolves to flow type phenomena, where the problem of changing from solid-like to fluid-like type of behaviour presents important difficulties. This is why we will deal separately with both phases, using different approaches for them.

This chapter is devoted to modelling of the initiation phase. After presenting the mathematical and the numerical model, we will consider examples of landslides with two different mechanisms of failure: localized and diffuse. Localized failure presents important mathematical and numerical difficulties. Concerning the former, the interested reader will find in Ref. Vardoulakis and Sulem (1995) a detailed description, while details of the latter are given in Refs. Pastor and Tamagnini (2002) and Mira (2002). Many landslides presenting a clear failure surface, like rotational and traslational landslides, fall in this category. Diffuse failures do not present such clear surfaces, and the mass of soil is much larger. This mechanism of failure is characteristic of soils presenting very loose or metastable structures with a strong tendency to compact under shearing. One paramount feature is that effective stresses approach zero, and the material behaves like a viscous fluid in which buildings can sink, as it happened during the 1966 earthquake of Niigata in Japan. When this failure mode takes place in a slope, the mass of mobilized soil can propagate downhill, evolving into flow slides or mudflows. Therefore, using a proper constitutive model is crucial.

Of course, there will be mixed types where an initial localized mechanism evolves into a diffuse mode as masses of soil liquefy.

2 Mathematical Model for soil skeleton-pore fluid coupling

2.1 Introduction

Soils and rocks are geomaterials with voids which can be filled with water, air, and other fluids. They are, therefore, multiphase materials, exhibiting a mechanical behaviour governed by the coupling between all the phases. Pore pressures of fluids filling the voids play a paramount role in the behaviour of a soil structure, and indeed, their variations can induce failure.

The purpose of this section is to describe the coupling between solid and fluid phases, following an approach which can be applied to both initiation and propagation phases.

The first mathematical model describing this coupling was proposed by Biot (1941), Biot (1955) for linear elastic materials. This work was followed by further development at Swansea University, where Zienkiewicz et al. (1980), Zienkiewicz and Shiomi (1984), Zienkiewicz et al. (1990a), Zienkiewicz et al. (1990b), Zienkiewicz et al. (2000) extended the theory to non-linear materials and large deformation problems. It is also worth mentioning the work of Lewis and Schrefler (1998), Coussy (1995) and de Boer (2000).

Among the different alternative ways which can be used to describe the coupling between solid skeleton and pore fluids, we have chosen an approach closer to mixture theories than to the more classical approach used in Computational Geotechnique, and it provides a more general description which can be used both for initiation of failure and for propagation of catastrophic landslides.

The soil skeleton consists of particles of density ρ_s having a porosity n (volume percent of voids in the mixture). The voids can be filled with air and water, but in some cases there will be mixtures of water and very fine particles which can be considered as a fluid phase. We will assume that fluid phases (α) are not miscible, defining S_α as the degree of saturation for the phase α, i.e., the volume fraction of voids occupied by it. In the case of air and water, we will define S_w and S_a as

$$S_w = \frac{\text{volume of water}}{\text{volume of voids}} \tag{2.1}$$

$$S_a = \frac{\text{volume of air}}{\text{volume of voids}} \tag{2.2}$$

with

$$S_w + S_s = 1 \tag{2.3}$$

We will introduce $\rho^{(\alpha)}$ to denote the density of the phase (α). If the fluid density is ρ_α,

$$\rho^{(\alpha)} = S_\alpha n \rho_\alpha \tag{2.4}$$

while for the solid we can write

$$\rho^{(s)} = (1-n)\rho_s \tag{2.5}$$

Density of the mixture is therefore obtained by adding the densities of all constituents in the mixture.

Concerning the kinematics, we will use u for displacements and v for velocities, with superindexes referring to the specific constituents. In this way, $u^{(s)}$ will be the displacement field of the solid phase, and $v^{(\alpha)}$ the velocity of pore fluid (α).

In most geotechnical applications, the movement of the fluid phases is described by the velocity relative to the soil skeleton. The so-called Darcy's velocity of phase (α) is defined as

$$w^{(\alpha)} = nS_\alpha \left(v^{(\alpha)} - v^{(s)} \right) \tag{2.6}$$

2.2 Effective and partial stresses

If we denote by σ_s and σ_α the Cauchy stress tensors acting on the solid particles and on the phase (α), we can define the partial stresses $\sigma^{(s)}$ and $\sigma^{(\alpha)}$ as

$$\begin{aligned} \sigma^{(s)} &= (1-n)\,\sigma_s \\ \sigma^{(\alpha)} &= nS_\alpha\sigma_\alpha \end{aligned} \tag{2.7}$$

The total stress tensor acting on the mixture is obtained as

$$\sigma = \sigma^{(s)} + \sum_{\alpha=1}^{\text{nphases}} \sigma^{(\alpha)} \tag{2.8}$$

The partial stresses $\sigma^{(\alpha)}$ can be decomposed into hydrostatic and deviatoric components as $\sigma^{(\alpha)} = -nS_\alpha p_\alpha I + nS_\alpha\,s_\alpha$ where $s_\alpha = dev(\sigma_\alpha)$ is the deviatoric part of σ_α, and I is the identity tensor of second order. It is important to note that if the saturating fluid is water, the deviatoric stresses can be neglected, but in other cases where viscous contributions are important, this term has to be taken into account. In above we have assumed that tractions are positive.

It is convenient to introduce an averaged pore pressure $\bar{p}$ as

$$\bar{p} = \sum_{\alpha=1}^{\text{nphases}} p^{(\alpha)} = \sum_{\alpha=1}^{\text{nphases}} S_\alpha\,p_\alpha \tag{2.9}$$

Two important particular cases are those of (i) non saturated soils, and (ii) saturated soils, where pore fluids are (i) water and air, and (ii) water. Above relations specialize to:

$$\begin{aligned} \sigma &= \sigma^{(s)} + \sigma^{(w)} + \sigma^{(a)} \\ \sigma^{(s)} &= (1-n)\,\sigma_s \\ \sigma^{(w)} &= -nS_w p_w I \\ \sigma^{(a)} &= -nS_a p_a I \\ \bar{p} &= S_w p_w + S_a p_a \end{aligned} \tag{2.10}$$

and

$$\begin{aligned} \sigma &= \sigma^{(s)} + \sigma^{(w)} \\ \sigma^{(s)} &= (1-n)\,\sigma_s \\ \sigma^{(w)} &= -np_w I \\ \bar{p} &= p_w \end{aligned} \tag{2.11}$$

respectively. Another important case is that of non saturated soils having $p_a = 0$, for which

$$\begin{aligned} \sigma &= \sigma^{(s)} + \sigma^{(w)} \\ \sigma^{(s)} &= (1-n)\,\sigma_s \\ \sigma^{(w)} &= -nS_w p_w I \\ \bar{p} &= S_w p_w \end{aligned} \tag{2.12}$$

This approach is valid for many engineering cases of interest. A more complete description can be found in Ref. Lewis and Schrefler (1998)

The effective stress σ' is defined as

$$\sigma' = \sigma + I\,\bar{p} \tag{2.13}$$

In order to relate effective and partial stresses, we can introduce the effective stress for the solid phase σ_s^e as

$$
\begin{aligned}
\sigma &= \sigma^{(s)} - n\bar{p}I \\
\sigma^{(s)} &= (1-n)\left(\sigma_s^e - \bar{p}I\right)
\end{aligned}
\tag{2.14}
$$

from where we obtain

$$\sigma = (1-n)\,\sigma_s^e - \bar{p}I \tag{2.15}$$

It follows that

$$\sigma' = (1-n)\,\sigma_s^e$$

In the case of viscous pore fluids where deviatoric stresses are important, the effective stress is given by

$$\sigma = \sigma' - I\,\bar{p} + n \sum_{\alpha=1}^{\text{nphases}} s^{(\alpha)}$$

Another point which is worth mentioning is that the effective stress above introduced has to be modified to account for compressibility of soil grains relative to that of solid skeleton. The total increment of strain $d\varepsilon$ can be decomposed into three parts: $d\varepsilon^g$ which accounts for the deformation of the solid grains, $d\varepsilon^0$ which describes changes of strain caused by other mechanisms than mechanical (thermal, for instance), and $d\varepsilon'$ which corresponds to the soil skeleton. It is possible then to write the constitutive equation as:

$$d\sigma' = D' : \left(d\varepsilon - d\varepsilon^0 - d\varepsilon^g\right)$$

The strain induced in the grains by a change in the pore pressure $\bar{p}$ is

$$d\varepsilon^g = -\frac{1}{3K_s}I\,d\bar{p}$$

where K_g is the volumetric stiffness of the soil grains. From here, we obtain

$$d\sigma' = D' : \left(d\varepsilon - d\varepsilon^0\right) + (D' : I)\frac{d\bar{p}}{3K_s}$$

The increment of total stress (Cauchy) will be given by:

$$d\sigma = D' : \left(d\varepsilon - d\varepsilon^0\right) - \left[I - \frac{D' : I}{3K_s}\right]d\bar{p} \tag{2.16}$$

As far as the term $D' : I$ is concerned, if we assume elastic behaviour of the soil skeleton, we will have

$$
\begin{aligned}
D'_{ijkl}\delta_{kl} &= \left(\lambda\delta_{ij}\delta_{kl} + \mu\delta_{ik}\delta_{jl} + \mu\delta_{il}\delta_{jk}\right)\delta_{kl} \\
&= \left(3\lambda' + 2\mu'\right)\delta_{ij} \\
&= 3K_T\delta_{ij}
\end{aligned}
$$

where K_T is the volumetric stiffness of the skeleton. Therefore,

$$
\left[I - \frac{D' : I}{3K_s}\right]\,d\bar{p} = \left(1 - \frac{K_T}{K_s}\right)\,d\bar{p} = (1 - \alpha)\,d\bar{p}
$$

where we have introduced the stiffness ratio $\alpha = K_V/K_s$. Using this result in the expression 2.16 we obtain

$$
\begin{aligned}
d\sigma &= D' : \left(d\varepsilon - d\varepsilon^0\right) - \alpha I\,d\bar{p} \\
d\sigma' &= D' : \left(d\varepsilon - d\varepsilon^0\right) + (1 - \alpha)\,I\,d\bar{p}
\end{aligned}
\tag{2.17}
$$

The value of α ranges from very small values in the case of soils, $\alpha \sim 0$, to values close to 0.7 in the case of rocks. Therefore, the definition of effective stress has to be modified in the case of soils having a stiff skeleton introducing the parameter α. Here we will make the assumption that α is equal to unity.

2.3   Balance of mass

We will introduce the material derivative of a magnitude ϕ following phase α as

$$
\frac{D^{(\alpha)}\phi}{Dt} = \frac{\partial\phi}{\partial t} + v^{(\alpha)}.grad\,\phi
\tag{2.18}
$$

One important relation which will be used later is

$$
\begin{aligned}
\frac{D^{(\alpha)}\phi}{Dt} &= \frac{D^{(s)}\phi}{Dt} + \left(v^{(\alpha)} - v^{(s)}\right)grad\,\phi \\
&= \frac{D^{(s)}\phi}{Dt} + \frac{w^{(\alpha)}}{nS_a}grad\,\phi
\end{aligned}
\tag{2.19}
$$

where we have introduced the Darcy's velocity of phase (α) relative to soil skeleton as

$$
w^{(\alpha)} = nS_a\left(v^{(\alpha)} - v^{(s)}\right)
\tag{2.20}
$$

The balance of mass equation for the solid phase is given by

$$
\frac{D^{(s)}\rho^{(s)}}{Dt} + \rho^{(s)}div\,v^{(s)} = 0
\tag{2.21}
$$

and for the fluid phases (α)

$$
\frac{D^{(\alpha)}\rho^{(\alpha)}}{Dt} + \rho^{(\alpha)}div\,v^{(\alpha)} = 0
\tag{2.22}
$$

where $\rho^{(s)} = (1-n)\rho_s$ and $\rho^{(\alpha)} = nS_\alpha\rho_a$. In the case of non saturated soils, we can write

$$\frac{D^{(s)}\{(1-n)\rho_s\}}{Dt} + (1-n)\rho_s div\ v^{(s)} = 0 \qquad (2.23)$$

$$\frac{D^{(\alpha)}}{Dt}(nS_\alpha\rho_a) + nS_\alpha\rho_a div\ v^{(\alpha)} = 0$$

with $\alpha = \{a, w\}$.

From here, we obtain

$$\frac{D^{(s)}n}{Dt} = \frac{1-n}{\rho_s}\frac{D^{(s)}\rho_s}{Dt} + (1-n)\ div\ v^{(s)} \qquad (2.24)$$

$$-\frac{D^{(\alpha)}n}{Dt} = \frac{n}{S_\alpha\rho_a}\frac{D^{(\alpha)}}{Dt}(S_\alpha\rho_a) + n\ div\ v^{(\alpha)}$$

If we now add both equations and use 2.19 we arrive to

$$div\left(\frac{w^{(\alpha)}}{S_\alpha}\right) + \frac{1-n}{\rho_s}\frac{D^{(s)}\rho_s}{Dt} + \frac{n}{S_\alpha\rho_a}\frac{D^{(\alpha)}}{Dt}(S_\alpha\rho_a) + div\ v^{(s)} = 0$$

for $\alpha = \{a, w\}$. If we neglect now thermal effects, we can elaborate further the second terms as follows:

$$\frac{1-n}{\rho_s}\frac{D^{(s)}\rho_s}{Dt} = \frac{1-n}{K_s}\frac{D^{(s)}\bar{p}}{Dt} \qquad (2.25)$$

In the case of a non saturated soil with $p_a = 0$ and $\bar{p} = S_w p_w$ we obtain

$$\frac{1-n}{K_s}\frac{D^{(s)}\bar{p}}{Dt} = \frac{1-n}{K_s}\left\{S_w\frac{D^{(s)}p_w}{Dt} + p_w\frac{D^{(s)}S_w}{Dt}\right\} \qquad (2.26)$$

The third term (water phase) can be obtained for non saturated soils with $p_a = 0$ as:

$$\frac{n}{S_w\rho_w}\frac{D^{(w)}}{Dt}(S_w\rho_w) = \left\{\frac{n}{K_w}\frac{D^{(w)}p_w}{Dt} + \frac{n}{S_w}\frac{D^{(w)}S_w}{Dt}\right\} \qquad (2.27)$$

Therefore, the balance of mass for the water (non saturated, $p_a = 0$) can be written as

$$div\left(\frac{\mathbf{w}}{S_w}\right) + \frac{n}{K_w}\frac{D^{(w)}p_w}{Dt} + \frac{(1-n)}{K_s}\frac{D^{(s)}p_w}{Dt} +$$

$$\frac{n}{S_w}\frac{D^{(w)}S_w}{Dt} + \frac{1-n}{K_s}p_w\frac{D^{(s)}S_w}{Dt} + div\ \mathbf{v}^{(s)} = 0 \qquad (2.28)$$

Sometimes, it is introduced for convenience the specific storage coefficient of the soil C_s as

$$C_s = n\frac{\partial S_w}{\partial p_w} \qquad (2.29)$$

From here we can obtain some particular cases of interest:

- Saturated soil

$$div \ w + \frac{n}{K_w} \frac{D^{(w)} p_w}{Dt} + \frac{(1-n)}{K_s} \frac{D^{(s)} p_w}{Dt} + div \ v^{(s)} = 0 \qquad (2.30)$$

This equation can be further simplificated by taking into account the relationship between material derivatives following the fluid and the solid phases 2.19

$$div \ w + \frac{n}{K_w} \frac{D^{(s)} p_w}{Dt} + \frac{(1-n)}{K_s} \frac{D^{(s)} p_w}{Dt} + div \ v^{(s)} + \frac{w}{K_w} grad \ p_w = 0$$

The last term is usually neglected in geotechnical analysis, and above equation reduces to

$$div \ w + \frac{1}{Q} \frac{D^{(s)} p_w}{Dt} + div \ v^{(s)} = 0 \qquad (2.31)$$

where

$$\frac{1}{Q} = \left(\frac{n}{K_w} + \frac{(1-n)}{K_s} \right) \qquad (2.32)$$

- Flow of dry geomaterials

In some circumstances, there may exist an important coupling between the pore air and the soil skeleton. Should this be the case, the balance of mass equation for the pore gas can be written in a similar way as

$$div \ w + \frac{1}{Q_a} \frac{D^{(s)} p_a}{Dt} + div \ v^{(s)} = 0 \qquad (2.33)$$

with

$$\frac{1}{Q_a} = \left(\frac{n}{K_a} + \frac{(1-n)}{K_s} \right) \qquad (2.34)$$

where have introduced the volumetric stiffness of the air K_a.

- The balance of mass equation of the mixture can be obtained by adding the equations for all constituents:

$$\frac{D^{(s)} \rho^{(s)}}{Dt} + \rho^{(s)} div \ v^{(s)} + \frac{D^{(\alpha)} \rho^{(\alpha)}}{Dt} + \sum_{\alpha=1}^{\text{nphases}} \rho^{(\alpha)} div \ v^{(\alpha)} = 0$$

from where, using (2.19) we get

$$\frac{D^{(s)} \rho}{Dt} + \rho \ div \ v^{(s)} + \sum_{\alpha=1}^{\text{nphases}} \left(\frac{w^{(\alpha)}}{n S_\alpha} \right) grad \ \rho^{(\alpha)} = 0 \qquad (2.35)$$

If all $w^{(\alpha)}$ are negligible, above equation reduces to

$$\frac{D^{(s)} \rho}{Dt} + \rho \ div \ v^{(s)} = 0 \qquad (2.36)$$

A particular case of interest is that of very fluid slurries, for which the mixture density can be assumed to be constant. The balance of mass reduces to:

$$div \ v^{(s)} = 0 \qquad (2.37)$$

2.4 Balance of momentum of pore fluid

The balance of momentum for a fluid phase (α) reads

$$\rho_\alpha \frac{D^{(\alpha)} v^{(\alpha)}}{Dt} = \rho_\alpha b + div\ \sigma_\alpha + R_{s\alpha} \qquad (2.38)$$

where b is the body forces vector, and $R_{s\alpha}$ is a force representing the interaction between the solid skeleton and the pore fluid (α), which can be assumed to follow Darcy's law

$$R_{s\alpha} = -k_\alpha^{-1} w^{(\alpha)} \qquad (2.39)$$

In above, k_α is the permeability matrix for phase (α),

$$k_\alpha = \frac{k_\alpha^{rel}.K_\alpha^{intr}}{\mu_\alpha} \qquad (2.40)$$

where k_α^{rel} is a relative permeability which depends on S_α, μ_α is the dynamic viscosity of fluid (α), and K_α^{intr} is the intrinsic permeability matrix which has dimensions of L^2. If the material is isotropic, then the matrix reduces to a constant. The permeability k_α has dimensions of $L^3.T\ /\ M$. It should be noted that in geotechnical analysis the permeability it is often defined in a slightly different way,

$$\bar{k}_\alpha = \rho_\alpha b\ k_\alpha \qquad (2.41)$$

with dimensions of LT^{-1}.

If we take into account the relation 2.19

$$\frac{D^{(\alpha)}\phi}{Dt} = \frac{D^{(s)}\phi}{Dt} + \frac{w^{(\alpha)}}{nS_a}\ grad\ \phi$$

we can write the balance of momemtum as

$$\rho_\alpha \left(\frac{D^{(s)} v^{(s)}}{Dt} + c.t. \right) = \rho_\alpha b + div\ \sigma_\alpha - k_\alpha^{-1} w^{(\alpha)} \qquad (2.42)$$

where the correcting terms $c.t.$ are given by

$$c.t. = \frac{D^{(s)}}{Dt}\left(\frac{w^{(\alpha)}}{nS_\alpha} \right) + \frac{w^{(\alpha)}}{nS_\alpha} grad\ v^{(s)} + \frac{w^{(\alpha)}}{nS_\alpha} grad\left(\frac{w^{(\alpha)}}{nS_\alpha} \right) \qquad (2.43)$$

In many geotechnical problems the correcting terms can be neglected , and $w^{(\alpha)}$ is obtained as

$$w^{(\alpha)} = k_\alpha \left(-\rho_\alpha \frac{D^{(s)} v^{(s)}}{Dt} + \rho_\alpha b + div\ \sigma_\alpha \right) \qquad (2.44)$$

This situation arises when (i) acceleration of fluid phases are small, and (ii) when cross products of $w^{(\alpha)}$ and $v^{(s)}$ can be neglected.

If the soil is saturated by water, and no viscous effects are present in the pore water, we can write

$$w = k_w \left(-\rho_w \frac{D^{(s)} v^{(s)}}{Dt} + \rho_w b - grad\ p_w \right) \tag{2.45}$$

and in the case of dry soils we will have

$$w = k_a \left(-\rho_a \frac{D^{(s)} v^{(s)}}{Dt} + \rho_a b - grad\ p_a \right) \tag{2.46}$$

2.5   Balance of momentum of the mixture

We will obtain the balance of linear momentum for the mixture combining the balance equations of the solid and fluid phases. Concerning the solid, we can write

$$\rho_s \frac{D^{(s)} v^{(s)}}{Dt} = \rho_s b + div\ \sigma_s + R_s \tag{2.47}$$

where R_s stands for the drag forces due to all fluid constituents.

Multiplying by $(1 - n)$, and taking into account the definition of partial stress $\sigma^{(s)} = (1 - n)\ \sigma_s$, we obtain

$$\rho_s (1 - n) \frac{D^{(s)} v^{(s)}}{Dt} = (1 - n)\ \rho_s b + (1 - n)\ div\ \sigma_s + (1 - n)\ R_s$$

and

$$\rho_s (1 - n) \frac{D^{(s)} v^{(s)}}{Dt} = (1 - n)\ \rho_s b + (1 - n)\ div\ \sigma^{(s)} + grad\ (n)\ div\ \sigma_s + (1 - n)\ R_s \tag{2.48}$$

The balance of momentum equations of all fluid phases can be cast as

$$\rho_\alpha n S_\alpha \left(\frac{D^{(s)} v^{(s)}}{Dt} + c.t. \right) = n S_\alpha \rho_\alpha b + n S_\alpha div\ \sigma_\alpha - n S_\alpha k_\alpha^{-1} w_\alpha$$

and taking into account

$$div\ \sigma^{(\alpha)} = (grad\ n S_\alpha)\ \sigma_\alpha + n S_\alpha\ div\ \sigma_\alpha$$

we arrive at

$$\rho_\alpha n S_\alpha \left(\frac{D^{(s)} \mathbf{v}^{(s)}}{Dt} + c.t. \right) = n S_\alpha \rho_\alpha \mathbf{b} + div\ \sigma^{(\alpha)} - (grad\ n S_\alpha)\ \sigma_\alpha - n S_\alpha k_\alpha^{-1} \mathbf{w}_\alpha \tag{2.49}$$

These equations are now added to 2.48, obtaining the balance of momentum for the mixture as:

$$\rho \frac{D^{(s)} \mathbf{v}^{(s)}}{Dt} + n S_\alpha \rho_\alpha \left\{ \frac{D^{(s)}}{Dt} \left(\frac{\mathbf{w}^{(\alpha)}}{n S_\alpha} \right) + \frac{w^{(\alpha)}}{n S_\alpha} grad\ \mathbf{v}^{(s)} + \frac{w^{(\alpha)}}{n S_\alpha} grad\ \left(\frac{\mathbf{w}^{(\alpha)}}{n S_\alpha} \right) \right\} =$$
$$\rho \mathbf{b} + div\ \sigma + \sigma_s\ grad\ n - \sum_\alpha \sigma_\alpha\ (grad\ n S_\alpha) \tag{2.50}$$

This equation can be simplified as

$$\rho \frac{D^{(s)} v^{(s)}}{Dt} = \rho b + div\ \sigma \qquad (2.51)$$

which is the form usually encountered in geotechnics.

3 Some useful forms in Computational Soil Mechanics: the u-pw formulation

In the preceding Section, we have presented the balance of mass and momentum equations for a mixture of solid particles and n_{phases} immiscible fluid phases occupying the pores, with some possible simplifications. The equations have been cast in a eulerian formulation, from which simple lagrangian forms involving small deformations can be immediately derived by neglecting the convective components of the material derivatives. The approach we propose in this work consists on modelling catastrophic landslides using a small deformation approach for the initiation phase followed by an eulerian approach for the propagation. Of course, in both phases it is of paramount importance taking into account the coupling between the soil particles and the pore fluids. Here we will present some useful forms for the initiation phase, and leave for the companion chapter the analysis of the propagation.

Therefore, we will not consider convective terms, and we will further assume that the accelerations of the pore fluids relative to the soil skeleton can be neglected. Validity of this assumption has been studied by Zienkiewicz et al. (1980), Zienkiewicz et al. (2000). The two cases we will study next are those of (i) non saturated soils with air at atmospheric pressure, and (ii) saturated soils. We will use displacements of the soil skeleton u instead of velocities.

In the former case, the system of equations describing the coupling are the following:

- Balance of mass for pore water

$$grad\ w + S_w \frac{\partial \varepsilon_v}{\partial t} + C_s \frac{\partial p_w}{\partial t} + \frac{1}{Q^*} \frac{\partial p_w}{\partial t} = 0 \qquad (3.1)$$

where ε_v is the volumetric strain

$$div\ v^{(s)} = \frac{\partial \varepsilon_v}{\partial t}$$

and

$$\frac{1}{Q^*} = \left\{ n \frac{S_w}{K_w} + \frac{1-n}{K_s} \left(S_w + p_w \frac{C_s}{n} \right) \right\}$$

- Balance of linear momentum (pore water)

$$-grad\ p_w + \rho_w b - k_w^{-1} w - \rho_w \left[\frac{\partial^2 u}{\partial t^2} + \frac{\partial}{\partial t} \left(\frac{w}{n S_w} \right) \right] = 0 \qquad (3.2)$$

- Balance of linear momentum (mixture)

$$div \ \sigma + \rho b - \rho \frac{\partial^2 u}{\partial t^2} - n\rho_w S_w \frac{\partial}{\partial t}\left(\frac{w}{nS_w}\right) = 0 \qquad (3.3)$$

- Constitutive equation

$$d\sigma' = D' : d\varepsilon$$

- Kinematics

$$d\varepsilon_{ij} = \frac{1}{2}\left(\frac{\partial u_i}{\partial x_j} + \frac{\partial u_j}{\partial x_i}\right) = S.du$$

Under certain conditions Zienkiewicz et al. (1980), we can neglect the relative accelerations of the fluid phase, and eliminate the relative displacements of the fluid phase w. This can be done by substituting the value of $div \ w$ obtained from 3.2

$$div \ w = \nabla^T \left[k_w\left(-\nabla p_w + \rho_w b - \rho_w \frac{\partial^2 u}{\partial t^2}\right)\right]$$

into the balance of mass of the fluid phase 3.1

$$div \ w + S_w \frac{\partial \varepsilon_v}{\partial t} + C_s \frac{\partial p_w}{\partial t} + \frac{1}{Q^*}\frac{\partial p_w}{\partial t} = 0$$

from where we obtain

$$\nabla^T\left\{k_w\left(-\nabla p_w + \rho_w b - \rho_w\frac{\partial^2 u}{\partial t^2}\right)\right\} + \left(C_s + \frac{1}{Q^*}\right)\frac{\partial p_w}{\partial t} + S_w \frac{\partial \varepsilon_v}{\partial t} = 0 \qquad (3.4)$$

The main advantage of this formulation is than the field variables reduce to two, displacements u and pressures p_w.

Therefore, the basic equations of this $u - p_w$ formulation are 3.3 and 3.4, together with the constitutive equation 2.17 and the kinematic relations between strains and displacements.

The simplification for saturated soils is immediate. The equation 3.4 reduces to:

$$\nabla^T\left\{k_w\left(-\nabla p_w + \rho_w b - \rho_w\frac{\partial^2 u}{\partial t^2}\right)\right\} + \frac{1}{Q}\frac{\partial p_w}{\partial t} + \frac{\partial \varepsilon_v}{\partial t} = 0 \qquad (3.5)$$

3.1 Boundary and Initial Conditions

The equations presented so far have to be complemented by suitable boundary and initial conditions for the problem variables. In the $u - p_w$ the conditions are the following:

(i) u prescribed on Γ_u

(ii) tractions prescribed on Γ_t, $\sigma.n = \bar{t}$, $\Gamma_u \cup \Gamma_t = \Gamma$, $\Gamma_u \cap \Gamma_t = \{\emptyset\}$

(iii) p_w prescribed on Γ_{pw}

(iv) Flux $-k_w\nabla p_w n$ prescribed along Γ_q, $\Gamma_{pw} \cup \Gamma_q = \Gamma$, $\Gamma_{pw} \cap \Gamma_q = \{\emptyset\}$

In the case of a non-saturated slope under rain, the boundary conditions to be applied can be simplified to a prescribed pressure (atmospheric) on the surface, but care should be taken as the inflow cannot be larger than the amount of percolating water. In this way, the saturation will increase within the material, and the effective stresses will decrease.

The initial conditions will be:

(i) Solid displacements and velocities at $t = 0$

(ii) Pore pressures at $t = 0$

4 Numerical Model: Discretization of the u-p_w model

The mathematical model consists on equations 3.3, 3.4,

$$div\ \sigma + \rho b - \rho \frac{\partial^2 u}{\partial t^2} - n\rho_w S_w \frac{\partial}{\partial t} \left(\frac{w}{nS_w} \right) = 0$$

$$\nabla^T \left\{ k_w \left(-\nabla p_w + \rho_w b - \rho_w \frac{\partial^2 u}{\partial t^2} \right) \right\} + \left(C_s + \frac{1}{Q^*} \right) \frac{\partial p_w}{\partial t} + S_w \frac{\partial \varepsilon_v}{\partial t} = 0$$

$$d\sigma' = D' : d\varepsilon$$

$$d\varepsilon_{ij} = \frac{1}{2} \left(\frac{\partial u_i}{\partial x_j} + \frac{\partial u_j}{\partial x_i} \right) = S.du$$

The system of partial differential equations can be discretized using standard Galerkin techniques, as described in Zienkiewicz et al. (1990a). After approximating the fields u and p_w as: $u = \mathbf{N}_u \bar{u}$, $p_w = \mathbf{N}_p \bar{p}_w$, it results in two ordinary differential equations:

$$\mathbf{M} \frac{d^2 \bar{u}}{dt^2} + \int_\Omega \mathbf{B}^T . \sigma' d\Omega - \mathbf{Q}\bar{p}_w - \mathbf{f}_u = 0 \tag{4.1}$$

where $\mathbf{B} = \mathbf{S}.\mathbf{N}_u$, and

$$\mathbf{Q}^T . \frac{d\bar{u}}{dt} + \mathbf{H}.\bar{p}_w + \mathbf{C}.\frac{d\bar{p}_w}{dt} - \mathbf{f}_p = 0 \tag{4.2}$$

The matrices given above are defined as:

$$\mathbf{M} = \int_\Omega \rho \mathbf{N}_u^T \mathbf{N}_u d\Omega$$

$$\mathbf{C} = \int_\Omega \frac{1}{Q^*} \mathbf{N}_p^T \mathbf{N}_p d\Omega$$

$$\mathbf{Q} = \int_\Omega S_w \alpha \mathbf{B}^T \mathbf{m} \mathbf{N}_p d\Omega \tag{3.3}$$

$$\mathbf{H} = \int \nabla \mathbf{N}_p^T k_w \nabla \mathbf{N}_p d\Omega$$

and

$$\mathbf{f}_u = \int_\Omega \mathbf{N}_u^T b d\Omega + \int_{\Gamma_t} \mathbf{N}_u^T \bar{t} d\Gamma$$

$$\mathbf{f}_p = \int_{\Gamma_q} \mathbf{N}_p^T k_w \frac{\partial p}{\partial n} d\Gamma + \int_\Omega \nabla \mathbf{N}_p^T k_w \rho_w b d\Omega - \left(\int_\Omega \nabla \mathbf{N}_p k_w \mathbf{N}_u d\Omega \right) \ddot{u} - \int_\Omega \mathbf{N}_p^T s_0 d\Omega \tag{3.4}$$

The term including accelerations in $\mathbf{f}_p$ is usually disregarded.

The time derivatives of $\mathbf{u}$ and $\mathbf{p}_w$ are approximated in a typical step of computation using the Generalized Newmark GN22 scheme for displacements and a GN11 for the water pressure Katona and Zienkiewicz (1985), Zienkiewicz and Taylor (1991).

If we introduce the following notation Katona and Zienkiewicz (1985):

$$\Delta\ddot{\mathbf{u}}^n = \ddot{\mathbf{u}}^{n+1} - \ddot{\mathbf{u}}^n$$

$$\Delta\dot{p}_w^n = \dot{p}_w^{n+1} - \dot{p}_w^n$$

$$\dot{\mathbf{u}}^{n+1} = \dot{\mathbf{u}}^{p,n+1} + \beta_1\Delta t\Delta\ddot{\mathbf{u}}^n$$

$$\mathbf{u}^{n+1} = \mathbf{u}^{p,n} + \tfrac{1}{2}\beta_2\Delta t^2\Delta\ddot{\mathbf{u}}^n \tag{3.5}$$

$$p_w^{n+1} = p_w^{p,n+1} + \theta\Delta t\Delta\dot{p}_w^n$$

where

$$\dot{\mathbf{u}}^{p,n+1} = \dot{\mathbf{u}}^n + \Delta t\ddot{\mathbf{u}}^n$$

$$\mathbf{u}^{p,n+1} = \mathbf{u}^n + \Delta t\dot{\mathbf{u}}^n + \tfrac{1}{2}\Delta t^2\ddot{\mathbf{u}}^n \tag{3.6}$$

$$p_w^{p,n+1} = p_w^n + \Delta t\dot{p}_w^n$$

we obtain the discretized system of equations valid in each time step:

$$\mathbf{M}\Delta\ddot{\mathbf{u}}^n + \int \mathbf{B}^T \sigma'^{n+1} - \theta\Delta t\mathbf{Q}\Delta\dot{\mathbf{p}}_w^n - \mathbf{F}_u^{n+1} = \mathbf{\Phi}_u = 0 \tag{3.7}$$

$$\beta_1\Delta t\mathbf{Q}^T\Delta\ddot{\mathbf{u}}^n + (\Delta t\theta\mathbf{H} + \mathbf{C})\Delta\dot{\mathbf{p}}_w^n - \mathbf{F}_p^{n+1} = \mathbf{\Phi}_p = 0 \tag{3.8}$$

where the unknown values are $\Delta\ddot{\mathbf{u}}^n$ and $\Delta\dot{\mathbf{p}}_w^n$

If this system is non-linear, it can be solved by using a Newton Raphson method with a suitable jacobian matrix:.

$$\begin{bmatrix} \dfrac{\partial\mathbf{\Phi}_u}{\partial\Delta\ddot{\mathbf{u}}} & \dfrac{\partial\mathbf{\Phi}_u}{\partial\Delta\dot{\mathbf{p}}} \\ \dfrac{\partial\mathbf{\Phi}_p}{\partial\Delta\ddot{\mathbf{u}}} & \dfrac{\partial\mathbf{\Phi}_p}{\partial\Delta\dot{\mathbf{p}}} \end{bmatrix}^{(i)} \begin{bmatrix} \delta(\Delta\ddot{\mathbf{u}}) \\ \delta(\Delta\dot{\mathbf{p}}) \end{bmatrix}^{(i+1)} = - \begin{bmatrix} \mathbf{\Phi}_u \\ \mathbf{\Phi}_p \end{bmatrix}^{(i)} \tag{3.9}$$

Using equations (3.7) and (3.8) we can write the above step as:

$$\begin{bmatrix} \mathbf{M} + \tfrac{1}{2}\Delta t^2\beta_2\mathbf{K}_T & -\theta\Delta t\mathbf{Q} \\ \beta_1\Delta t\mathbf{Q}^T & \Delta t\theta\mathbf{H} + \mathbf{S} \end{bmatrix}^{(i)} \begin{bmatrix} \delta(\Delta\ddot{\mathbf{u}}) \\ \delta(\Delta\dot{\mathbf{p}}) \end{bmatrix}^{(i+1)} = - \begin{bmatrix} \mathbf{\Phi}_u \\ \mathbf{\Phi}_p \end{bmatrix}^{(i)} \tag{4.3}$$

where $\mathbf{K}_T$ is the tangent stiffness matrix. $\mathbf{K}_T = \int \mathbf{B}^T \mathbf{D}_{ep} \mathbf{B} d\Omega$

5 Localized mechanisms of failure

One main fundamental difficulty of modelling localized failure problems is the ill posed-ness nature of the problem if the constitutive behaviour of the material is described by an elastoplastic model. As a direct consequence, the width of the shear band will depend on the mesh size h, tending to zero with it. In the limit $h \to 0$, the failure takes place in a surface of zero width. To overcome this problem, the description of the continuum has to be enriched by using non-local, gradient, Cosserat, or viscoplastic models Vardoulakis and Sulem (1995).

In addition to this, other difficulties are the following:

(i) Low order elements such as linear triangles and tetrahedra tend to present locking, and failure loads are overestimate. To solve this problem, mixed pressure and displace-ment elements or enhanced strain formulations can be used.

(ii) Low order elements are strongly affected by mesh alignment, and in some patho-logical cases the failure mechanism computed can be unrealistic. The solution consists on using more accurate elements, such as those based on enhanced strain or mixed stress-displacement or velocities, Mira (2002), Mabssout and Pastor (2003).

(iii) The interpolation spaces used for displacements and pore pressures have to fulfill the so called Babuska-Brezzi conditions, which prevents using the same shape functions for displacements and pore pressures unless some stabilization technique is used. This is the case of the enhanced strain formulations which can be stabilized by the technique proposed by Mira (2002).

(iv) The failure surface is smeared by the elements, resulting on a less fragile computed behaviour. A possible solution is to use mesh adaptivity, as proposed in Refs. Peraire et al. (1987) and Pastor et al. (1999).

Here we will present two examples. The first is used to show an application of adap-tive mesh refinement together with the stabilization technique proposed in Pastor et al. (1999), while the second uses a more classical element which can be found in commer-cial finite element codes, the 8 noded quadrilateral with bilinear pore pressures, where reduced integration (2 x 2) with control of hourglassing.

5.1 Failure of a vertical slope

The first problem we will consider is that of the failure of a 45° slope under undrained loading. The geometry and boundary conditions are given in Fig. 1. In the quasi-static limit, this is a simple Boundary Value Problem for which an analytical solution of collapse load exists Chen (1975):

$$\frac{F}{BC_u} = 3.57$$

It is important to note that this problem becomes ill-posed from a mathematical point of view which results in non-uniqueness and mesh dependence. We will not focus on these issues, but on how simple T3P3 are able to provide an accurate description of both limit load and failure mechanism.

We will use adaptive remeshing to improve resolution of the shear band as introduced by Peraire et al. (1987) and described in Ref. Zienkiewicz et al. (1995).

The soil has been assumed to be an elastoplastic material with a Tresca yield surface and no hardening.

Applied external forces F and vertical displacement δ at the point at which load is applied are described in terms of the non-dimensional parameters $\frac{F}{BC_u}$ and $\alpha = \frac{E\delta}{BC_u}$.

The soil domain is discretized with an initial mesh of linear T3P3, elements which can be seen in Fig. 2. The problem is solved using adaptive remeshing, and Fig.2 shows the first and third meshes together with the plastic strain contours at $\alpha = 15$. Finally, the load vs. displacement curves are given in Fig. 3.

The error in the computed limit load is below 5%, with a failure surface clearly defined.

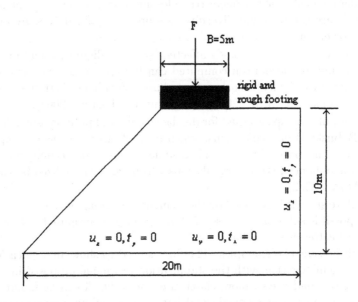

Figure 1. Slope stability problem: geommetry and boundary conditions

5.2   Failure of a cut slope under rain action

The first example we will present concerns the progressive failure of a non saturated clay cut slope due to rainfall. The height has been chosen as 10 meters and the slope is 2:1. Initial conditions have been simulated by assuming an initial horizontal soil layer which has been excavated to obtain the cut slope. The finite element mesh can be seen in Fig. 4. The interpolation functions are quadratic for displacements and bilinear for pore pressures. A reduced integration rule of 2x2 points has been used.

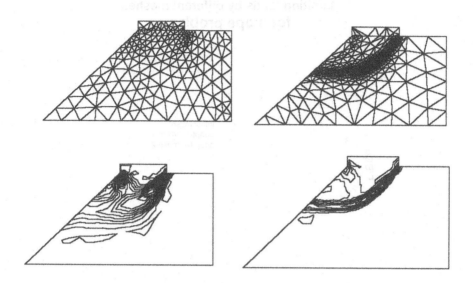

Figure 2. Adaptive meshes and plastic strain contours

Concerning boundary conditions, we have used $\partial_n p_w = 0$ and $u.n = 0$, where n is the normal to the artificial boundary. In the surface, we have assumed traction free conditions and a suction of $20\,kPa$.

The material has been assumed elasto-plastic, with a non associated Drucker–Prager yield criterion and a softening law characterized by $\varepsilon^p_{peak} = 0.05$, $\phi_{peak} = 20°$, $c_{peak} = 7\,kPa$, $\varepsilon^p_{res} = 0.20$, $\phi_{res} = 13°$ and $c_{res} = 2\,kPa$. The Young modulus varies with the effective confining pressure as

$$E = \min\left(400\,kPa,\ 25\left(p' + 100\right)\right) \qquad \nu = 0.2$$

After checking equilibrium steady state conditions after excavation (Fig. 5), the rain has been applied in two simplified ways:

1. Flooding, $p_w = 0$

2. Rain intensity known, flow $= 6$ mm/day.

In both cases failure is reached. Fig. 5 shows the plastic strain contours at different times. The failure occurs naturally and progressively through a localized surface. Fig. 6 displays the time evolution of the horizontal displacement of the midslope point.

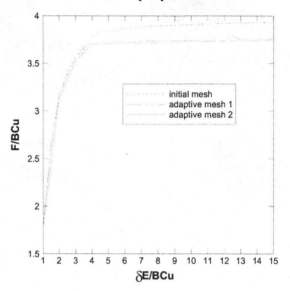

Figure 3. Slope stability problem: Force vs. displacement plot

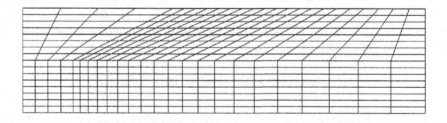

Figure 4. Finite element mesh

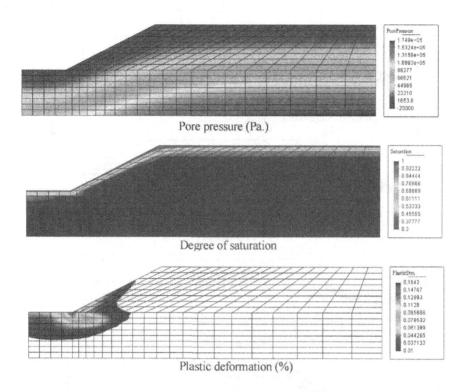

Pore pressure (Pa.)

Degree of saturation

Plastic deformation (%)

Figure 5. Equilibrium steady state conditions after excavation

6 Diffuse mechanisms of failure

If the soil has a low relative density, and tends to compact when sheared, failure mechanism can be different from the localized mode described in the preceding section. It affects not only the material on a narrow band or failure surface, but a much larger mass of soil. An example is shown in Fig.8 , where part of a vertical soil slope collapsed. Failure can be described as of a flow type. Indeed, flowslides use to involve loose, poorly compacted materials. Pore pressures developed may cause liquefaction of the soil, and liquefaction -or conditions close to it- may have played an important role on failure of slopes such as those of Aberfan, Jupille and Rocky Mountains coal mine dumps which will be described in next chapter.

Therefore, the key to properly model this type of landslides is the ability of the constitutive model to reproduce liquefaction.

This type of mechanism has been described as "diffuse" Darve and Laouafa (2002), Pastor et al. (2002), and will be illustrated with two examples.

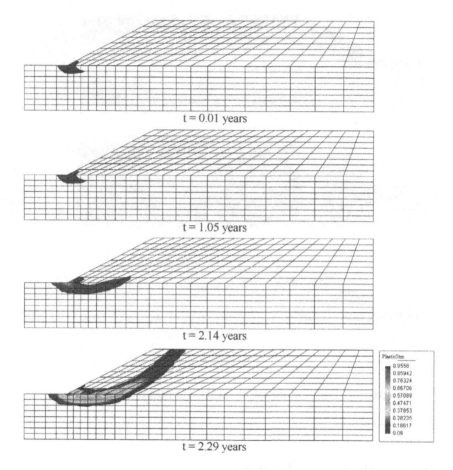

t = 0.01 years

t = 1.05 years

t = 2.14 years

t = 2.29 years

PlasticStrn
0.9558
0.85942
0.76324
0.66706
0.57089
0.47471
0.37853
0.28235
0.18617
0.09

Figure 6. Plastic strain contour

6.1   Liquefaction of a sand layer under earthquake loading

Fig. 9 shows a fully saturated sand layer which is subject to a horizontal earthquake. The seismic input is the accelerogram given in Fig. 10. The sand layer is modelled by a sand column with both sides and the bottom assumed impermeable. Pore pressures are assumed to be zero at the surface of the sand layer. Repeated boundary conditions are assumed for the lateral nodes, which implies that the displacement of a right hand side node is to be equal to that of the corresponding left hand side node.

The finite element mesh is shown in 9 and consists of stabilized four node quadrilateral elements, where bilinear shape functions are used both for displacements and pressures.

Concerning the material, it has been assumed to be a very loose sand. Constitutive behaviour has been modelled using the Pastor-Zienkiewicz model Mk3 described in Zienkiewicz et al. (2000). Material properties and other relevant data used in the

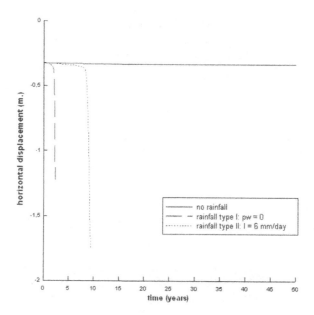

Figure 7. Horizontal displacement at middle height of slope

Figure 8. Failure of diffuse type on a vertical slope of loose saturated soil

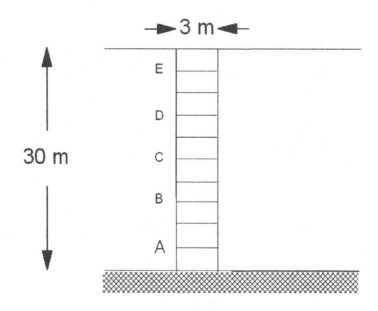

Figure 9. Soil layer problem: finite element mesh

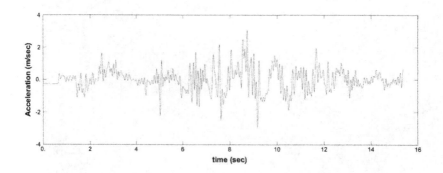

Figure 10. Seismic input for the horizontal layer problem

analysis are listed in Table 1.

Table 1. Data used in sand layer analysis

Soil data:	
Solid density, ρ_s	$= 2720 kg/m^3$
Fluid density, ρ_w	$= 980 kg/m^3$
Solid grain bulk modulus, K_s	$= 1.0E + 20Pa$
Fluid grain bulk modulus, K_w	$= 1.092E + 9Pa$
Passion's ratio, ν	$= 0.2875$
Porosity, n	$= 0.363$
Permeability, k	$= 2.1E - 3m/s$
Acceleration due to gravity, g	$= 9.81m/s^2$
Pastor-Zienkiewicz soil model parameters:	
K_{evo}	$= 45.0$
K_{eso}	$= 67.5$
M_g	$= 1.5$
M_f	$= 0.4$
$\alpha_f = \alpha_g$	$= 0.45$
β_0	$= 4.2$
β_1	$= 0.2$
H_0	$= 350.$
H_{u0}	$= 6.0E + 6Pa$
$\gamma_u = \gamma_{DM}$	$= 2.0$

The results can be seen in Figs. 11 and 12, where we have depicted the short and long term evolution of the pore pressure at stations A to E. It is interesting to note that pore pressure reaches a plateau between 5 and 15 seconds, depending on the depth, and then dissipate slowly.

Fig. 13 provides further insight in the phenomenon of liquefaction. We have plotted profiles of pore pressure at several instants together with the vertical stress. In this way, it is possible to see how liquefaction extends from the surface to a maximum depth of 18 m. approximately.

6.2 Liquefaction failure of a dyke under earthquake action

The case we will considered next is that of an earthquake induced flowslide in very loose saturated sand. The problem consists on a dike 10 m in height with slopes 2:1, founded on a sand layer which extends 10 m in depth and lies on a rigid rockbed. The material of both the dike and the foundation is a very loose saturated sand.

Initial conditions correspond to geostatic equilibrium under gravity forces. Pore pressure at the surface has been assumed to be equal to -20 kPa. The finite element mesh can be seen in Fig. 14 and consists on 500 quadrilaterals with 8 nodes for displacements and 4 for pore pressures. A reduced integration rule has been used in the solid part to avoid locking. The number of nodes is 1611, with 3535 degrees of freedom.

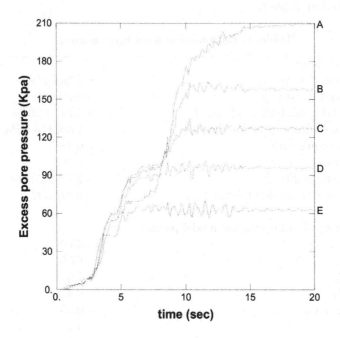

Figure 11. Short term evolution of pore pressure at several stations

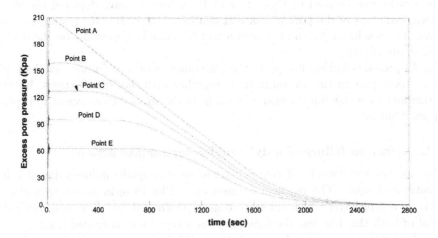

Figure 12. Long term evolution of pore pressures at several depths

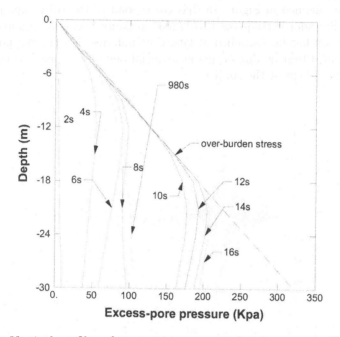

Figure 13. Vertical profiles of pore water pressure showing extent of liquefaction

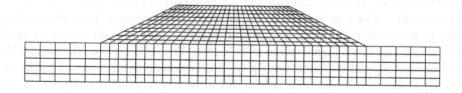

Figure 14. Finite element mesh

Loading is applied by prescribing horizontal accelerations at the base. We have used the input motion defined in Figure 15 that correspond to the NS component accelerogram of the El Salvador earthquake 13/01/2001 in Santa Tecla. A simplified absorbing boundary condition has been applied at lateral boundaries. Concerning pore pressures, it has been assumed that no flux occurs at artificial boundaries, and the constant value of -20 kPa has been kept at the surface.

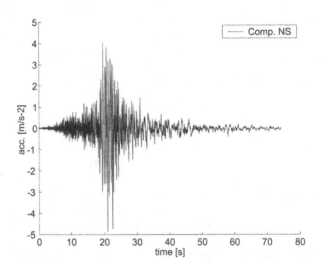

Figure 15. Seismic input: NS component accelerogram of the El Salvador earthquake 13/01/2001 in Santa Tecla

The behaviour of the loose material, $D_R = 27^o$, is represented, using the Pastor-Zienkiewicz model for sand. To better understand this behaviour we simulate a loading-unloading-reloading undrained triaxial test. The material loose progressively its resistant capacity, the effective stress decrease and the pore pressure increase, Fig. 16.

The results can be seen in Fig. 17, 18, 19, where the contours of pore pressure, plastic strain, p'/p'_0 and displacements are given at different times. Plastic strain accumulates in two zones, with much higher values at the outer. The ratio between the mean effective confining pressure and its initial value p'/p'_0 has been used as an indicator of the extent of the liquefied zones. From these results, it can be concluded that failure of the dike is caused by liquefaction of the outer liquefied zone.

6.3   A note on air liquefaction failure

If the soil is partially saturated, its behaviour depends on the coupling between the solid skeleton and the pore air and water. In the limit case of a dry soil, the air has to flow out from the pores for the material to consolidate, but typical air consolidation times are much smaller than those of the water. Therefore, in practical cases, the role of air pressures is neglected, as the characteristic time of loading is much larger than consolidation time. However, it is possible to imagine situations with much smaller

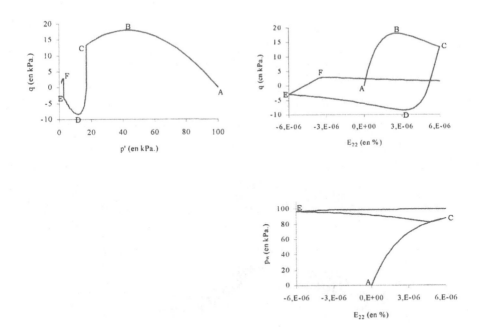

Figure 16. Nondrained triaxial simulation of the loose sand, loading, unloading, reloading

loading times, where coupling between pore air and soil skeleton plays a paramount role. This is the case of fluidized granular beds, just to mention a particular example of industrial interest.

Bishop (1973) describes the case of Jupille flowslide, which happened in Belgium in February 1961. A tip of uncompacted fly ash located in the upper part of a narrow valley collapsed, and the subsequent flowslide travelled for about 600 m at very high speeds (130 km/h) until it stopped. Triggering mechanism was suggested to be "collapse due to undermining of a steep, partly saturated and slightly cohesive face". Bishop referred to Calembert and Dantinne (1964), who pointed out the role of the entrapped air. The mechanism of pore air consolidation can explain the "sort of fog" which formed above the flowing material, as warmer air met the colder winter air in the exterior. The "honeycomb" like structure of soil in the tip was responsible of its sudden collapse when movement started.

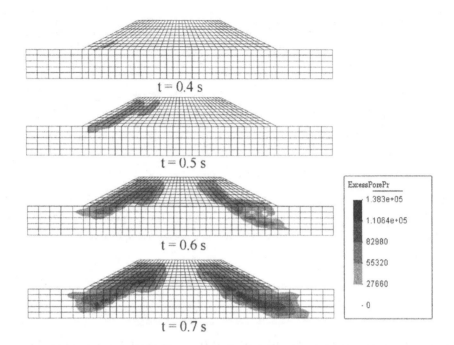

Figure 17. Evolution of excess pore pressure contour (Pa.)

7 Conclusions

We have proposed in this paper a theoretical framework for modelling the initiation mechanisms of catastrophic landslides. First, the equations describing the coupling have been presented following an eulerian approach close to the mixture theory. The main advantage is that they provide a unified formulation for both initiation and propagation phases. The system of partial Differential Equations is then discretized using the classical Galerkin Finite Element Method.

Finally, applications to localized and diffuse failure have been presented.

Bibliography

M.A.Biot, General theory of three-dimensional consolidation. **J.Appl.Phys. 12**, 155-164, 1941.

M.A.Biot, Theory of elasticity and consolidation for a porous anisotropic solid. **J.Appl.Phys. 26**, 182-185, 1955.

A.W.Bishop, The stability of tips and spoil heaps. **Quart.J.Eng.Geol. 6**, 335-376, 1973.

L.Calembert & R.Dantinne, The avalanche of ash at Jupille (Liege) on February 3rd, 1961. From: **The commemorative volume dedicated to Professeur F.Campus**, pp. 41-57, Liége, Belgium, 1964.

W.F.Chen, **Limit Analysis and Soil Plasticity**, Elsevier, 1975.

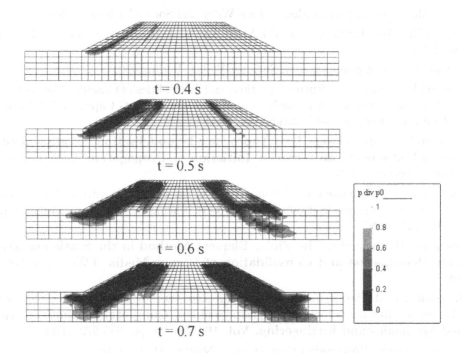

Figure 18. Evolution of p'/p'_0 contour

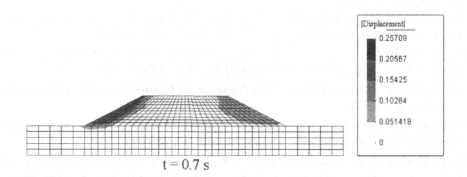

Figure 19. Deformation contour (m)

O.Coussy, Mechanics of Porous Media, John Wiley and Sons, Chichester, 1995.

R.Dikau, D.Brundsen,L.Schrott and M.L.Ibsen, Landslide Recognition, John Wiley and Sons, 1996.

R. de Boer, Theory of porous media, Springer-Verlag, Berlin,2000.

F.Darve and F.laouafa, Constitutive Equations and Instabilities of Granular Materials, in Modeling and Mechanics of Granular and Porous Materials, G.Capriz, V.N.Ghionna and P.Giovine (Eds), pp.3-44, Birkhäuser, 2002.

Jose Antonio Fernández-Merodo, **Une Approche a la modelisation des glissements et des effonfrements de terrains: Initiation et Propagation**, Ecole Centrale de Paris, December 2001.

M.G.Katona and O.C.Zienkiewicz, "A unified set of single-step algorithms, Part 3: the beta-m method, a generalisation of the Newmark scheme", **Int.J.Num.Meth:eng.**, **21**, 1345-1359, 1985.

R.L.Lewis and B.A.Schrefler, **The Finite Element Method in the Static and Dynamic Deformation and Consolidation of Porous Media**, J.Wiley and Sons, 1998.

M. Mabssout and M. Pastor, A Taylor-Galerkin algorithm for shock wave propagation and strain localization failure of viscoplastic continua, **Computer Methods in Applied Mechanics and Engineering, Vol. 192**, No.7-8, pp. 955-972, 2003.

P.Mira, Ph.D.Thesis, Universidad Politécnica de Madrid, Madrid, 2002.

M.Pastor, T.Li,X.Liu and O.C.Zienkiewicz, "Stabilized Low order Finite Elements for Failure and Localization Problems in Undrained Soils and Foundations", **Comp.Meth.Appl.Mech.Eng. 174**, 219-234, 1999.

M.Pastor, M.Quecedo, J.A.Fernández Merodo, M.I.Herreros, E.González, and P.Mira, Modelling Tailing dams and mine waste dumps failures, **Geotechnique, Vol.52**, No.8, pg.579-591, 2002.

M.Pastor and C.Tamagnini (Eds), Numerical Modelling in Geomechanics, Revue Française du génie civil, Vol.6, 2002, Hermes-Lavoisier, Paris 2002.

J.Peraire, M.Vahdati, K.Morgan and O.C.Zienkiewicz, "Adaptive remeshing for compressible flow computations", **J.Comp.Phys.**, **72**, 449-466, 1987.

I.Vardoulakis and J.Sulem, **Bifurcation Analysis in Geomechanics**, Blakie academic & professional, 1995.

O.C.Zienkiewicz, C.T.Chang and P.Bettess. Drained, undrained, consolidating dynamic behaviour assumptions in soils. **Geotechnique 30**, 385-395, 1980.

O.C.Zienkiewicz, A.H.C.Chan, M.Pastor, D.K.Paul and T.Shiomi. Static and dynamic behaviour of soils: a rational approach to quantitative solutions. I. Fully saturated problems **Proc.R.Soc.Lond. A 429**, 285-309, 1990.

O.C.Zienkiewicz, Y.M.Xie, B.A.Schrefler, A.Ledesma and Bicanic.N . Static and dynamic behaviour of soils: a rational approach to quantitative solutions. II. Semi-saturated problems **Proc.R.Soc.Lond. A 429**, 311-321, 1990.

O.C.Zienkiewicz, A.H.C.Chan, M.Pastor, B.Schrefler and T.Shiomi, **Computational Geomechanics**, J.Wiley and Sons, 2000.

O.C.Zienkiewicz, M. Huang and M. Pastor. "Localization problems in plasticity using finite elements with adaptive remeshing", **Int.J.Num.Anal.Meth.Geomechs. 19**, 127-148, 1995.

O.C.Zienkiewicz and T.Shiomi. Dynamic behaviour of saturated porous media: The generalised Biot formulation and its numerical solution. **Int.J.Num.Anal.Meth.Geomech.,8**, 71-96, 1984.

O.C.Zienkiewicz and R.L.Taylor, The Finite Element Method (4th. edition), Vol.2, McGraw-Hill, 1991.

Modelling of Landslides: (II) Propagation[*]

M.Pastor [*†] , M.Quecedo [††] , E.Gonzalez [*†] , M.I.Herreros [*†] ,
J.A.Fernandez Merodo [*†] and P.Mira [*†]

[*] Centro de Estudios y Experimentación de Obras Públicas, Madrid, Spain
[†] M2i (Math.Model.Eng.Group), Department of Applied Mathematics,
ETS de Ingenieros de Caminos, UPM Madrid, Spain
[‡] ENUSA, Madrid, Spain

Abstract This paper presents a numerical model wich can be used to simulate phenomena such as flowslides, avalanches, mudflows and debris flows. The proposed approach is eulerian, and balance of mass and momentum equations are integrated on depht. Depending on the material involved, the model can implement rheological models such as Bingham or frictional fluids. A simple dissipation law allows the approximation of pore pressure dissipation in the sliding mass. Finally some applications are presented.

1 Introduction

The term "landslide" embraces a wide variety of phenomena, from the very slow reptations with velocities in the rang of cm/yr, to the fast catastrophic flowslides and debris flows, where velocities can be of the order of 60 Km/h. The causes, origin, and nature of these phenomena are also very different from each other. Among the causes we can mention the following: (i) changes in soil or rock properties due to weathering, (ii) changes of effective stresses induced by variations in pore pressures caused in turn by the rain, (iii) earthquakes, and (iv) loading or change of geometry due to human action. The slopes can be natural, or man made, as in the case of tailing dams and mine waste dumps.

We will focus here in the propagation phase of fast landslides, specially flowslides, debris flow and mudflows. They are dangerous because they can propagate over long distances in very short times without warning, reaching zones which were considered previously "safe".

We will consider here three **main types** of fast landslides: (i) avalanches of rocks and granular materials, (ii) flowslides, and (iii) debris flows. Concerning avalanches, they consist on flows of dry materials, with a high permeability to the air, so the pore pressure effects can be neglected.

From an engineering point of view, it is important to:

[*]The authors gratefully acknowledge the financial support provided by the Spanish Agency for International Cooperation (AECI), by the Spanish Ministry of Science and Technology (Projects Andes and Ramón y Cajal contract) and by the European Union (Projects Diga and Lamé).

(i) Assess the risk of flowslides occurrence, providing remediation measures such as soil reinforcement or improvement.

(ii) Predict the characteristics of the flowslide (mass of material involved, velocity, runout distance and path of propagation), in order to design suitable protection and channeling structures.

Unfortunately, classical engineering tools such as slip circles or finite element codes with Mohr-Coulomb like models are unable to provide accurate predictions even for the initiation phase. Which is why more accurate models are needed to describe the propagation phase.

One important characteristic common to the three types is that the moving mass of soil behaves in a fluid-like manner. Once failure has been triggered, the soil mass will move with a velocity which will depend largely on the type of problem. In some cases (flow slides, mudflows,...) the material will flow in a "fluid-like" manner. There exists a strong coupling between the solid and the fluid phases, and the equations describing the movement are similar to those of the initiation phase. However, a first simplification of assuming a single phase can be made in two limit cases (i) very permeable or dry materials, and (ii) materials for which the time scale of consolidation is much larger than that of propagation. Examples of both situations are rock avalanches and mudflows. Care should be taken when selecting constitutive material properties, as the apparent friction angle can be much smaller than the effective, because of the pore pressures. In fact, liquefied material can flow over slopes much smaller than its effective friction angle.

If some assumptions are made about the vertical structure of the flow, it is possible to integrate the balance equations in depth, arriving to the so-called "depth-integrated" equations or "shallow water equations" in coastal and hydraulic engineering. The equations can be further simplified by integrating on cross sections, arriving to simple 1D models.

The depth integrated equations can be cast either in an eulerian or a lagrangian form. The second approach has been used mainly in one dimensional situations. The models proposed by Savage and Hutter (1991), Hutter and Koch (1991), Hungr (1995) and Rickenmann and Koch (1997) incorporate features such as introducing a K_0 coefficient relating the horizontal and vertical stress components, or formulating the equations in a curvilinear coordinate system following the terrain. However, lagrangian formulations present some important difficulties in the 2D case. For instance, we can mention the problem of bifurcation of the flow when arriving to an obstacle, or merging after it has been passed. Because of this problem, other authors have chosen eulerian formulations (Vulliet and Hutter (1988); Laigle and Coussot (1997); Laigle (1997), Pastor et al. (2002)).

In real situations it is necessary to predict also the sliding mass and the initial conditions, including pore pressures. Moreover, it is important to know whether the slide will evolve or not into a flowslide. Nowadays, it is difficult to provide accurate answers to these questions. The fundamental difficulty is the "phase transition" between the solid and fluidized states (and viceversa), for which a consistent model is required.

Most of the examples presented in the literature deal with flowslides which have been produced in the past. Therefore, the sliding mass is known, and both material parameters and the basal pore pressure factor can be determined by back analysis.

The approximation proposed by the authors consists on using non-linear coupled models implementing suitable constitutive equations to determine the extent of the mobilized mass of soil together with the initial distribution of pore pressures. The modelling method will consist on determining the initial conditions (stresses and pore pressures), and then simulating the failure mechanism (changes in pore pressures induced by rain, external loading, earthquake,..). Even if the external loading is not of dynamic nature, it is highly recommended to perform the analysis taking into account possible accelerations which may develop when failure approaches. Otherwise, loss of convergence of non-linear iterations will stop the process. This is the case, for instance, of an applied load higher than the limit load. Indeed, these computations are usually carried out under displacement control.

1.1   Avalanches

When a rock slope fails, large blocks can disaggregate into smaller particles, which behaviour will be closer to that of a fluid. Another example is the discharge of granular materials from silos. In the solid state, when the material is at rest, the stress field can be obtained using suitable constitutive equations for granular media and numerical models such as non-linear finite elements. Two neighbour particles will continue being so when the stresses change. However, once the trap opens, the paths followed by them may diverge, and the behaviour is of fluid type.

Velocities of rock avalanches can reach tens of meters per second, propagating over kilometers. Total mass of rock varies, but can be of the order of million cubic meters.

As examples we can mention the rock avalanches of Randa (Switzerland, 1991) and Valtellina (Italy, 1987) (Fig. 1).

1.2   Debris Flows

The main difference between flowslides and debris flows is the amount an origin of the water. In the case of flowslides, they can be of course triggered by heavy rains, but the mechanism of fluidification is the tendency of the soil to compact which causes pore pressure to increase.

On the contrary, the origin of debris flows is different. A typical case consists on a zone where rainfall erodes the slopes and concentrates into steep, narrow canyons (ravines or gullies). The material is transported along the valley and it is deposited in the fan which develops when the velocity falls below a critical value due to a widening of the valley or a reduction of the slope.

This mechanism can be combined with others. For instance, Fig. 2 Du et al. (1987) shows several landslides which have provided additional material in the ravine.

Depending on topography, rain intensity, etc., debris flows can also be produced in larger valleys, where tributary ravines arrive. In this case, material provided by secondary debris flows can be eroded by the main stream and become part of a larger flow. The slope of the main valley can be much smaller, but this does not prevents large velocities of debris flows to occur (up to 60 Km/h). After the debris flow stops, the current can wash out finer particles, and erode the deposit forming a new bed. Patterns can become

Figure 1. Rock avalanche of Valtellina (Italy,1987)

complex at junctions. Fig. 3 shows the JiangJia ravine, at a point close to Dongchuan Observation Station.

Debris flows are quite often of catastrophic nature. They cause:

- Destruction of farmland and crops
- Looses of human lives, houses and equipment in cities close to active ravines, as it happens, for instance in Dongchuan (Yunnan, China).
- Partial destruction of lifelines such as roads, railroads, etc. In China, the 2400 Km long road from Chengdu (Sichuan) to Lhasa (Tibet) is cut 2 months per year, and in the south province of Yunnan railroads are often interrupted. Fig. 4 shows the result of an event which took place on September the 9th of 1999, burying some 600 m of railroad.

Another example of debris flow is that of Venezuela (Dec.1999), where a previous period of rain had softened slopes. Heavy rain then caused general landslides,

Figure 2. Landslides provide material for debris flows (Du et al 1987)

providing the solid material of the debris flows. Hugh rock blocks were transported, causing in some cases severe damage to buildings. Fig. 5 shows some an example of the damage caused.

1.3 Flowslides

A paramount aspect of flowslides is the fluidization or liquefaction process, in which the soil mass is transformed into a fluid like material. Most of the soils involved in flowslides are very loose and metastable, with a strong tendency to compact under shear. If time of loading is much smaller than consolidation time, pore pressures (of water or air) will make the effective stresses path to approach failure conditions. In the limit, the mean hydrostatic confining pressure can become very close to zero, and the soil will behave as a viscous fluid in which buildings can sink, as it happened in Niigata during the 1966 earthquake.

It has to be mentioned here that this mode of failure can be exhibited also by non-saturated soils such as those of volcanic origin. Indeed, collapse of the mate-

Figure 3. Junction of debris flows

Figure 4. Railroad cut by debris flow)

Figure 5. Damage to buildings in Venezuela (Photograph by Prof.Zhang)

rial under the loading induced by an earthquake can make the pore air pressure to increase. The phenomenon is controlled by two characteristic time scales, a characteristic time for the consolidation and a characteristic time of loading. If the former is much larger than the latter, there will be not enough time for dissipation of air pore pressures, and the material will arrive to a "dry" liquefaction. Bishop describes the failure of a fly ash tip at Jupille in Belgium and refers to the explanation provided by Calembert and Dantinne (1964). Some of the catastrophic landslides caused in El Salvador by the 13th of January 2001 earthquake can be explained by this mechanism. Fig. 6 shows an aerial view of the dry flowslide triggered as a consequence of the 13th February earthquake.

Figure 6. Flow slide at Santa Tecla, El Salvador (2001)

One classical example of flowslide is that of the failure of Tip No.7 flowslide at Aberfan (1966). In October 21st 1966, a flow slide developed at a tip of loose colliery waste in Aberfan. It propagated downhill and into the village of Aberfan, causing 144 deaths. The material was loose coal mine waste, which was deposited by end-tipping. The height of the tip was 67 m. from the toe, and the natural slope of natural terrain was 12%. The failure mechanism and geotechnical properties have been described by Bishop et al. (1969) and summarized in "The stability of tips and spoil heaps" Bishop (1973) and "A sliding- consolidation model for flow slides" Hutchinson (1986), where the interested reader can find a detailed description. Here we will recall some aspects which will are of interest in the analysis.

First of all, failure was caused by artesian pore pressures at the toe, which probably saturated the lower part of the tip, while the upper part remained unsaturated. Bishop reports that "...the rescue work was complicated by water which flowed, after the slip, from the sandstone at the base of the rotational slip, where the boulder clay and head was stripped off...". Once failure was triggered, the flowslide propagated downhill 275 m., then divided into two a north and a south lobes. It was the larger south lobe which run into the village, while the northern lobe stopped after reaching an embankment.

Aberfan flowslide had a runout of 600 m., although it is believed that it could have been larger if no obstructions were within his path. The velocity has been estimated from witnesses as 4.5-9 m/s.

2 Rheological models

2.1 Introduction

Once downhill movement has started, material behaviour becomes more and more "fluid-like". The problem of modelling such fluid is complex, and it will depend on the type of mixture. In the simplest case of a flow of a granular material without interstitial water, it could be thought of being composed of a single phase. However, the flow presents inverse segregation, and the coarsest fraction will move upwards. Therefore, this phenomenon cannot be modelled if the assumption of a single phase material is made. The problem gets more complicated if an interstitial fluid such as water of mud is present.

A first simplification commonly found consists in studying the overall mixture behaviour, formulating ad-hoc models for it. This approach precludes the relative movement of the interstitial fluid relative to the solid fraction, and does not allow modelling of stabilization of the flow using bottom drainage systems.

Next refinement consists on considering two phases, a granular skeleton with voids filled with either water of mud. If the shear resistance of the fluid phase can be neglected, the stress tensor in the mixture can be decomposed into a "pore pressure" and a effective stress.

The purpose of this chapter is to present some of the models which have been used in the past, providing values of their material parameters whenever they are available. Most of them have been formulated in terms of total stresses for one phase, but can be generalized including the stresses of the fluid. It seems reasonable to consider one phase in the case of mudflows, and two in the case of debris flows.

The difference could be the ability of the fluid phase to percolate trough the solid. First of all, we will assume that the stress tensor has two parts, a thermodynamic pressure p and an "extra stress" σ^* which depends on the rate of deformation tensor $\mathbf{D}$. Using the representation theorems of tensor functions of a tensor variable we arrive at:

$$\sigma = -p\mathbf{I} + \Phi_0\mathbf{I} + \Phi_1\mathbf{D} + \Phi_2\mathbf{D}^2 \tag{2.1}$$

where $\mathbf{I}$ is the second order identity tensor, and $\Phi_k, k = 0..3$, are functions of the three invariants of the rate of deformation tensor.

If the fluid is incompressible, dev $\mathbf{D} = 0$. From here on, we will assume that material is incompressible, unless otherwise specified. This approach is consistent with the well known fact that soils fail at constant volume.

The stress tensor can be decomposed into hydrostatic and deviatoric components, as usually done in Continuum Mechanics:

$$\sigma = -\tilde{p}\mathbf{I} + \tau \tag{2.2}$$

where it can be seen that the deviatoric component τ depends on the rate of deformation $\mathbf{D}$, and the hydrostatic is given by:

$$\tilde{p} = p - \left(\Phi_0 + \frac{1}{3}\Phi_2 \mathrm{tr}\, \mathbf{D}^2 \right) \tag{2.3}$$

$$= p - \left(\Phi_0 + \frac{2}{3}\Phi_2 I_{2D} \right) \tag{2.4}$$

where we have used the definition of the second invariant of $\mathbf{D}$:

$$I_{2D} = \frac{1}{2}\,\mathrm{tr}\, \mathbf{D}^2$$

One interesting case is that of a simple shear flow, which is a two dimensional flow (in x and z directions, for instance), with velocities v depending only on z. It can be shown easily that tensor $\mathbf{D}$ reduces to

$$\mathbf{D} = \begin{pmatrix} 0 & 0 & \frac{1}{2}\frac{\partial v}{\partial z} \\ 0 & 0 & 0 \\ \frac{1}{2}\frac{\partial v}{\partial z} & 0 & 0 \end{pmatrix}$$

and $\mathbf{D}^2$ is

$$\mathbf{D}^2 = \begin{pmatrix} \frac{1}{4}\left(\frac{\partial v}{\partial z}\right)^2 & 0 & 0 \\ 0 & 0 & 0 \\ 0 & 0 & \frac{1}{4}\left(\frac{\partial v}{\partial z}\right)^2 \end{pmatrix}$$

The second invariant I_{2D} of the rate of deformation tensor is now:

$$I_{2D} = \frac{1}{2}\left(\frac{\partial v}{\partial z}\right)^2$$

and the components of the stress tensor are:

$$\sigma_{xx} = \sigma_{zz} = -\bar{p} + \tau_{xx} \tag{2.5}$$

$$= -p + \Phi_0 + \frac{1}{4}\Phi_2 \left(\frac{\partial v}{\partial z}\right)^2$$

$$\sigma_{yy} = -p + \Phi_0 \tag{2.6}$$

$$\sigma_{xz} = \frac{1}{2}\Phi_1 \left(\frac{\partial v}{\partial z}\right) \tag{2.7}$$

It is important to know that Φ_k are functions of the invariants of $\mathbf{D}$ I_{2D} and I_{3D} as $I_{1D} = 0$.

The normal stresses depend on the rate of deformation through Φ_0 and $\Phi_2 \left(\frac{\partial v}{\partial z}\right)^2$. If Φ_0 were a constant, it could be included with p in a single term. Other possibilities can be dealt with by modifying Φ_2 accordingly. For instance, for a linear dependence on $\left(\frac{\partial v}{\partial z}\right)$, Φ_2 can be proportional to $\left(\frac{\partial v}{\partial z}\right)^{-1}$. Therefore, the pressure caused by the intergranular stress can be accounted for by the term in $\mathbf{D}^2$, which only causes normal stresses.

In the case of plane, iconmpressible flow, i.e., only the velocities along X and Z are not zero, $\mathbf{D}$ and $\mathbf{D}^2$ are:

$$\mathbf{D} = \begin{pmatrix} D_{xx} & 0 & D_{xz} \\ 0 & 0 & 0 \\ D_{xz} & 0 & D_{zz} \end{pmatrix}$$

and

$$\mathbf{D}^2 = \begin{pmatrix} D_{xx}^2 + D_{xz}^2 & 0 & D_{xz}(D_{xx} + D_{zz}) \\ 0 & 0 & 0 \\ D_{xz}(D_{xx} + D_{zz}) & 0 & D_{xx}^2 + D_{xz}^2 \end{pmatrix}$$

from which the stress components are:

$$\sigma_{xx} = -p + \Phi_0 + \Phi_1 D_{xx} + \Phi_2 \left(D_{xx}^2 + D_{xz}^2\right)$$

$$\sigma_{yy} = -p + \Phi_0 \tag{2.8}$$

$$\sigma_{zz} = -p + \Phi_0 + \Phi_1 D_{zz} + \Phi_2 \left(D_{xx}^2 + D_{xz}^2\right) \tag{2.9}$$

$$\sigma_{xz} = \Phi_1 D_{xz} + \Phi_2 D_{xz} (D_{xx} + D_{zz}) \tag{2.10}$$

Again, the shear stress depends only on $\Phi_1 \mathbf{D}$, and the normal stresses on the flow plane are increased by the extra-stress components.

2.2 Phenomenological Aspects

To develop suitable constitutive relations it is necessary to have experimental results obtained under controlled conditions. It is convenient to use simple stress and strain rate fields to obtain the basic features of the model. At a latter stage, more complex tests can be introduced to improve our understanding.

Testing apparatuses used to reproduce simple flows are called rheometers. Most of them aim to obtain fluid viscosity -or relations between shear stress and rate of shear strain-. Simplest conditions are those of simple shear flows, where the fluid is sheared on a plane XZ, the velocity being parallel to X (Fig. 7).

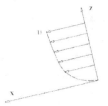

Figure 7. Simple shear flow

Debris flows and mudflows are terms covering a wide range of phenomena such as flows of dry boulders or granular materials in a mudlike fluid.

Sometimes, due to the size of particles, the size of a representative volume element will differ greatly from one case to another. Therefore, it will be impossible to reproduce a flow of large boulders in the laboratory because of the large size of the apparatus needed for this task.

The study of the flow of granular materials of very small diameter (in the mm range) in a viscous fluid can provide insight into the basic features of flow. From here, constitutive equations for larger grains can be developed, provided they are validated against data obtained either in large scale laboratory flumes or in real debris or mudflows.

The first rheometer built to study the flow of granular particles is due to Bagnold (1954), which is sketched in Fig. 8.

Bagnold used the annular space between two drums to create a simple shear Couette flow when the outer drum rotated keeping the inner stationary.

The granular material consisted on 0.132 cm. diameter spherical droplets of a mixture of paraffin wax and lead stearate, with a density equal to that of water.

Other experimental devices which are worth mentioning are those of Hanes and Inman (1985) and the rolling sleeve rheometer of Johnson and Martosudarmo (1997).

In Bagnold's rheometer, it is possible to study the influence of grains concentration, shear strain rate and fluid viscosity. The main results obtained in his pioneering work are:

- For a given mixture, there are two regimes:
  * The inertia regime, where the effects of collisions between grains are dominant and the shear stress is proportional to the square of the shear strain rate.

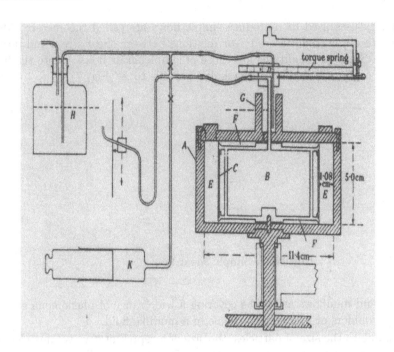

Figure 8. Bagnold's Rheometer (Bagnold 1954)

* The macro viscous regime, where viscosity effects dominate and the relation between shear stress and strain rate is linear.
- Shearing causes an extra pressure of dispersive nature, which was found to be proportional to the shear stress.

Fig. 9 and 10 show the dependence of the shear and normal stresses on the rate of shear strain. In the Figure, taken from Bagnold (1954), σ is the density, D the particle diameter, and T and P the shear and normal stresses respectively. λ is a linear concentration defined as:

$$\frac{1}{\lambda} = (\frac{C_{\max}}{C})^{\frac{1}{3}} - 1 \qquad (2.11)$$

and C and $C_{\max}$ are the concentration and the maximum possible concentration, which for spheres is $\pi/3\sqrt{2} = 0.74$.
Bagnold (1966) suggested two critical values of λ, 17 for general shearing and 14 for fluidification (with zero residual stress at $\mathbf{D} = \mathbf{0}$).

2.3 Newtonian Fluids

The simplest case is that of a Newtonian Fluid, for which

$$\sigma = -p\,\mathbf{I} + \tau = -p\,\mathbf{I} + 2\,\mu\,\text{dev}\mathbf{D} \qquad (2.12)$$

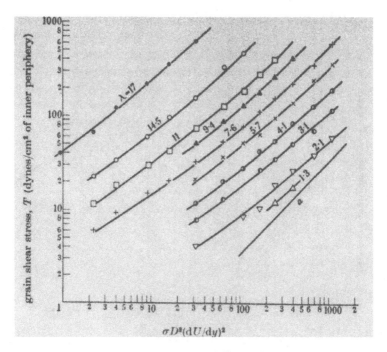

Figure 9. Shear stress vs. shear strain rate at different linear concentrations

where p is the pressure and μ is the viscosity coefficient.

In the case of a simple shear flow, the shear stress τ_{xz} which will be called τ for the sake of simplicity is given by

$$\tau = \mu \frac{\partial v}{\partial z} \qquad (2.13)$$

Typical values of the viscosity coefficient are:

material	μ (Pa.s)
Air	10^{-6}
Water	10^{-3}
Mud	10^{-2}

where it should be remembered that the values depend on temperature.

The experimental results obtained by Bagnold (1954) displayed in Figs. 9 and 10 show that the behaviour at low rates of strain in the macro-viscous region can be reproduced with a Newtonian fluid model,

$$\tau = (1 + \lambda)\left(1 + \frac{1}{2}\lambda\right)\mu \frac{\partial v}{\partial z} \qquad (2.14)$$

The role of the linear concentration consists, therefore, in increase the effective viscosity of the mixture.

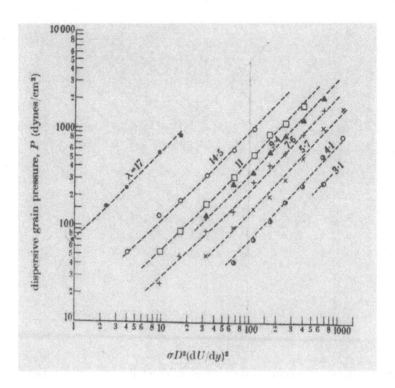

Figure 10. Dispersive pressure for different linear concentrations

It will be shown later that newtonian fluids do not allow the formation of "plug regions" in the flow, which are featured by debris and mud flows.

Another limitation of newtonian fluids, as far as modelling of debris and mud flows are concerned is that the fluid will not stop when the slope decreases.

2.4 Non-Newtonian Flows: Bagnold's dilatant fluid

So far we have considered the simplest case of a linear relation between shear stress and rate of shear deformation. Bagnold's experiments in the inertia regime showed that the relation between both should be quadratic (See Figs. 9 and 10). Bagnold proposed a relation based on the frequency of shocks between particles:

$$\sigma_{xz} = \sigma_{xz}^{turb} + a_i \sin \alpha_i \, \rho \lambda^2 d^2 \left(\frac{\partial v}{\partial z} \right)^2 \tag{2.15}$$

$$\sigma_{zz} = -p + a_i \cos \alpha_i \, \rho \lambda^2 d^2 \left(\frac{\partial v}{\partial z} \right)^2$$

where a_i and α_i where model parameters for which he suggested that typical values could be $a_i = 0.042$ and $\tan \alpha_i = 0.4$. Above relations were valid for linear concentrations $\lambda > 12$.

The terms σ^{turb} and p refer to shear stress caused by turbulence and fluid pressure, respectively.

Fig. 11 shows the relation between the dimensionless numbers, N, G_T and G_P defined as:

$$N = \frac{\text{Inertia Force}}{\text{Viscous Force}} = \frac{\lambda^2 \rho d^2 \left(\frac{\partial v}{\partial z}\right)^2}{\lambda^{3/2} \mu \left(\frac{\partial v}{\partial z}\right)} = \frac{\lambda^{1/2} \rho d^2 \left(\frac{\partial v}{\partial z}\right)}{\mu} \qquad (2.16)$$

$$G_T = \frac{\sigma_{xz} \rho d^2}{\lambda \mu^2} \quad \text{and} \quad G_P = \frac{\sigma_{zz} \rho d^2}{\lambda \mu^2}$$

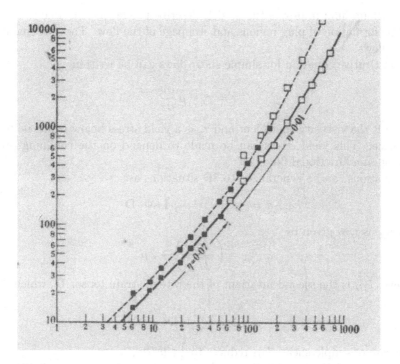

Figure 11. Relation between dimensionless numbers N and G (From Bagnold 1954)

The experiments were carried out using two different viscosities and the same linear concentration $\lambda = 11$. It can be seen there that there exist two segments, $N < 40$ and $N > 450$ which correspond to the macro-viscous and the inertia regimes respectively. Therefore, the Bagnold's number N can be used to discriminate the most suitable model to be used.

Bagnold's dilatant fluid model can be generalized to 3D as:

$$\sigma = -p\mathbf{I} + 4a_i \sin \alpha_i \, \rho \lambda^2 d^2 \, |I_{2D}|^{1/2} \, \mathbf{D} - 4a_i \cos \alpha_i \, \rho \lambda^2 d^2 \mathbf{D}^2 \qquad (2.17)$$

where we have not considered stress components caused by turbulence.

It can be seen that above expression is a particular case of 2.1, having chosen

$$\Phi_0 = 0 \qquad (2.18)$$
$$\Phi_1 = 4a_i \sin\alpha_i \, \rho\lambda^2 d^2 \, |I_{2D}|^{1/2}$$
$$\Phi_2 = -4a_i \cos\alpha_i \, \rho\lambda^2 d^2$$

Bagnold's model is the base of many advanced models used today, as it will be shown later.

2.5   Bingham fluid

The Bingham fluid improves the models described so far as it is able to reproduce both the formation of plug regions and stoppage of the flow. They are mainly used for mudflows.

The constitutive equation for simple shear flows can be written as:

$$\tau = \tau_y + \mu \frac{\partial v}{\partial z} \qquad (2.19)$$

where μ is the viscosity coefficient and τ_y is a yield stress below which no flow will take place. This yield stress can be made to depend on the confining pressure, resulting on a "frictional fluid".

This equations can be generalized to 3D situations as:

$$\sigma = -p\,\mathbf{I} + \tau = -p\,\mathbf{I} + \Phi_1\,\mathbf{D}$$

where Φ_1 is now given by

$$\Phi_1 = \left(\frac{\tau_y}{\sqrt{I_{2D}}}\right) + 2\mu \qquad (2.20)$$

In above, I_{2D} is the second invariant of the rate of strain tensor $\mathbf{D}$, which is given by:

$$I_{2D} = \frac{1}{2}\operatorname{tr}\mathbf{D}^2$$

In the case of simple shear, I_{2D} reduces to $\frac{1}{4}\left(\partial v/\partial z\right)^2$.

Rickenmann and Koch (1997) have taken τ_y and μ in the ranges 100-800 Pa and 400-800 Pa.s in simulations of debris flows in Kamikamihori (Japan) and Saas (Switzerland) valleys events. From their simulations it can be concluded that Bingham models predict higher velocities than other models. Jan (1997) provides both values used by himself and by other researchers.

Author	ρ (Kg/m^3)	τ_y(Pa)	μ(Pa.s)
Jan		100-160	40-60
Johnson	2000,2400	60,170-500	45
Sharp & Nobles	2400		20-60
Pierson	2090	130-240	210-810

It is also interesting to reproduce the values selected by Jin and Fread (1997) for several sites:

Site	ρ (Kg/m^3)	τ_y(Pa)	μ(Pa.s)
Anhui Debris Dam	1570	38	2.1
Aberfan	1764	4794	958
Rudd Creek	15750	956	958

2.6   Advanced Models

The simple models presented so far can be further improved by combining features of them. For instance, the Bingham Fluid can be generalized to a Non-Newtonian-like fluid by expressing the constitutive law as:

$$\tau = \tau_y + \mu \left(\frac{\partial v}{\partial z}\right)^{\eta} \tag{2.21}$$

where η is a new material parameter. Of course, the Bingham model is recovered by taking $\eta = 1$.
This law can be generalized to 3D situations as:

$$\sigma = -p\,\mathbf{I} + \tau = -p\,\mathbf{I} + \Phi_1\,\mathbf{D} \tag{2.22}$$

where Φ_1 is

$$\Phi_1 = \left(\frac{\tau_y}{\sqrt{I_{2D}}}\right) + 2\,\mu\,|\,4\,I_{2D}|^{\frac{\eta-2}{2}}$$

An interesting improvement is that proposed by Julien and Lan (1991) for simple shear flows as follows:

$$\tau = \tau_y + \mu\frac{\partial v}{\partial z} + \nu\left(\frac{\partial v}{\partial z}\right)^2 \tag{2.23}$$

which has been proved to provide good agreement with laboratory experiments of several authors.
The parameter ν is given by:

$$\nu = \rho_m\,l_m^2 + \hat{a}_i\,\rho_s\lambda^2 d_s^2$$

where:
ρ_m and ρ_s are densities of the mixture and of the solid phase
l_m is the mixing length, which can be approximated by $0.4h$, where h is the flow depth
$\hat{a}_i$ is a material parameter taken as 0.01 (It corresponds to $a_i\sin\alpha_i$ in Bagnold's model)
λ is the linear concentration
d_s is the particle diameter
Julien and Lan's model can be written in 3D by taking :

$$\Phi_1 = \left(\frac{\tau_y}{\sqrt{I_{2D}}}\right) + 2\,\mu + 2\,\nu\,|\,4\,I_{2D}|^{1/2} \tag{2.24}$$

Several authors have proposed assumptions concerning the form of functions Φ_1 and Φ_2. The approaches generally made are based on obtaining functions which reduce to laws developed for simple shear flows. The Generalized Viscoplastic Fluid proposed by Chen and Ling (1996) aims to generalize the laws

$$\sigma_{xz} = s + \mu_1 \left(\frac{\partial v}{\partial z}\right)^{\eta_1} \tag{2.25}$$

$$\sigma_{zz} = -\tilde{p} + \frac{1}{3}\mu_2 \left(\frac{\partial v}{\partial z}\right)^{\eta_2}$$

where μ_1, η_1, μ_2 and η_2 are model parameters for which they propose ad-hoc expressions, and s is a Mohr-Coulomb like yield stress:

$$s = c \cos\phi + \tilde{p}\sin\phi$$

with c and ϕ being the cohesion and friction angle. The expressions for Φ_1 and Φ_2 are the following:

$$\Phi_1 = \left(\frac{s}{\sqrt{I_{2D}}} + 2\mu_1 |4\,I_{2D}|^{\frac{\eta_1-1}{2}}\right) \tag{2.26}$$

$$\Phi_2 = 4\mu_2 |4\,I_{2D}|^{\frac{\eta_2-2}{2}}$$

The stress is therefore given by:

$$\sigma = -\tilde{p}\mathbf{I} + \left(\frac{s}{\sqrt{I_{2D}}} + 2\mu_1 |4\,I_{2D}|^{\frac{\eta_1-1}{2}}\right)\mathbf{D} + 4\mu_2 |4\,I_{2D}|^{\frac{\eta_2-2}{2}}\,\mathbf{D}^2$$

It can be seen that above expressions do not depend on the third invariant of the rate of deformation tensor $\mathbf{D}$ the reason being probably related to the fact that most experimental results have been obtained in rheometers for which it is zero. Another interesting fact which can be observed when particularizing above formulas for simple shear flows is that

$$\sigma_{xx} = \sigma_{zz}$$

It is also interesting to analyze the stresses predicted for the limit case $\mathbf{D} = 0$:

$$\sigma = -\tilde{p}\mathbf{I} + \frac{s}{\sqrt{I_{2D}}}\mathbf{D}$$

This stress is not dependent on the rate of deformation, but there exists an indetermination in the direction of $\mathbf{D}/\sqrt{I_{2D}}$. In the case of a simple shear flow, the stress components ($D_{xx} = D_{zz} = 0$) are given by:

$$\sigma_{xz} = c \cos\phi + \tilde{p}\sin\phi \tag{2.27}$$

$$\sigma_{zz} = \sigma_{xx} = -\tilde{p}$$

Above stresses do not fulfill Mohr-Coulomb criterion of failure unless $\phi = 0$. Moreover, if these results are to be linked to the classical solid (or soil) mechanics approach, some inconsistencies have to be removed from the present model. In the simple shear, the stress conditions will be such $\sigma_{xx} \neq \sigma_{zz}$. In some cases, an earth pressure coefficient K_0 could be assumed.

Chen and Ling propose to relax the condition $D_{xx} = D_{zz} = 0$ for the simple shear flow. If we consider a Mohr-Coulomb material in plane strain conditions, the stresses are given by:

$$\sigma_{xx} = -\bar{p} + (c \cos \phi + \bar{p} \sin \phi) \frac{D_{xx}}{\sqrt{D_{xx}^2 + D_{xz}^2}} \qquad (2.28)$$

$$\sigma_{zz} = -\bar{p} + (c \cos \phi + \bar{p} \sin \phi) \frac{D_{zz}}{\sqrt{D_{xx}^2 + D_{xz}^2}}$$

$$\sigma_{xz} = (c \cos \phi + \bar{p} \sin \phi) \frac{D_{xz}}{\sqrt{D_{xx}^2 + D_{xz}^2}}$$

where we have used $I_{2D} = D_{xx}^2 + D_{xz}^2$ and $D_{xx} = -D_{zz}$.
We will assume that

$$\lim_{D \to 0} \frac{D_{xx}}{\sqrt{D_{xx}^2 + D_{xz}^2}} = \cos \beta$$

$$\lim_{D \to 0} \frac{D_{xz}}{\sqrt{D_{xx}^2 + D_{xz}^2}} = \sin \beta$$

After substituting in 2.28 we obtain

$$\sigma_{xx} = -\bar{p} + (c \cos \phi + \bar{p} \sin \phi) \cos \beta \qquad (2.29)$$

$$\sigma_{zz} = -\bar{p} - (c \cos \phi + \bar{p} \sin \phi) \cos \beta$$

$$\sigma_{xz} = (c \cos \phi + \bar{p} \sin \phi) \sin \beta$$

which fulfill Mohr-Coulomb criterion as

$$(\sigma_{xx} + \bar{p})^2 + \sigma_{xz}^2 = (c \cos \phi + \bar{p} \sin \phi)^2 = s^2 \qquad (2.30)$$

This results can be interpreted from a different point of view. There exists a process of fluidification or liquefaction in which the solid is transformed into a fluid. Failure of the solid takes place at the Critical State, with zero volume change. Rate of deformation fulfills $\mathrm{tr} D = 0$, and is close to the limit $\|\mathbf{D}\| \to 0$. If velocity and rate of deformation increases, the stress will abandon the yield surface, and material behaviour will be governed by relations 2.29.

There are other models which can be mentioned within this framework, such as that of Johnson (1996).

It is also worth mentioning the work of Hutter et al. (1996), who propose to decompose the stress tensor of the granular phase into static and dynamic components, given by:

$$\sigma^s = -p_s \mathbf{I} + 2\mu^s \mathbf{D}$$

$$\sigma^d = -p_d \mathbf{I} + 2\mu^d \mathbf{D} + 4\eta \mathbf{D}^2$$

where p_s and p_d are the static and dynamic grain pressures.

3 Mathematical models for landslide propagation

3.1   Introduction

Mathematical and numerical models for fast landslide propagation can be classified in two main groups:

- Full 3D formulations where the equations of balance of mass and linear momentum are solved on a domain which changes with time. The position of the interface between the soil and the air is tracked using special techniques such as the MAC algorithm applied by Sousa and Blight (1991), or the pseudoconcentration function implemented in the model proposed by Frenette et al. (1997). The pseudoconcentration function is an indicator which takes two different values depending on whether the point being considered is occupied by soil or by air. The pseudoconcentration function is advected by the velocity field, presenting the well known difficulties of the advection of a discontinuous function. In consequence, the solution can be corrupted as time advances. As an effective alternative, it is possible to use the "level set" technique, which consists on transforming the indicator function on a distance function Quecedo and Pastor (2001). These models are expensive in terms of computer effort and therefore their use is restricted to cases in which it is necessary to know the fine structure of the flow, forces against structures, etc.
- If some assumptions are made about the vertical structure of the flow, it is possible to integrate the balance equations in depth, arriving at the so-called "depth-integrated" equations or "shallow water equations" in coastal and hydraulic engineering. The equations can be further simplified by integrating on cross sections, arriving at simple 1D models.

The depth integrated equations can be cast either in an Eulerian or a Lagrangian form. The second approach has been used mainly in one dimensional situations. The models proposed by Savage and Hutter (1991), Hutter and Koch (1991), Hungr (1995) and Rickenmann and Koch (1997) incorporate features such as introducing a K_0 coefficient relating the horizontal and vertical stress components, or formulating the equations in a curvilinear coordinate system following the terrain. However, lagrangian formulations present some important difficulties in the 2D case. For instance, we can mention the problem of bifurcation of the flow when arriving to an obstacle, or merging after it has been passed. Because of this problem, other authors have chosen eulerian formulations (Vulliet and Hutter (1988); Laigle and Coussot (1997); Laigle (1997), and Pastor et al. (2002)). This approach has been followed also in Hydraulic Engineering to model the propagation of flood waves, dam break problems, etc. (Laigle and Coussot (1997); Jin and Fread (1997)).

In the past years, there has been a continuously increasing interest in the so-called Riemann solvers, where a Riemann problem is solved at the boundaries between cells or elements at every time step. (See, for instance the work of Laigle and Coussot (1997) and Laigle (1997) for mudflows or the more recent of Zoppou and Roberts (1999) or Toro (2001) in Hydrodynamics). The use of such specialized algorithms introduces an additional computational load, as a non-linear Riemann problem has to be solved at boundaries between cells or elements at every time step. To circumvent this difficulty, approximate Riemann solvers have been introduced, but care must be taken to avoid entropy violating solutions.

As a simple yet effective alternative, the numerical model proposed here is based on the classical Taylor-Galerkin algorithm introduced by Peraire et al. (1984) and Donea (1984), which has been modified to deal with propagation over dry regions and which is able to provide accurate enough solutions. It is based on the balance of mass and momentum equations for the mixture of soil and interstitial fluid.

3.2   An eulerian approach to soil dynamics

Following the conclusions of the preceding Section, we will describe here the eulerian approach which we will use to model the propagation phase of catastrophic landslides. We will consider the soil as a mixture of a solid phase -the grains- and the pore fluids, water and air in the general case.

The basic equations are, therefore:

(**A.i**) Balance of mass for solid and fluid phases:

$$\frac{D^{(s)}\rho^{(s)}}{Dt} + \rho^{(s)} div\ v^{(s)} = 0 \tag{3.1}$$

$$\frac{D^{(\alpha)}\rho^{(\alpha)}}{Dt} + \rho^{(\alpha)} div\ v^{(\alpha)} = 0 \tag{3.2}$$

where $\rho^{(s)} = (1-n)\rho_s$ and $\rho^{(\alpha)} = nS_\alpha\rho_a$.

(**A.ii**) Balance of momentum for the pore fluids

$$\rho_\alpha \frac{D^{(\alpha)}v^{(\alpha)}}{Dt} = \rho_\alpha b + div\ \sigma_\alpha - k_\alpha^{-1}w^{(\alpha)} \tag{3.3}$$

and for the solid skeleton

$$\rho_s \frac{D^{(s)}v^{(s)}}{Dt} = \rho_s b + div\ \sigma_s + k_\alpha^{-1}w^{(\alpha)}$$

(**A.iii**) Constitutive equations for all constituents relating stresses and rate of deformation tensors

(**A.iv**) Kinematic relations between velocities and rate of deformation tensors

$$D^{(\alpha)} = \frac{1}{2}\left(\frac{\partial v_i^{(\alpha)}}{\partial x_j} + \frac{\partial v_j^{(\alpha)}}{\partial x_i}\right) \tag{3.4}$$

It was shown in the preceding Chapter that some simplifications can be done. If we assume that the **soil is non-saturated**, with $p_a = 0$, and we neglect the assume that relative velocities of pore fluids relative to soil skeleton are small, the problem can be cast in terms of soil grains velocity v, relative velocity of the water phase w and pore water pressures. The system of equations will reduce to:

(**B.i**) Balance of mass and momentum for the pore water

$$div \ v + div \left(\frac{w}{S_w} \right) + \frac{1}{Q} \frac{D \ p_w}{Dt} + \frac{1}{G} \frac{D \ S_w}{Dt} = 0 \tag{3.5}$$

with

$$\frac{1}{Q} = \left(\frac{n}{K_w} + \frac{1-n}{K_s} \right) \tag{3.6}$$

$$\frac{1}{G} = \left(\frac{n}{S_w} + \frac{1-n}{K_s} p_w \right)$$

and

$$w = k_w \left(-\rho_w \frac{Dv}{Dt} + \rho_w b - grad \ p_w \right) \tag{3.7}$$

Combining equations 3.1 and 3.7 we arrive to

$$div \ v + div \left\{ \frac{k_w}{S_w} \left(-\rho_w \frac{D \ v}{Dt} + \rho_w b - grad \ p_w + div \ S_w \right) \right\} + \frac{1}{Q} \frac{D \ p_w}{Dt} + \frac{1}{G} \frac{D \ S_w}{Dt} = 0 \tag{3.8}$$

(**B.ii**) Balance of momentum of the mixture

$$\rho \frac{D^{(s)} v^{(s)}}{Dt} = \rho b + div \ \sigma \tag{3.9}$$

(**B.iii**) and (**B.iv**) remain the same than (A.iii) and (A.iv). This formulation is the eulerian version of Zienkiewicz's "$p_w - u$" formulation Zienkiewicz and Shiomi (1984).

In the case of **saturated soils**, the system is:

(**C.i**)

$$div \ v + div \left\{ k_w \left(-\rho_w \frac{D \ v}{Dt} + \rho_w b - grad \ p_w \right) \right\} + \frac{1}{Q} \frac{D \ p_w}{Dt} = 0 \tag{3.10}$$

with (**C.ii**), (**C.iii**) and (**C.iv**) being the same than above. Divergence of accelerations and body forces is usually neglected in geotechnical analysis.

An important case is that of dry soils where the time scale of air pore pressure dissipation is similar to that of propagation of the landslide. In this case, pore pressures of the pore air can play a paramount role, and in the limit it is possible to arrive at "dry liquefaction". The equations are the same than in the case (C), but substituting now density, permeability and volumetric stiffness of the water by those of the air.

So far, we have considered coupling between the soil grains and the pore pressures. There are two limit cases where the soil can be approximated as a single phase material:

(i) Flow of dry granular materials with high permeability. The permeability is high enough so that the consolidation time is much smaller than the time of propagation, the material behaves as "drained".

(ii) Flow of slurries with a high water content, where the time of dissipation is much higher than that of propagation. The behaviour can be assumed to be of undrained type. Material behaviour can be approximated using rheological models such as Bingham, etc.

In both cases, it is usually assumed that material density does not change. This approach can be justified in the case (i) by the existence of a "critical void ratio".

3.3   Propagation and Consolidation

Following the approaches described in the preceding Section, the system of equations for a saturated soil is:

$$\rho \frac{D^{(s)} v^{(s)}}{Dt} = \rho b + div\ \sigma \tag{3.11}$$

$$div\ v + div\ (-k_w\ grad\ p_w) = 0$$

where we have neglected the term $\frac{1}{Q}\frac{D\ p_w}{Dt}$, together with the rheological law relating stresses and velocities.

One important aspect is that fast landslides involve two physical phenomena which appear in above equations: (i) consolidation and dissipation of pore pressures, and (ii) propagation. In order to gain insight on the relative importance of all terms, we will express above equations in non-dimensional form Hutter and Koch (1991), introducing the a characteristic length of the landslide L and H a characteristic depth of the sliding mass, and the ratio $\varepsilon = H/L$. In typical cases, L will be of the order of 10^2m and H of the order of 5 m. Therefore, ε will be small.

Horizontal velocities will be scaled using $\sqrt{gL}$, vertical velocities with $\varepsilon\sqrt{gL}$, time with $\sqrt{L/g}$ and pressures and stresses with $\rho_0 gH$, ρ_0 being a typical density of the mixture. The new nondimensional magnitudes will be denoted using " ˆ ". The second equation in 3.11 is written as

$$div\ \hat{v} = \theta\ \left\{ \varepsilon^2 \frac{\partial^2 \hat{p}_w}{\partial \hat{x}_1^2} + \varepsilon^2 \frac{\partial^2 \hat{p}_w}{\partial \hat{x}_2^2} + \frac{\partial^2 \hat{p}_w}{\partial \hat{x}_3^2} \right\} \tag{3.12}$$

where

$$\theta = \left(\frac{\sqrt{L/g}}{H/(k_w \rho_0 g)} \right) \tag{3.13}$$

Assuming the same typical values for L and H, and a permeability $k_w \rho_0 g$ of the order of $10^{-n}\ ms^{-1}$, it can be seen that θ will be close to 10^{-n}. Typical values are 10^{-9} for clays, 10^{-7} for silts, 10^{-5} for fine sands and 10^{-1} for gravels. Therefore, equation 3.12 can be approximated by

$$div\ \hat{v} = \theta\ \frac{\partial^2 \hat{p}_w}{\partial \hat{x}_3^2} \tag{3.14}$$

The component along x_3 of the balance of momentum equation for the mixture results in

$$\varepsilon \left\{ \frac{\partial \hat{u}_3}{\partial \hat{t}} + grad \ \hat{u}_3 . \hat{u} \right\} = -1 + \left\{ \varepsilon \ \frac{\partial \hat{\sigma}_{13}}{\partial \hat{x}_1} + \varepsilon \ \frac{\partial \hat{\sigma}_{23}}{\partial \hat{x}_2} + \frac{\partial \hat{\sigma}_{33}}{\partial \hat{x}_3} \right\} \qquad (3.15)$$

where we have assumed that gravity acts along the axis x_3, and therefore, $b_3 = -\hat{g} = -1$. If we assume that ε is small, above equation reduces to

$$-1 + \frac{\partial \hat{\sigma}_{33}}{\partial \hat{x}_3} = 0$$

or, in terms of effective stresses,

$$-1 + \frac{\partial \hat{\sigma}'_{33}}{\partial \hat{x}_3} - \frac{\partial \hat{p}_w}{\partial \hat{x}_3} = 0 \qquad (3.16)$$

The dimensional equations are then

$$div \ v = \frac{\partial}{\partial x_3} \left(- k_w \frac{\partial \ p_w}{\partial x_3} \right) \qquad (3.17)$$

and

$$-\rho g + \frac{\partial \sigma_{33}}{\partial x_3} = 0 \qquad (3.18)$$

$$-\rho g + \frac{\partial \sigma'_{33}}{\partial x_3} - \frac{\partial p_w}{\partial x_3} = 0 \qquad (3.19)$$

It can be observed that the distribution of σ_{33} is hydrostatic along x_3. However, neither the effective stress nor the pressure p_w are necessary hydrostatic.
We will assume next that the velocity field can be decomposed as

$$\hat{v} = \hat{v}_0 + \theta \ \hat{v}_1 \qquad (3.20)$$

which may be viewed as a perturbation in the parameter θ. The pore pressure field will be assumed to of the form

$$\hat{p}_w = \hat{p}_{w0} + \hat{p}_{w1} \qquad (3.21)$$

where p_{w0} is a hydrostatic field varying linearly from zero at the surface to $\rho g h$ at the bottom. Substituting into 3.17 we obtain

$$div \ \hat{v}_0 + \theta \ div \ \hat{v}_1 = \theta \left(\frac{\partial^2 \hat{p}_{w0}}{\partial \hat{x}_3^2} + \frac{\partial^2 \hat{p}_{w1}}{\partial \hat{x}_3^2} \right) = \theta \ \frac{\partial^2 \hat{p}_{w1}}{\partial \hat{x}_3^2}$$

as $\frac{\partial^2 \hat{p}_{w0}}{\partial \hat{x}_3^2} = 0$. From here,

$$div \ \hat{v}_0 = 0 \qquad (3.22)$$

$$div \ \hat{v}_1 = \frac{\partial^2 \hat{p}_w}{\partial \hat{x}_3^2}$$

In this way, we can identify the perturbed field $\theta \, \hat{v}_1$ as the velocity field corresponding to consolidation and $\hat{v}_0$ as the velocity field corresponding to propagation. This result is of paramount importance, and clarifies the assumptions which should be made when modelling these phenomena. First of all, incompressibility is not a feature of rheological soil behaviour, but a consequence of the coupled behaviour between the pore fluid and the soil skeleton, which is only valid when the parameter θ is small enough. Indeed, this will explain the "undrained" behaviour in simple shear devices, where pressures depend on shear strain rate. We will come back now to the dimensional equations and apply the results obtained above.

Considering equation 3.18, we will decompose the effective vertical stress into two components, $\sigma'_{33}{}^0$ which will be assumed to vary in a hydrostatic manner and $\sigma'_{33}{}^1$:

$$\left(-\rho g + \frac{\partial \sigma'_{33}{}^0}{\partial x_3} - \frac{\partial p_{w0}}{\partial x_3} \right) + \left(\frac{\partial \sigma'_{33}{}^1}{\partial x_3} - \frac{\partial p_{w1}}{\partial x_3} \right) = 0 \qquad (3.23)$$

Of the two components, the former can be interpreted as the balance of momentum along x_3 once the pore pressures have been dissipated, while the latter allow us to assume that $\sigma'_{ij}{}^1$ is an hydrostatic tensor. The changes in p_{w1} due to consolidation will result in changes in $\sigma'_{33}{}^1$ which in turn will produce a volumetric strain in the soil skeleton given by

$$div \; v_1 = \frac{1}{K_T} \frac{Dp_w}{Dt} \qquad (3.24)$$

K_T is the volumetric stiffness of the soil skeleton. In consequence, the consolidation equation will be

$$\frac{1}{K_T} \frac{Dp_w}{Dt} = \frac{\partial}{\partial x_3} \left(-k_w \frac{\partial \, p_w}{\partial x_3} \right) \qquad (3.25)$$

where p_w depends on x_1, x_2, x_3 and t.

3.4 Depth Integrated equations for propagation

The model describing the propagation phase consists on the following set of PDE's

$$\rho \frac{D^{(s)} v_0^{(s)}}{Dt} = \rho b + div \; \sigma \qquad (3.26)$$

$$div \; v_0^{(s)} = 0 \qquad (3.27)$$

$$\frac{1}{K_T} \frac{Dp_w}{Dt} = \theta \frac{\partial^2 p_{w1}}{\partial x_3^2} \qquad (3.28)$$

In what follows, we will drop the subindexes "0" in the velocity field and "1" in the pressures. Taking into account the incompressibility condition eqn 3.26 can be written in a conservative form

$$\rho \frac{\partial v_i}{\partial t} + \frac{\partial}{\partial x_j} (\rho v_i v_j) = \frac{\partial}{\partial x_j} \sigma_{ij} + \rho b_i$$

In certain cases, as in the case of debris flows, there exist turbulent fluctuations over averaged states. The equations are then averaged over a representative time length T, and an extra term σ_{ij}^R

$$\sigma_{ij}^R = -\frac{1}{T} \int_t^{t+T} \left(\rho u_i' u_j' \right) dt \tag{3.29}$$

is added to the stresses. In above, u_i' represents the fluctuations of u_i over the average $\bar{u}_i$. We will assume that the stress tensor σ_{ij} includes Reynolds stresses, and, for the sake of simplicity drop the overbar, assuming that all magnitudes are the averaged values.

The equations of the depth averaged model are obtained integrating along X_3 the balance of mass and momentum equations, and taking into account Leibniz's rule. The details can be found in standard text books Zienkiewicz and Taylor (2000), Toro (2001).

We will use the reference system given in Fig. 12, where we have depicted some magnitudes of interest which will be used in this section.

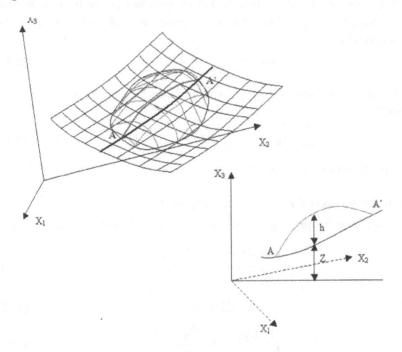

Figure 12. Reference system and notation used in the analysis

Integration along X_3 of the balance of mass equation results in

$$\frac{\partial h}{\partial t} + \frac{\partial}{\partial x_j} \left(h \bar{v}_j \right) = 0 \quad \text{with } j = (1, 2) \tag{3.30}$$

where $\bar{v}_j$ is the depth-averaged velocity along X_j.

Integration along X_i $(i = 1, 2)$ of the equations of balance of linear momentum results in:

$$\rho \frac{\partial}{\partial t}(h\bar{v}_i) + \rho \frac{\partial}{\partial x_j}(h\bar{v}_i\bar{v}_j) = \rho b_i h + \frac{\partial}{\partial x_j}(h\,\bar{\sigma}_{ij}) + t_i^A + t_i^B \quad i, j = 1, 2 \quad (3.31)$$

where t_i^A and t_i^B are the surface forces at the surface and the bottom respectively,

$$t_i^A = \sigma_{ij}.n_j|_{Z+h}$$
$$t_i^B = \sigma_{ij}.n_j|_{Z}$$

and $\bar{\sigma}_{ij}$ is the stress tensor averaged over depth plus a new contribution similar to Reynolds stresses given by:

$$-\rho \int_Z^{Z+h} u_i^* u_j^* dx_3 = -\rho h \overline{u_i^* u_j^*}$$

where

$$u_i = \bar{u}_i + u_i^*$$

Concerning the equation of balance of linear momentum along X_3:

$$\rho \frac{\partial v_3}{\partial t} + \rho \frac{\partial}{\partial x_j}(v_j v_3) = \frac{\partial}{\partial x_j}\sigma_{3j} + \rho b_3 \quad (3.32)$$

we will assume that vertical accelerations (total) are zero. Vertical velocity will vary linearly between the bottom and the surface, and it is interesting to note that the average rate of deformation D_{33} is given by

$$\bar{D}_{33} = \frac{1}{h}\frac{\partial h}{\partial t} \quad (3.33)$$

After integrating along X_3 we arrive at:

$$t_{3B} = -\rho b_3 h - \frac{\partial}{\partial x_j}(h\,\bar{\sigma}_{3j})|_Z - t_{3A} \quad (3.34)$$

This expression differs from the classical used in Shallow Water computations in the second term of the right hand side. Integration can also be performed between the bottom and a given height $\xi = x_3 - Z$ above the bottom:

$$\sigma_{33}(\xi) = \rho b_3 (h - \xi) + \frac{\partial}{\partial x_j}\int_{\xi=0}^h \sigma_{3j}(\xi)\,d\xi + t_{3A} \quad (3.35)$$

To conclude, the depth integrated equations can be cast in conservative form as:

$$\frac{\partial U}{\partial t} + \frac{\partial F_j}{\partial x_j} = S \quad (3.36)$$

where
$$U^T = (h, h\bar{v}_1, h\bar{v}_2)$$

$$
\begin{aligned}
F_1^T &= \left(h\bar{v}_1, h\bar{v}_1^2, h\bar{v}_1\bar{v}_2\right) \\
F_2^T &= \left(h\bar{v}_1, h\bar{v}_1\bar{v}_2, h\bar{v}_2^2\right)
\end{aligned}
$$

and

$$
S = \begin{pmatrix} 0 \\ b_1 h \\ b_2 h \end{pmatrix} + \frac{1}{\rho}\frac{\partial}{\partial x_1}\begin{pmatrix} 0 \\ h\bar{\sigma}_{11} \\ h\bar{\sigma}_{12} \end{pmatrix} + \frac{1}{\rho}\frac{\partial}{\partial x_2}\begin{pmatrix} 0 \\ h\bar{\sigma}_{12} \\ h\bar{\sigma}_{22} \end{pmatrix} + \frac{1}{\rho}\begin{pmatrix} 0 \\ t_{1A} + t_{1B} \\ t_{2A} + t_{2B} \end{pmatrix}
$$

3.5   Depth Integrated model for Consolidation

The equation governing vertical consolidation is 3.25

$$\frac{1}{K_T}\frac{Dp_w}{Dt} = \frac{\partial}{\partial x_3}\left(k_w \frac{\partial p_w}{\partial x_3^2}\right)$$

which can be written as

$$\frac{\partial p_w}{\partial t} + v_i \frac{\partial p_w}{\partial x_i} = \frac{\partial}{\partial x_3}\left(c_v \frac{\partial p_w}{\partial x_3^2}\right) \tag{3.37}$$

where the coefficient c_v is the consolidation coefficient.
Integrating this equation along x_3 we arrive to

$$\frac{\partial}{\partial t}\left(\bar{p}_w\, h\right) + \frac{\partial}{\partial x_k}\left(\overline{v_k\, p_w}h\right) = c_v \left.\frac{\partial p_w}{\partial x_3^2}\right|_{Z+h} - c_v \left.\frac{\partial p_w}{\partial x_3^2}\right|_{Z} \tag{3.38}$$

In above, the overbar refers to vertical averaged magnitudes.
In order to obtain the averaged values, we will assume that pore pressure can be approximated by

$$p_w\left(x_1, x_2, x_3, t\right) = \sum_{j=1}^{nf} N_j^{(3)}\left(x_3\right)\, P_{wj}\left(x_1, x_2, t\right) \tag{3.39}$$

where $N_j^{(3)}\left(x_3\right)$ are shape functions used to approximate the variation of pore pressure along x_3. We will choose harmonic functions satisfying boundary conditions. If we assume that pore pressure is zero at the surface and the bottom is impervious,

$$N_j^{(3)}\left(x_3\right) = \cos\left(\frac{2j-1}{2h}\pi\left(x_3 - Z\right)\right) \qquad j = 1..nf \tag{3.40}$$

The averaged value of p_w is given by

$$
\begin{aligned}
\bar{p}_w &= \frac{1}{h}\int_Z^{Z+h}\sum_{j=1}^{nf} N_j^{(3)}\left(x_3\right)\, P_{wj}\left(x_1, x_2, t\right) dx_3 \tag{3.41}\\[2mm]
&= \sum_{j=1}^{nf}\frac{2}{(2j-1)\,\pi}(-1)^{j+1}\, P_{wj}\left(x_1, x_2, t\right)
\end{aligned}
$$

This expression is substituted in eqn.3.38, resulting in

$$\frac{\partial}{\partial t}\left\{\sum_{j=1}^{nf}\frac{2}{(2j-1)\,\pi}\,(-1)^{j+1}\,P_{wj}\,h\right\}+ \tag{3.42}$$

$$\frac{\partial}{\partial x_k}\left\{\sum_{j=1}^{nf}\frac{2}{(2j-1)\,\pi}\,(-1)^{j+1}\,\bar{v}_k P_{wj}h\right\}$$

$$=-\sum_{j=1}^{nf}\frac{2}{(2j-1)\,\pi}\,(-1)^{j+1}\,P_{wj}$$

If we take one term in the series, above equation reduces to

$$\frac{\partial}{\partial t}\left(P_{w1}\,h\right)+\frac{\partial}{\partial x_k}\left(\bar{v}_k P_{w1}h\right)=-\frac{\pi^2}{4h^2}c_v P_{w1} \tag{3.43}$$

where P_{jw} depends on x_1, x_2 and t. This equation is incorporated into the general model 3.36 adding one more term to the unknowns, fluxes and sources. Therefore, the system can be cast as

$$\frac{\partial U}{\partial t}+\frac{\partial F_j}{\partial x_j}=S \tag{3.44}$$

where

$$U^T=(h, h\bar{v}_1, h\bar{v}_2, hP_{w1})$$

$$F_1^T = \left(h\bar{v}_1, h\bar{v}_1^2, h\bar{v}_1\bar{v}_2, hP_{w1}\bar{v}_1\right)$$
$$F_2^T = \left(h\bar{v}_1, h\bar{v}_1\bar{v}_2, h\bar{v}_2^2, hP_{w1}\bar{v}_2\right)$$

and

$$S=\begin{pmatrix}0\\b_1 h\\b_2 h\\0\end{pmatrix}+\frac{1}{\rho}\frac{\partial}{\partial x_1}\begin{pmatrix}0\\h\bar{\sigma}_{11}\\h\bar{\sigma}_{12}\\0\end{pmatrix}+\frac{1}{\rho}\frac{\partial}{\partial x_2}\begin{pmatrix}0\\h\bar{\sigma}_{12}\\h\bar{\sigma}_{22}\\0\end{pmatrix} \tag{3.45}$$

$$+\frac{1}{\rho}\begin{pmatrix}0\\t_{1A}+t_{1B}\\t_{2A}+t_{2B}\\0\end{pmatrix}+\begin{pmatrix}0\\0\\0\\-\frac{\pi^2}{4h^2}c_v P_{w1}\end{pmatrix}$$

It is important to notice that consolidation results in a source term which is incorporated into the system.

4 Depth Integrated Rheological models

4.1   Introduction

Once the depth integrated equations have been obtained, it is necessary to provide suitable expressions for the averaged stress tensor $\bar{\sigma}_{ij}$ and for the tractions acting on the boundaries (which are given by the product $\sigma_{ij}n_j$).

We will assume a decomposition of the stress tensor of the type

$$\sigma = -p\mathbf{I} + \sigma^* \tag{4.1}$$

where p is the hydrostatic pressure $p = -\rho b_3 (h - \xi)$ and $\mathbf{I}$ the identity tensor of second order. It is important to note that σ^* is not a deviatoric tensor in the sense $\mathrm{tr}(\sigma^*) = 0$.

If the material is isotropic, the second component will be assumed to be given by:

$$\sigma^* = \Phi_1 \mathbf{D} + \Phi_2 \mathbf{D}^2 \tag{4.2}$$

where $\mathbf{D}$ is the rate of deformation tensor

$$D_{ij} = \frac{1}{2} \left(\frac{\partial v_i}{\partial x_j} + \frac{\partial v_j}{\partial x_i} \right)$$

and Φ_1 and Φ_2 are scalar functions of the second and third invariants of $\mathbf{D}$, $I_{2D} = \frac{1}{2}\mathrm{tr}\left(\mathbf{D}^2\right)$, and $I_{3D} = \frac{1}{3}\mathrm{tr}\left(\mathbf{D}^3\right)$.

The implementation of rheological laws in the depth integrated models present the difficulty of obtaining the rate of deformation tensor along the x_3 axis at any given point, and then integrating the stresses given by the law. Here, it is proposed to use "integrated" rheological laws relating averaged stresses and rates of deformation. Concerning the latter, we will introduce

$$\bar{D}_{ij} = \frac{1}{2} \left(\frac{\partial \bar{v}_i}{\partial x_j} + \frac{\partial \bar{v}_j}{\partial x_i} \right) \qquad i,j = 1,2 \tag{4.3}$$

As the material has been assumed to be incompressible, $\bar{D}_{33}$ is simply given by $-\left(\bar{D}_{11} + \bar{D}_{22}\right)$. To obtain $\bar{D}_{i3}$, it is possible to assume that the structure of the flow is that of a steady state, uniform, simple shear flow, and then relate the value of $\bar{D}_{i3}$ to that of the average velocity. The invariant $\bar{I}_{2D}$ is then obtained as

$$\bar{I}_{2D} = \frac{1}{2}\bar{D}_{ij}\bar{D}_{ij} \tag{4.4}$$

Therefore, we will assume the the relation between the averaged stress $\bar{\sigma}_{ij}$ $(i,j = 1,2)$ and the rate of deformation is given by the general expression:

$$\bar{\sigma} = -\bar{p}I + \Phi_1\bar{\mathbf{D}} + \Phi_2\bar{\mathbf{D}}^2 \tag{4.5}$$

where Φ_1 and Φ_2 depend on $\bar{I}_{2D}$, and the mean pressure $\bar{p}$ is given by $-\frac{1}{2}\rho b_3 h$. We will analyze here some particular cases of the rheological law 4.5.

4.2   Newtonian Fluid.

The newtonian fluid is obtained taking $\Phi_2 = 0$ and $\Phi_1 = 2\mu$, where μ is the fluid viscosity. The averaged stress tensor is given by

$$\bar{\sigma} = -\bar{p}I + \bar{\sigma}^*$$
$$\bar{\sigma}^* = 2\mu\bar{D}$$

Concerning the pressure term, it can be seen that

$$\bar{p} = -\frac{1}{2}\rho b_3 h$$

The averaged stresses are:

$$\bar{\sigma}_{11} = -\bar{p} + 2\mu\,\bar{D}_{11} \tag{4.6}$$
$$\bar{\sigma}_{22} = -\bar{p} + 2\mu\,\bar{D}_{22}$$
$$\bar{\sigma}_{12} = 2\mu\,\bar{D}_{12}$$

The balance of momentum equation can be written as:

$$\rho\frac{\partial}{\partial t}(h\bar{v}_i) + \rho\frac{\partial}{\partial x_j}(h\bar{v}_i\bar{v}_j) = \rho b_i h + \frac{\partial}{\partial x_j}\left(\frac{1}{2}\rho b_3 h^2 \delta_{ij} + h\,\bar{\sigma}^*_{ij}\right) + t_i^A + t_i^B \quad i,j = 1,2$$

The pressure term is then incorporated to the fluxes $h\bar{u}_i\bar{u}_j$

$$\rho\frac{\partial}{\partial t}(h\bar{v}_i) + \rho\frac{\partial}{\partial x_j}\left(h\bar{v}_i\bar{v}_j - \frac{1}{2}b_3 h^2 \delta_{ij}\right) = \rho b_i h + \frac{\partial}{\partial x_j}(2\mu h\,\bar{D}_{ij}) + t_i^A + t_i^B \quad i,j = 1,2$$

The term $\frac{\partial}{\partial x_j}(2\mu h\,\bar{D}_{ij})$ is usually neglected in depth integrated models, as is much smaller than the pressure term. Concerning the tractions on the boundaries (free surface and bottom), it is usually assumed that $t_i^A = 0$. The bottom friction can be approximated by that of a uniform simple shear flow. The resulting expression is:

$$t_i^B = \frac{3\mu\rho\,\bar{v}_i}{h}$$

In the case of turbulent flows, it is possible to use Chezy-Manning formula:

$$t_i^B = \frac{\rho b_3 \, n^2 \,|\mathbf{V}|\,\bar{v}_i}{h^{1/3}} \tag{4.7}$$

The experimental results obtained by Bagnold (1954) show that the behaviour at low rates of strain in the macro-viscous region can be reproduced with a Newtonian fluid model, where the viscosity of the fluid is corrected by a factor which accounts for the linear concentration of the material

$$\mu^* = (1 + \lambda)\left(1 + \frac{1}{2}\lambda\right)\mu \tag{4.8}$$

where λ is the linear concentration defined as:

$$\frac{1}{\lambda} = (\frac{C_{\text{max}}}{C})^{\frac{1}{3}} - 1 \tag{4.9}$$

and C and C_{max} are the concentration and the maximum possible concentration, which for spheres is $\pi/3\sqrt{2} = 0.74$.

If we consider a 2D flow over an horizontal surface, steady state conditions are obtained from the balance of momentum along X_1, assuming that accelerations are zero:

$$\frac{\partial}{\partial x_1} (\bar{\sigma}_{11}) + t_{B1} = 0 \tag{4.10}$$

where $\bar{\sigma}_{11} = -\rho g h$. It follows that the slope of the free surface has to be zero.

4.3 Bingham fluids

The general rheological law is obtained taking Φ_1 as:

$$\Phi_1 = \left(\frac{\tau_y}{\sqrt{\bar{I}_{2D}}}\right) + 2\mu \tag{4.11}$$

where τ_y is a yield stress.

The averaged stresses involved in the balance of momentum equations along X_1 and X_2 are given explicitly by:

$$\bar{\sigma}_{11} = -\bar{p} + \tau_y \frac{\bar{D}_{11}}{\sqrt{\bar{I}_{2D}}} + 2\mu \bar{D}_{11} \tag{4.12}$$

$$\bar{\sigma}_{22} = -\bar{p} + \tau_y \frac{\bar{D}_{22}}{\sqrt{\bar{I}_{2D}}} + 2\mu \bar{D}_{22}$$

$$\bar{\sigma}_{12} = \tau_y \frac{\bar{D}_{12}}{\sqrt{\bar{I}_{2D}}} + 2\mu \bar{D}_{12}$$

where $\bar{p} = -\frac{1}{2}\rho b_3 h$ and $\bar{D}_{ij} = \frac{1}{2}\left(\frac{\partial \bar{v}_i}{\partial x_j} + \frac{\partial \bar{v}_j}{\partial x_i}\right)$. This form differs from the classic Bingham models used in the context of depth integrated models which assume

$$\bar{\sigma}_{11} = \bar{\sigma}_{22} = -p$$

and $\bar{\sigma}_{12} = 0$.

Concerning bottom friction, we will assume that it can be approximated by the value under simple shear flow conditions. The shear stress at the bottom τ_B can be related to the modulus of depth averaged velocity $\bar{u}$ by

$$\bar{v} = \frac{\tau_B}{6\mu} h \left(1 - \frac{\tau_y}{\tau_B}\right)^2 \left(2 + \frac{\tau_y}{\tau_B}\right) \tag{4.13}$$

which can be written as

$$P_3(\xi) := \xi^3 - (3+a)\xi + 2 = 0 \tag{4.14}$$

where $\xi = h_P/h = \tau_y/\tau_B$ and the parameter $a = (6\mu\,\bar{u})\,/\,(h\,\tau_y)$. Alternatively, it is possible to use the Bingham number B defined as:

$$B = \frac{4h\tau_y}{\mu\,\bar{u}} = \frac{24}{a}$$

The solution of equation (4.14) lies in the interval $[0,1]$. Once ξ has been obtained, the value of the bottom friction is obtained as $\tau_B = \tau_y/\xi$.

The roots of a 3rd order polynomial $\{\xi_1, \xi_2, \xi_3\}$ can be obtained using Cardano's formula once the equation has been reduced to the canonical form

$$\xi^3 + p\xi + q = 0 \qquad (4.15)$$

and are given by

$$\xi_1 = \sqrt[3]{-\frac{q}{2} + \sqrt{R}} + \sqrt[3]{-\frac{q}{2} - \sqrt{R}} \qquad (4.16)$$

where the discriminant R is

$$R = \left(\frac{q}{2}\right)^2 + \left(\frac{p}{3}\right)^3 \qquad (4.17)$$

Once ξ_1 has been obtained, the remaining roots are given by

$$\xi_{2,3} = -\frac{\xi_1}{2} \pm \sqrt{\left(\frac{\xi_1}{2}\right)^2 + \frac{q}{\xi_1}} \qquad (4.18)$$

In the particular case considered here, $R = (2/2)^2 - \left(1 + \frac{a}{3}\right)^3 < 0$, and there exist three complex solutions for ξ_1. Because of this complexity the friction term has been obtained either finding the root between 0 and 1 using an iterative method such as Newton Raphson, or by using approximate solutions:

 - Jin and Fread (1997) proposed the semi-empirical law

$$\xi = \frac{\tau_y}{\tau_0} = \left(1 + \frac{(m+1)\,(m+2)}{(0.74 + 0.656m)}\,\frac{\bar{u}}{h\left(\frac{\tau_y}{\mu}\right)^m}\right)^{-\frac{1}{m+0.15}}$$

which for $m = 1$ represents a Bingham fluid:

$$\xi = \left(1 + \frac{1}{1.396}a\right)^{-\frac{1}{1.15}} \qquad (4.19)$$

 - The approximation used in NWS DAMBREAK (Fread, 1988), which can be written as

$$\xi = \left(\frac{3}{2} + \frac{3\,\bar{u}\mu}{\tau_y h}\right)^{-1} = \frac{2}{3+a} \qquad (4.20)$$

- The approximation proposed by Jeyapalan, Duncan and Seed (1983), where the relative depth of the plug is given by:

$$\xi = \frac{3}{3+a} \tag{4.21}$$

However, a simple yet effective and accurate approximation can be obtained using a polynomial economization technique, in which we approximate the third order polynomial $P_3(\xi)$ by a second order interpolating polynomial $P_2(\xi)$ in the interval $0 \leq \xi \leq 1$ where the root we are interested lies. It can be seen that the error E

$$P_3 = P_2 + E \tag{4.22}$$

is given by

$$E = \frac{1}{3!} \frac{d^3 P_3(\xi)}{d\xi^3} \Pi(\xi) = \Pi(\xi) \tag{4.23}$$

where $\Pi(\xi) = (\xi - \xi_0)(\xi - \xi_1)(\xi - \xi_2)$, with ξ_0, ξ_1 and ξ_3 being the points used in the interpolation. The maximum error can be minimized if ξ_0, ξ_1 and ξ_3 are chosen as the roots of the third Tchebychef polynomial

$$T_3(t) = 4t^3 - t^2 \qquad t \in [-1, 1] \tag{4.24}$$

or

$$T_3(\xi) = 32\xi^3 - 48\xi^2 + 18\xi - 1 \qquad \xi \in [0, 1] \tag{4.25}$$

Therefore, the error will be given by

$$E = \xi^3 - \frac{3}{2}\xi^2 + \frac{9}{16}\xi - \frac{1}{16} \tag{4.26}$$

and the second order polynomial $P_2(\xi)$ which is the best uniform or minimax approximation of $P_3(\xi)$ is obtained as

$$\begin{aligned} P_2(\xi) &= P_3(\xi) - E & (4.27) \\ &= \left(\xi^3 - (3+a)\xi + 2\right) - \left(\xi^3 - \frac{3}{2}\xi^2 + \frac{9}{16}\xi - \frac{1}{16}\right) & (4.28) \\ &= \frac{3}{2}\xi^2 - \left(\frac{57}{16} + a\right)\xi + \frac{65}{32} & (4.29) \end{aligned}$$

with a maximum error of $1/32$ in the interval $[0, 1]$.
Therefore, the proposed approximation consists of solving the second order equation

$$\frac{3}{2}\xi^2 - \left(\frac{57}{16} + a\right)\xi + \frac{65}{32} = 0$$

or

$$\frac{3}{2}\xi^2 - \left(\frac{384 + 57B}{16}\right)\xi + \frac{65}{32} = 0$$

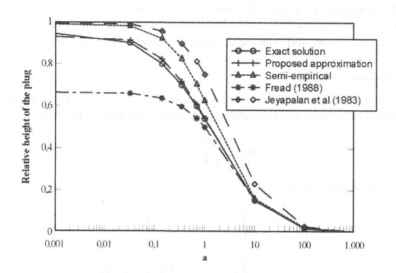

Figure 13. Comparison of exact and approximate solutions

Figure 13 presents a comparison of the exact and approximate values of the relative plug weight ξ for different values of the parameter a.

It can be seen how the proposed approach provides reasonable values for the whole range of values of a studied.

The equilibrium conditions for a 1D flow over a horizontal bottom are obtained substituting the averaged horizontal stress in the balance of momentum equation 4.10:

$$\frac{\partial}{\partial x_1}\left(-\frac{1}{2}\rho gh^2 + \tau_y \cos\beta\right) + \tau_y = 0 \tag{4.30}$$

Therefore, the equilibrium slope dh/dx is $\tau_y/(\rho gh)$, which depends on h. Integration of the slope results in a parabolic equilibrium profile.

4.4 Generalized Viscoplastic Models

The Generalized Viscoplastic Model ? proposed a general rheological model for incompressible fluids, where the stresses are given by:

$$\sigma = -p\mathbf{I} + \left(\frac{s}{\sqrt{I_{2D}}} + 2\mu_1 |4\,I_{2D}|^{\frac{n_1-1}{2}}\right)\mathbf{D} + 4\mu_2 |4\,I_{2D}|^{\frac{n_2-2}{2}}\mathbf{D}^2 \tag{4.31}$$

The depth averaged rheological model is:

$$\bar{\sigma}_{ij} = -\bar{p}\delta_{ij} + \left(\frac{\bar{s}}{\sqrt{\bar{I}_{2D}}} + 2\mu_1 |4\,\bar{I}_{2D}|^{\frac{n_1-1}{2}}\right)\bar{D}_{ij} + 4\mu_2 |4\,\bar{I}_{2D}|^{\frac{n_2-2}{2}}(\bar{D}^2)_{ij} \qquad (i,j=1,2) \tag{4.32}$$

A possible simplified version of this model consists on taking $\bar{\sigma}_{11} = \bar{\sigma}_{22} = -\bar{p}$, and $\bar{\sigma}_{12} = 0$. The bottom friction should be obtained assuming that is the same than in a simple shear flow, and relating its value to the averaged velocity.

4.5   Frictional Fluids

The Frictional Fluid without cohesion is a particular case of the Bingham fluid for which the yield stress is given by $p \sin \phi$. Therefore, Φ_1 will be given by

$$\Phi_1 = \frac{p \sin \phi}{\sqrt{I_{2D}}} + 2\mu$$

The depth averaged stresses are: ,

$$\bar{\sigma}_{11} = -\bar{p}\left(1 - \sin\phi\frac{\bar{D}_{11}}{\sqrt{\bar{I}_{2D}}}\right) + 2\mu\,\bar{D}_{11} \tag{4.33}$$

$$\bar{\sigma}_{22} = -\bar{p}\left(1 - \sin\phi\frac{\bar{D}_{22}}{\sqrt{\bar{I}_{2D}}}\right) + 2\mu\,\bar{D}_{22}$$

$$\bar{\sigma}_{12} = \bar{p}\sin\phi\frac{\bar{D}_{12}}{\sqrt{\bar{I}_{2D}}} + 2\mu\,\bar{D}_{12}$$

Concerning the bottom friction term, it can be obtained using the Mohr-Coulomb criterion and the expression of t_{3B} obtained after integrating along X_3 the third component of the balance of momentum equation 3.34.

In the case of a 1D flow, the averaged stress components are given by:

$$\bar{\sigma}_{11} = -\bar{p}\left(1 - \sin\phi\cos\beta\right) + 2\mu\bar{D}_{11} \tag{4.34}$$

$$\bar{\sigma}_{33} = -\bar{p}\left(1 + \sin\phi\cos\beta\right) + 2\mu\bar{D}_{33}$$

$$\bar{\sigma}_{13} = \bar{p}\sin\phi\sin\beta + 2\mu\bar{D}_{13}$$

where $\sin\beta$ and $\cos\beta$ have been introduced. These stress components satisfy in the limit case $\bar{\mathbf{D}} \to \mathbf{0}$ the Mohr-Coulomb criterion. Another interesting aspect is that $|\bar{\sigma}_{11}| < |\bar{\sigma}_{33}|$ if $\cos\beta > 0$, i.e., $\bar{D}_{11} > 0$ (active state), and $|\bar{\sigma}_{11}| > |\bar{\sigma}_{33}|$ if $\cos\beta < 0$, i.e., $\bar{D}_{11} < 0$ (passive state).A further simplified version will be obtained as shown in the previous sections, assuming $\bar{\sigma}_{11} = \bar{\sigma}_{22} = -\bar{p}$, and $\bar{\sigma}_{12} = 0$.

The equilibrium slope in 1D can be obtained following the same procedure described above. Substituting $\bar{\sigma}_{11} = -\bar{p}\left(1 - \sin\phi\cos\beta\right)$ in the eqn.4.10, we arrive at

$$\rho g h\frac{\partial h}{\partial x}\left(1 - \sin\phi\cos\beta\right) = t_{B1}$$

If we assume an active state $(\cos\beta > 0)$, The Mohr-Coulomb condition at the bottom gives

$$t_{B1} = \sigma_{11}\tan\phi = -\rho g h\left(1 - \sin\phi\cos\beta\right)\tan\phi$$

Therefore, the equilibrium slope is $\tan\phi$.

5 Numerical Model

This system of conservation equations can be solved with the Finite Element Method using the Taylor-Galerkin algorithm introduced by Donea (1984) and Peraire et al (1984).

The Taylor-Galerkin algorithm can be considered as the FEM counterpart of the Lax-Wendroff procedure in the FDM. It basically consists in a higher order expansion of the time derivative, followed by the spatial discretization of the resulting equation using the conventional Galerkin weighting method.

However, following the general procedure described in Zienkiewicz and Taylor (2000) requires the calculation of the derivatives of the flux tensor, $\mathbf{F}$, and source vector, $\mathbf{S}$, relative to the vector of unknowns, ϕ, for each element in the mesh and performing a number of matrix multiplications.

To avoid this computer memory and time consuming operations Peraire et al (1984), developed a two step algorithm that can be regarded as the FEM implementation of the Richtmyer scheme. Globally, the Richtmyer scheme is of second order accuracy in space and time. Due to its accuracy and simplicity it has been used by the authors to discretize the depth integrated equations. the Taylor-Galerkin procedure for solving equation (3.36) starts from a second order expansion in time

$$\phi^{n+1} = \phi^n + \Delta t \left. \frac{\partial \phi}{\partial t} \right|^n + \frac{1}{2} \Delta t^2 \left. \frac{\partial^2 \phi}{\partial t^2} \right|^n \tag{5.1}$$

where the first order time derivative of the unknowns can be calculated using equation (3.36)

$$\left. \frac{\partial \phi}{\partial t} \right|^n = (S\text{-}div\,\underline{F})^n$$

To obtain the second order time derivative, the two-steps Taylor-Galerkin procedure considers an intermediate step between t^n and t^{n+1}. The aim of this first time step is to calculate the solution at a time $t^{n+1/2}$. This step is followed by a second one that brings the solution to t^{n+1}.

In this way, the first step results in

$$\phi^{n+1/2} = \phi^n + \frac{\Delta t}{2} (\mathbf{S}\text{-}div\,\underline{F})^n \tag{5.2}$$

which allows the calculation of $\underline{F}^{n+1/2}$ and $\mathbf{S}^{n+1/2}$.

Considering now a Taylor series expansion of the flux and source terms,

$$\underline{F}^{n+1/2} = \underline{F}^n + \left(\frac{\partial \underline{F}}{\partial t} \right)^n \frac{\Delta t}{2}$$

$$\mathbf{S}^{n+1/2} = \mathbf{S}^n + \left(\frac{\partial \mathbf{S}}{\partial t} \right)^n \frac{\Delta t}{2}$$

and the values of $\underline{F}^{n+1/2}$ and $S^{n+1/2}$, the flux and sources time derivatives are calculated as

$$\left(\frac{\partial \underline{F}}{\partial t}\right)^n = \frac{2}{\Delta t}\left(\underline{F}^{n+1/2} - \underline{F}^n\right)$$

$$\left(\frac{\partial S}{\partial t}\right)^n = \frac{2}{\Delta t}\left(S^{n+1/2} - S^n\right)$$

Incorporating these expressions into the second order time derivative

$$\left.\frac{\partial^2 \phi}{\partial t^2}\right|^n = \frac{\partial}{\partial t}(S\text{-}div\,\underline{F})^n$$

results in

$$\left.\frac{\partial^2 \phi}{\partial t^2}\right|^n = \frac{2}{\Delta t}\left(S^{n+1/2} - S^n - div\left(\underline{F}^{n+1/2} - \underline{F}^n\right)\right)$$

Now replacing the expressions for the first and second order time derivatives in the Taylor series expansion, (5.1), allows the determination of the unknowns at time t^{n+1}

$$\phi^{n+1} = \phi^n + \Delta t\left(S^{n+1/2} - div\,\underline{F}^{n+1/2}\right)$$

This equation is spatially discretized using conventional Galerkin weigthing to finally result in the system of equations to be solved to obtain the unknowns increment in the time step

$$\underline{M}\,\Delta\phi = \Delta t\left(\int_\Omega \underline{N}\,S^{n+1/2}d\Omega - \int_{\Gamma_N} \underline{N}\,\underline{F}^{n+1/2}d\Omega + \int_\Omega \underline{F}^{n+1/2}\,grad\,\underline{N}\,d\Omega\right) \quad (5.3)$$

6 Examples and Applications

6.1 Failure of Gypsum Tailings Impoundment

Bingham models have been widely applied to the analysis of flowslides, mudflows and debris flows. Rickenmann and Koch (1997) propose values of τ_y and μ in the ranges 100-800 Pa and 400-800 Pa.s in simulations of debris flows in Kamikamihori (Japan) and Saas (Switzerland) valleys events. From their simulations, it can be concluded that Bingham models predict higher velocities than other models, for the particular set of parameters chosen by the authors.. Table I summarizes the values used by Jan (1997) and other researchers. It is also interesting to reproduce the values selected by Jin and Fread (1997) for several sites, which are reproduced in Table II, as they provide information on the range of values of Bingham yield stress and viscosity for different sites.

It is interesting in order to assess the predictive performance of the model to compare its predictions against in situ data. Here we have chosen a well documented case of a failure of a tailings dam.

Jeyapalan et al. (1983) reports the flow of liquefied tailings from Gypsum Tailings Impoundment in East Texas which took place in 1966. The impoundment was rectangular, and had reached a height of 11 m. by the time failure took place. The slide was caused by seepage at the toe of the slope, and affected a length of 140 m. of the dyke, extending 110 m. into the impoundment lagoon. The released material travelled 300 m. before stopping, with an average velocity of 2.5 to 5 m/s. A section of the flow by a vertical plane perpendicular to the dyke at the centre of the breach is given in Fig. 14.

Properties of the tailings have been taken from Jeyapalan et al. (1983) who assumed that the behaviour of the flowing material could be described by a Bingham model. The tailings were classified as "non-plastic silts", with $D_{50} \approx 0.07$ mm, a uniformity coefficient close to 3, density of particles 2450 Kg/m^3 and an average water content of 30 %. The yield stress was determined from simple slope stability analysis as $\tau_y = 10^3$ Pa, and the viscosity was taken as $\mu = 50$ Pa.s. Finally, the density of the tailings was assumed to be $\rho = 1400$. Kg/m^3.

The finite element mesh used in the analysis is shown in Fig. 15. The model comprises 350 m. of the dyke length, removing the 140 m. long section which failed. Therefore, the model assumes that collapse of the dyke takes place instantaneously. The plain onto which the tailings flow has been limited for computational reasons, imposing absorbing conditions at the artificial boundaries.

Fig. 16 shows the contours of tailings depth on the plane at times 0, 30, 60, 90 and 120 s. The vertical scale has been enlarged by a factor of ten in order to better describe the properties of the flowslide. It is interesting to see the development of a transient jump at $t = 30s$ which propagates backwards. This jump is also seen in Fig. 17, where the free surface profiles along the vertical plane passing by the centerline of the breach are given.

The results agree well with observed values of inundation distance (300 m), freezing time $(60 - 120\,s)$ and mean velocity $(2.5 - 5.0$ m/s$)$. However, the flowslide progressed back into the pond about 110 m, while in the simulation reaches a longer distance.

In order to illustrate the performance of the proposed model, we have analyzed this case assuming that viscosity is zero, and, therefore, the relative height of the plug is $\xi = 1$.

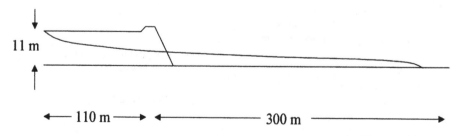

Figure 14. Sketch of the flowslide at Gypsum Tailings Impoundments

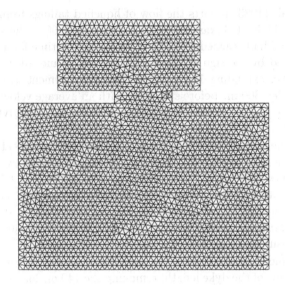

Figure 15. Finite element mesh used in the analysis of Gypsum Tailings flowslide

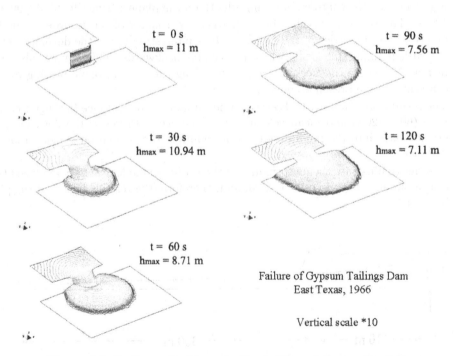

Figure 16. Isolines of tailings depth at different times after failure

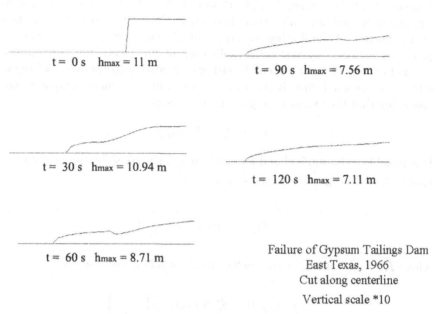

t = 0 s   hmax = 11 m

t = 30 s   hmax = 10.94 m

t = 60 s   hmax = 8.71 m

t = 90 s   hmax = 7.56 m

t = 120 s   hmax = 7.11 m

Failure of Gypsum Tailings Dam
East Texas, 1966
Cut along centerline
Vertical scale *10

Figure 17. Sections of the free surface at different times by a vertical plane passing through the centerline of the breach

6.2 *Tip No.7 flowslide at Aberfan (1966)*

One of the main difficulties when modelling Aberfan flowslide is the role of pore pressures. Coal debris were a material composed of solid, fluid and gas phases, with a strong interaction between them. However, models cast in terms of effective stresses, pore pressures, and velocities of all constituents have not been used so far, because of the problem of moving interfaces mentioned above. Depth integrated models can provide important information about runout, propagation paths and velocities, which is frequently sufficient to design protection structures. The main shortcoming of depth integrated models comes from the fact that pore fluid and solid particles are modelled as a single phase material, with properties that do not change with time. This is why Aberfan flowslide has been modelled assuming a Bingham fluid rheological law for the debris. For instance, Jeyapalan et al. (1983) and ? obtained results which fitted well the observations choosing $\tau_y = 4794$ Pa., $\mu = 958$ Pa.s and $\rho = 1760$ kg/m^3. Even if the results are good, it is possible to argue that waste coal was not fully saturated, and the material was frictional. Of course, the apparent angle of friction introduced above will be much smaller than ϕ', but vertical consolidation could have made it to change during the propagation phase. Hutchinson (1986) proposed a simple "sliding-consolidation" model in which it was clear that the combination of friction with basal pore pressures could provide accurate results of runout and velocities.

We will compare the method proposed in this work with the simplified analysis introduced by the authors in Ref. Pastor et al. (2002). This simplified method consists on we will assuming that there exist a layer of saturated soil of height h_s on the bottom of the flowing material Hutchinson (1986). The decrease in pore pressures is caused by vertical consolidation of this layer. Pore pressures on the top and bottom of this layer can be either estimated from the values of the vertical stresses or obtained directly from the results of finite element computations. Assuming that the excess pore pressure evolves as

$$p_w(x_3, t) = N(x_3)\,\bar{p}_w(t) \tag{6.1}$$

it is possible to obtain a closed form solution of the consolidation equation. In the case of $N(x_3) = \sin\left(\frac{\pi}{2}\frac{x_3}{h_s}\right)$, the solution is

$$\bar{p}_w(t) = \bar{p}_w^0 \exp\left(-\frac{t}{T_v}\right) \tag{6.2}$$

where $T_v = \frac{4h_s^2}{\pi^2 c_v}$ and c_v is the coefficient of consolidation. Finally,

$$p_w(x_3, t) = \bar{p}_w^0\, N(x_3) \exp\left(-\frac{t}{T_v}\right) \tag{6.3}$$

The coefficient r_u can be estimated as:

$$r_u = r_u^0 \exp\left(-\frac{t}{T_v}\right) \tag{6.4}$$

In above, we have used "pore pressures" instead of "pore water pressures". The reason is that dry or partially saturated materials can collapse generating high air pressures on their pores, which will cause a similar effect.

The information provided in the literature do not give enough data to perform a realistic analysis in two dimensions. Therefore, we have used a simple 1D model with the terrain profiles given by Jeyapalan et al. (1983). The main purpose of this example is to show that a depth integrated model using pore pressure dissipation can reproduce the basic patterns observed.

Density of the mixture ρ and friction angle ϕ', have been taken as 1740 kg/m^3 and 36° respectively. Consolidation time T_v has been estimated from the range of values proposed by Hutchinson (1986) for the depth of the saturated liquefied layer $h_s = 0.1$ m and the coefficient of consolidation $c_v = 6.4\,10^{-5}$ m^2/s. Assuming $T_v = \frac{4h_s^2}{\pi^2 c_v}$, we arrive to $T_v = 64$ s. The initial value of the pore pressure coefficient r_{u0} has been taken as 0.78.

The results obtained in the simulation are given in Fig. 18 where sections of the free surface of the flowslide are given at times 0., 5, 10, 20. and 60 s. On the left, we have depicted the results obtained with the simplified method. The column on the right correspond to the method proposed in this paper. The basic features of the flow obtained in the simulation (propagation distance$\approx$ 600 m, freezing time ≈ 40 s and average velocity 15 m/s) coincide reasonably well with those reported in the literature. Fig. 19 shows position of the front as a function of time. The differences between both of the approaches are small, and affect only the profile. With the advected pore pressure, the front depth can be seen to be larger.

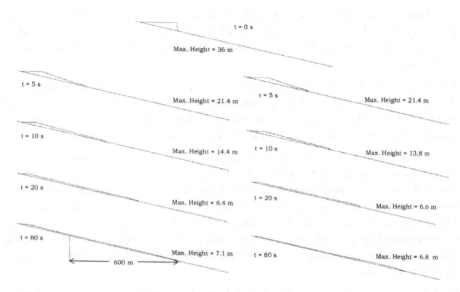

Figure 18. Vertical sections ofberfan flowslide at different times (i) Left, with simplified approach for consolidation (ii) Right, enhanced method

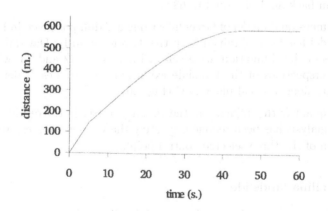

Figure 19. Position of head of Aberfan flowslide as a function of time

6.3 Greenhills Cougar 7 dump failure

Among of the large number of flowslides that have occurred due to liquefaction from the Rocky Mountains coal mine waste dumps, three interesting cases have been analyzed in detail by Dawson et al. (1998). Therefore, information on the characteristics of the flowslides, the properties of the involved materials, etc. is reasonably available.

This paper analyzed the case of Fording Greenhills, where in May 1992 the Cougar 7 dump failed. The mobilized mass of debris consisted on approximately 200,000 m^3, which slided off the 100 m high dump travelling a distance of 700 m before coming to rest.

The debris were mainly sandy gravel, and the foundation beneath the dump and the runout debris consisted on a sand and gravel colluvium layer with a depth varying between 0.3 m and 0.5 m beneath the dump. Wet fine grained layers were found at the foundation contact, near the crest, and in the debris.

According to the post failure analysis carried out by Dawson and coworkers, these layers played a paramount role in both the initiation and the propagation phases. The collapse model proposed by them is based on the existence of layers of finer sandy gravel materials of low permeability deposited parallel to the dump face. Triggering mechanism could have consisted on: (i) a redistribution of effective stresses caused by pore pressure changes, and (ii) liquefaction of these layers under quasi undrained conditions. The flowslide could have ridded over these layers of liquefied materials.

We have based the analysis of this flowslide on the geotechnical properties obtained by Dawson et al. (1998), and we have chosen accordingly a density of 1900 kg/m^3 and an effective friction angle $\phi' = 37^o$. The mobilized mass has been taken from the data given in by Dawson et al. (1998), and the consolidation time has been chosen from back analysis equal to 68 s.

The two dimensional model of Greenhills Cougar 7 dump is given in Figure 20. The terrain model has been obtained from the data given in by Dawson et al. (1998). The results of the simulation are given in Figures 21, 22 and 23, where it can be seen: (i) perspectives of the flowslide extension, (ii) isolines of debris depth and (iii) sections along vertical plane AA' (Fig. 20).

Finally, to quantify the influence of the viscous stresses in the results of the calculations, the analysis has been redone neglecting the viscous stresses. Fig. 20 presents the position of the three selected control points.

6.4 Valtellina landslide

So far, we have considered examples concerning flowslides which could be classified as "undrained" and "coupled". Here, we will consider one case of the "drained" type, the avalanche which take place in Valtellina (northern Italy) in 1987. The mobilized mass consisted on 40 million m^3 of rock and debris which covered 2.4 km^{2}of the valley with a layer reaching 90 m of depth. The material has been considered dry, with a dissipation time of the air pore pressures much smaller than the time of propagation. The results of the computations are shown in Fig. 24, together with predictions obtained with the code PFC2D.

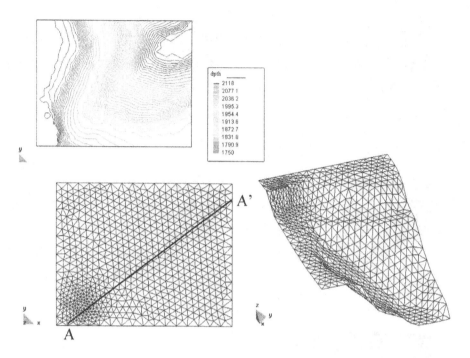

Figure 20. Terrain model of Greenhills Cougar 7 dump and control points

Bibliography

R.A.Bagnold, Experiments on a gravity-free dispersion of large solid spheres in a Newtonian fluid under shear, **Proc.Royal Society A**, 225, pp.49-63, 1954.

R.A.Bagnold, The shearing and dilatation of dry sand and the singing mechanism, **Proc.Royal Society A**, 295, pp.219-232, 1966.

A.W.Bishop, J.N.Hutchsinson, A.D.M.Penman, and H.E.Evans, Geotechnical Investigations into the causes and circumstances of the disaster of 21st October 1966. **A selection of technical reports submitted to the Aberfan Tribunal**, pp. 1-80, Welsh Office, H.M.S.O. London, England, 1969.

A.W.Bishop, The stability of tips and spoil heaps, **Quart.J.Eng.Geol. 6**, 335-376, 1973.

L.Calembert and R.Dantinne, The avalanche of ash at Jupille (Liege) on February 3rd, 1961. From: **The commemorative volume dedicated to Professeur F.Campus**, pp. 41-57, Liége, Belgium, 1964.

C.L.Chen and C.H.Ling, Granular-Flow Rheology: Role of Shear-Rate Number in Transition Regime, **J.Engng.Mech. ASCE**, 122, 5, 469-481, 1996.

R.F.Dawson, N.R.Morgenstern and A.W.Stokes, "Liquefaction flowslides in Rocky Mountain coal mine waste dumps", **Can.Geotech.J.**, **35**, 328-343, 1998.

J.Donea, A Taylor-Galerkin Method for Convective Transport Problems, **Int.J.Num.Meth.Engng. 20**, pp. 101-109, 1984.

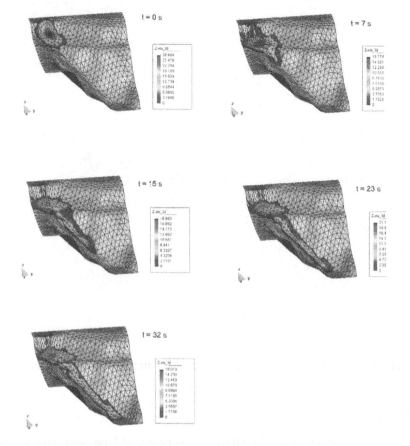

Figure 21. Propagation of flowslide at Greenhills: perspective

R.Du, Z.Kang and P.Zhu, **Debris Flow of Xiaojiang basin in photographs**, Sichuan Publishing House of Science and Technology, 1987.

R.Frenette, D.Eyheramendy and T.Zimmermann, Numerical Modelling of dambreak type problems for Navier-Stokes and Granular Flows, in C.L.Chen (Ed.), **Debris-Flow Hazards Mitigation: Mechanics, Prediction and Assessment, ASCE** , pp. 586-595, 1997.

D.M.Hanes and D.L.Inman, Observations of rapidly flowing granular fluid materials, **J.Fluid Mech. 150**, 357-380, 1985.

O.Hungr, A model for the runout analysis of rapid flow slides, debris flows and avalanches, **Can.Geotech.J. 32**, pp 610-623, 1995.

J.N.Hutchinson, Undrained loading, a fundamental mechanism of mudflows and other mass movements, **Géotechnique 21**, No.4, 353-358, 1971.

J.N.Hutchinson, A sliding-consolidation model for flow slides, **Can.Geotech.J., 23**, 115-126, 1986.

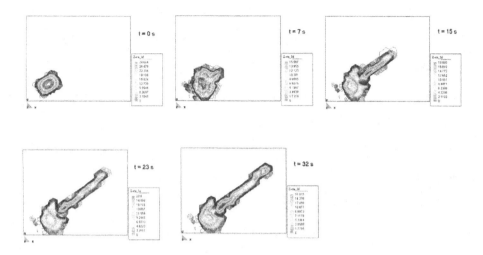

Figure 22. Propagation of flowslide at Greenhills: isolines of depth

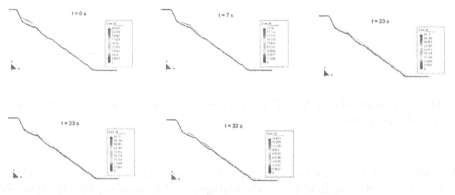

Figure 23. Propagation of flowslide at Greenhills: section along AA'

K.Hutter and T.Koch, Motion of a granular avalanche in an exponentially curved chute: experiments and theoretical predictions,**Phil.Trans.R.Soc.London, A 334**, pp 93-138, 1991.

K.Hutter, B.Svendsen and D.Rickenmann, Debris Flow modelling: A review. Continuum Mech.Thermodyn, 8, 1-35, 1996.

C.D.Jan, A study on the numerical modelling of debris flows, in **Debris-Flow Hazards Mitigation:Mechanics, Prediction and Assessment**, Proc.1st Int.Conference, ASCE, C-l.Chen (Ed.), pp.717-726, 1997.

J.K.Jeyapalan, J.M.Duncand and H.B.Seed, Investigation of flow failures of tailing dams, **J. Geotech. Engng ASCE 109**, , pp.172-189, 1983.

M.Jing and D.L.Fread, One-dimensional routing of mud/debris flows using NWS FLDWAV model, in **Debris-Flow Hazards Mitigation and Assessment**, Proc.1st Int.Conference, ASCE, C-l.Chen (Ed.), pp.687-696, 1997.

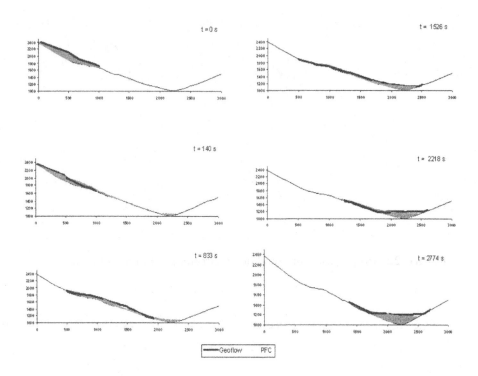

Figure 24. Rock avalanche at Valtellina: comparison between PFC2D and the depth integrated model proposed

A.M.Johnson, A model for grain flow and debris flow, U.S.Geological Survey Open-File Report 96-728, 1996 **Hazards Mitigation:Mechanics, Prediction and Assessment,** Proc.1st Int.Conference, ASCE, C-l.Chen (Ed.), pp 687-696, 1997.

A.M.Johnson and S.Y.Martosudarmo, Discrimination between inertial and macro-viscous flows of fine-grained debris with a rolling-sleeve viscometer, in **Debris-Flow Hazards Mitigation:Mechanics, Prediction and Assessment,** Proc.1st Int.Conference, ASCE, C-l.Chen (Ed.), pp.229-238, 1997.

P.Y.Julien and Y.Lan, Rheology of hyperconcentrations, **J. Hydr.Engng. ASCE, 117,** 3, 346-353, 1991.

D.Laigle and P.Coussot, Numerical Modelling of Mudflows, **J.Hydr.Engng. ASCE, Vol.123,** pp. 617-623, 1997.

D.Laigle, A two-dimensional model for the study of debris-flow spreading on a torrent debris fan, in C.L.Chen (Ed.), **Debris-Flow Hazards Mitigation: Mechanics, Prediction and Assessment, ASCE** , pp. 123-132, 1997.

J.S.O'Brien, P.Y.Julien and W.T.Fullerton, Two-dimensional water flow and mud-flow simulation, **J.Hydr.Engng. ASCE, Vol.119,** pp. 244-261, 1993.

M.Pastor, M.Quecedo, J.A.Fernández Merodo, M.I.Herreros, E.González, and P.Mira, Modelling Tailing dams and mine waste dumps failures, **Geotechnique, Vol.52**, No.8, pg.579-591, 2002.

J.Peraire, **A Finite Element Method for convection dominated flows**. PhD thesis. University of Wales, Swansea, 1986.

J.Peraire, O.C.Zienkiewicz and K.Morgan, A Taylor-Galerkin method for convective transport problems, **Int. J. Num. Methods in Engineering, 20**, 101-109, 1984.

M.Quecedo, and M.Pastor, Application of the level set method to the finite element solution of two-phase flows. **Int. J. Num. Methods in Engineering**, 50, 645-663, 2001.

D.Rickenmann and T.Koch, Comparison of debris flow modelling approaches, in **Debris-Flow Hazards Mitigation:Mechanics, Prediction and Assessment**, Proc.1st Int.Conference, ASCE, C-l.Chen (Ed.), pp.576-585, 1997.

S.B.Savage and K.Hutter, The dynamics of avalanches of granular materials from initiation to runout. Part I: Analysis, **Acta Mechanica 86**, pp 201-223, 1991.

J.Sousa and B.Blight, Continuum simulation of flow failures, **Géotechnique 41**, 4, pp. 515-538, 1991.

E.F.Toro, **Shock-Capturing Methods for Free-Surface Shallow Flows**, Sons, New York, 2001.

L.Vulliet and K.Hutter, Continuum model for natural slopes in slow movement, **Géotechnique 38**, pp. 199-217, 1988.

S.Zhang, A comprehensive Approach to the Observation and prevention of Debris Flows in China, **Natural Hazards**, 7, 1-23, 1993.

O.C.Zienkiewicz and R.L.Taylor, **The Finite Element Method, Vol.3, Fluid Dynamics**, Butterworth and Heinemann, Oxford, 2000.

O.C.Zienkiewicz and T.Shiomi. Dynamic behaviour of saturated porous media: The generalised Biot formulation and its numerical solution. **Int.J.Num.Anal.Meth.Geomech.,8**, 71-96, 1984.

C.Zoppou and S.Roberts, Catastrophic collapse of water supply reservoirs in urban areas, **J.Hydr.Engng. ASCE, Vol.125**, pp. 686-695, 1997.

Printed in the United States
By Bookmasters

Printed in the United States
By Bookmasters